BACTERIAL
PHYSIOLOGY AND BIOCHI

Progress in Biochemistry and Biotechnology

BACTERIAL PHYSIOLOGY AND BIOCHEMISTRY

Ivan Kushkevych

Department of Experimental Biology,
Faculty of Science, Masaryk University,
Brno, Czech Republic

Edited by
Josef Jampílek

ELSEVIER

ACADEMIC PRESS
An imprint of Elsevier

Academic Press is an imprint of Elsevier
125 London Wall, London EC2Y 5AS, United Kingdom
525 B Street, Suite 1650, San Diego, CA 92101, United States
50 Hampshire Street, 5th Floor, Cambridge, MA 02139, United States
The Boulevard, Langford Lane, Kidlington, Oxford OX5 1GB, United Kingdom

ISBN: 978-0-443-18738-4

For information on all Academic Press publications
visit our website at https://www.elsevier.com/books-and-journals

Working together
to grow libraries in
developing countries

www.elsevier.com • www.bookaid.org

Publisher: Stacy Masucci
Acquisitions Editor: Linda Versteeg-Buschman
Editorial Project Manager: Barbara L. Makinster
Production Project Manager: Stalin Viswanathan
Cover Designer: Matthew Limbert

Cover input from Igor Starunko
Typeset by STRAIVE, India

CONTENTS

ACKNOWLEDGEMENTS

Scientific reviewers:

Prof. PharmDr. **Josef Jampílek**, Ph.D., Professor of Medicinal Chemistry, Department of Chemical Biology, Faculty of Science, Palacky University Olomouc, Slechtitelu 27, 783 71 Olomouc, Czech Republic; Department of Analytical Chemistry, Faculty of Natural Sciences, Comenius University in Bratislava, Ilkovicova 6, 842 15 Bratislava, Slovakia

Prof. **Lorenzo Drago**, Ph.D., Professor of Clinical Microbiology, Department of Biomedical Sciences for Health, University of Milan, 20133 Milan, Italy

Prof. **Aidan Coffey**, Ph.D., Professor of Microbiology, Department of Biological Sciences, Cork Institute of Technology, Bishopstown, Cork, Ireland

This educational book contains the modern knowledge of bacterial physiology and biochemistry. The book includes seven chapters that describe the subject of bacterial physiology, chemical composition, functional cell structures, nutrition, growth, and the processes of cell differentiation. Special attention is paid to the bacterial metabolism, including the processes of catabolism and anabolism. The energy of biochemical reactions, carriers of hydrogen, the role of ATP in the bacterial cells, and types of phosphorylation are also discussed. The physiological role of bacteria in ecosystems, intercellular and internal population interactions, quorum-sensing regulation of gene expression, and luminescent bacteria and bioluminescence are described. The effect of environmental factors on bacteria, including physical and chemical factors, and their mechanisms are presented. The description of chemotherapeutics and antibiotics and mechanisms of their action is also provided. This book can be helpful for bachelor and master students who study General Microbiology, Medical Microbiology, Veterinary Microbiology, and Molecular Biology, for microbiologists, biochemists, biologists, experts in the cell and molecular biology, and for readers interested in the study of bacterial physiology and biochemistry.

Acknowledgments: The author sincerely thanks **Igor Starunko** from Ivan Franko National University of Lviv (Ukraine) for his help and the technical preparation of the illustrative material and layout of the book. The author also expresses special thanks to the scientific editor, Prof. **Josef Jampílek,** as well as the reviewers, Prof. **Lorenzo Drago** and Prof. **Aidan Coffey**, for their time and help, critical comments, and recommendations during preparation of this book.

Ivan Kushkevych

ABBREVIATIONS

ATP	Adenosine triphosphate
Abe	Abequose
ADP	Adenosine diphosphate
AMP	Adenosine monophosphate
APS	Adenosine-5′-phosphosulfate
BChl	Bacteriochlorophylls
BLAST	Basic local alignment search tool
CAP	Catabolite activator protein
Chl	Chlorophylls
CTP	Cytidine triphosphate
Cyt	Cytochrome
Da	Daltons
DAP	Diaminopimelic acid
dATP	Deoxyadenosine triphosphate
dCTP	Deoxycytidine triphosphate
dGTP	Deoxyguanosine triphosphate
DHB	2,3-Dihydroxybenzoate
DNA	Deoxyribonucleic acid
dTTP	Deoxythymidine triphosphate
dXDP	Deoxyribonucleotide diphosphate
EDP	Entner–Doudoroff pathway
EDTA	Ethylenediaminetetraacetic acid
EMBL	European Molecular Biology Laboratory
EMP	Embden–Meyerhof–Parnas pathway
FA	Fatty acids
FACS	Fluorescence activated cell sorting
FAD	Flavin adenine dinucleotide
Fd	Ferredoxin
Feb	Ferrienterochelin-binding protein
FMN	Flavin mononucleotide
Gal	Galactose
GDP	Guanosine diphosphate
Glu	Glucose
GluN	Glucosamine
GSB	Green sulfur bacteria
GSH	Reduced glutathione
GSSG	Oxidized glutathione

GTP	Guanosine triphosphate
GTP	Guanine triphosphate
Hase	Hydrogenase
HDHD	3-Hydroxydecanoyl-3-hydroxydecanic acid
Hep	Heptose
HM	Hydroxy muramic acid
HMP	Hexose-monophosphate pathway
HPLC	High-performance liquid chromatography
KDO	2-Keto-3-deoxy-octonoic acid
KEGG	Kyoto Encyclopedia of Genes and Genomes
LPS	Lipopolysaccharide
LTPP	Lipothiamine pyrophosphate
Man	Mannose
MR-test	Methyl red test
NAD	Nicotinamide adenine dinucleotide
NAG	*N*-Acetylglucosamine
NAM	*N*-Acetylmuramic acid
NRPS	Nonribosomal peptide synthetases
OAc	*O*-acetyl
ORF	Open reading frame
PAPS	3′-Phosphoadenosine-5′-phosphosulfate
PAS	4-Aminosalicylic acid
PCR	Polymerase chain reaction
PEP	Phosphoenolpyruvate
QS	Quorum-sensing
Rha	Rhamnose
RNA	Ribonucleic acid
SDS	Sodium dodecyl sulfate
TCA	Tricarboxylic acid
TDP-rha	Thymidine diphosphate rhamnose
THF	Tetrahydrofolic acid
TPP	Thiamine pyrophosphate
UMP	Uridine monophosphate
UTP	Uridine triphosphate
UV	Ultraviolet
VP-test	Voges–Proskauer test
XDP	Ribonucleotide diphosphate

FOREWORD

All things are hidden, obscure and debatable if the cause
of the phenomena is unknown, but everything is clear if its cause be known.

Louis Pasteur

Bacterial Physiology and Biochemistry is an educational book designed for use in advanced bachelor's and master's courses in biology, including microbiology, biochemistry, and molecular biology. This book contains curriculum taught to biology students specializing in microbiology. The knowledge of this subject is required within the above-mentioned specialization in partial and state exams. The content of the text and the way of organization are based on the relevant microbiology curriculum, corresponding to the requirements of the new educational system at the faculties of science. The necessary knowledge of biochemistry and organic chemistry and mastering the basics of general microbiology are assumed. In a clear form defined by the relevant curriculum, the text can also be used to prepare students of professional and teacher biology for the partial exam in the physiological part of general microbiology. This book provides the most current, authoritative, and relevant presentation of bacterial physiology and biochemistry and includes seven chapters about the subject of the book in general, bacterial chemical composition and functional cell structure, nutrition and growth, processes of cell differentiation, metabolism, and the influence of environmental factors.

The first chapter describes the subject of bacterial physiology and biochemistry, bacteria in the phylogeny of living organisms and their shape diversity and evolution. This chapter also presents methods for studying bacterial properties. Detailed information on the chemical composition and functional cell structure of bacteria is described in the second chapter. It includes the element and compound composition, and characterizes the bacterial nucleus, cytoplasm, plasma membrane, cell wall, flagella, pilus and fimbriae, bacterial capsule, endospores, and pigments.

An integral part of bacterial physiology is microbial nutrition and growth as well as basic sources of nutrition, including sources of carbon and nitrogen. Therefore, the third chapter of this book presents mineral nutrition, growth factors, energy sources, and transport of substances across the plasma membrane. Furthermore, the growth and multiplication of bacteria, their growth under static culture conditions, growth constants, and deviations from the normal growth curve are described. This important chapter concludes with the multiplication of bacterial populations under conditions of continuous (dynamic) cultivation, synchronous multiplication, and the bacterial cell cycle.

The text of the fourth chapter presents the process of cell differentiation, characteristics of differentiation processes, polar differentiation, differentiation of photosynthetic membranes in facultative phototrophic bacteria, and the formation of heterocysts in cyanobacteria under bound nitrogen deficiency.

Special attention is paid to bacterial metabolism, energy of biochemical reactions, hydrogen carriers, the role of ATP in cells, and types of phosphorylation. The chapter describing metabolic processes is divided according to two metabolic pathways into catabolism and anabolism sections. The processes of catabolism of carbon compounds and fermentation of ethanol, lactic acid, pentose sugars, propionic acid, butyric acid with solvent formation, mixed acids, and sugars and the fermentation pathway of polysaccharides are described in detail. Then anaerobic respiration is shown, including nitrate reduction and denitrification, sulfate reduction, and reduction of carbon dioxide to methane. The process of aerobic respiration in chemolithotrophic bacteria, including oxidation of ammonia, reduced sulfur and iron compounds, hydrogen, and methane, is represented in sections of this chapter. Aerobic respiration in chemoorganotrophic bacteria is also described, especially incomplete oxidation of alcohols and glucose as well as complete oxidation of substrates. Nitrogen catabolism, protein, and amino acid dissimilation are characterized. Anaerobic degradation and aerobic metabolism of amino acids and heterocyclic compounds as well as fermentation and oxidation of heterocyclic compounds complete the section about bacterial catabolism. Special attention is paid to the processes of anabolism, including carbohydrate and lipid biosynthesis, carbon dioxide consumption by heterotrophs, and molecular nitrogen fixation. The biosynthesis of amino acids, nucleotides, nucleic acids, and proteins is shown. In addition, the regulation of the metabolic process and the metabolism of phototrophic bacteria are summarized in this chapter, which is one of the largest chapters of the book.

Physiological and biochemical processes depend on the environment. Therefore, the next chapter characterizes microbial growth in nature, bacterial populations as part of the ecosystem, and their physiological role in ecosystems. In addition, intercellular and internal population interactions, quorum-sensing regulation of gene expression, luminescent bacteria, and bioluminescence are presented. The last chapter of this book describes the influence of environmental factors on bacteria, including physical and chemical factors, and chemotherapeutics. Antibiotics and their mechanisms of action are also demonstrated. This book concludes with the description of antibiotics that inhibit cell wall synthesis, disrupt the plasma membrane, and affect nucleic acid metabolism, protein synthesis, and the phosphorylation process.

Prof. Lorenzo Drago, Ph.D.,
Professor of Clinical Microbiology,
Department of Biomedical Sciences for Health,
Faculty of Medicine and Surgery at University of Milan, Italy

CHAPTER 1

INTRODUCTION INTO BACTERIAL PHYSIOLOGY AND BIOCHEMISTRY

In this chapter, the studies of bacterial physiology and biochemistry and their main goals are characterized. Bacteria in the phylogeny of living organisms and diversity of cell shapes are described. Special attention is paid to the comparison of cell structures and their metabolic properties in different organisms, including bacteria, archaea, and eukaryotes. Moreover, main features of bacterial evolution and methods of modern bacterial physiology and biochemistry are presented.

1.1. Subject of study

The subject of bacterial physiology is the study of the functions of bacterial cells, that is, the study of all physical, chemical, and biological processes occurring in the cell as well as physical, chemical, and biological transformations caused by bacteria in the environment during their development. Such studies are impossible without familiarizing with the organization of morphology and functional structures of the microbial cell. For performing biological functions, bacterial cells are differentiated into architectural and functional structures.

An important part of bacterial physiology is the study of the chemical composition of bacterial cells, which, albeit similar to the chemical composition of the cells of higher organisms, has certain peculiarities. These features are related to the ability of bacteria to adapt to the environment. Due to such adaptation, depending on the environment, different variants of the same microorganism by morphological and physiological characteristics can be observed. The process of adaptation is explained not only by the existence of various types of bacterial metabolism, but also by the transformation of the saprophytic state into parasitic and the appearance of pathogenic microorganisms.

The following two types of living organisms are distinguished by the cell structure:
- **prokaryotic** cells (bacterial and archaeal cells that do not have a formed nucleus);
- **eukaryotic** (cells with the nucleus).

Bacterial Physiology and Biochemistry
http://doi.org/10.1016/B978-0-443-18738-4.50001-2,

Prokaryotic and eukaryotic cells differ in structural organization. Prokaryotes, unlike eukaryotes, do not have membrane-bound compartments called organelles that perform specialized functions. As mentioned above, prokaryotic cells do not have a formed nucleus, and hereditary information (DNA) in prokaryotes is not separated from other components of the cell. This feature is the most important difference between prokaryotic and eukaryotic cells. A comparison of the cellular structure of eukaryotic and prokaryotic cells is presented in **Table 1.1.**

Table 1.1. Comparison of bacteria, archaea, and eukaryotes

Properties	Bacteria	Archaea	Eukaryotes
Nucleus with a nuclear membrane around the DNA	Absent	Absent	Present
Complex of internal membrane organelles	Absent	Absent	Present
Cell wall	Have peptidoglycan, muramic acid	Do not have muramic acid	Do not have muramic acid
Membrane lipids	Esterified, not branched	Esterified, branched, aliphatic chains	Esterified, not branched
Gas vesicles	Present	Present	Absent
RNA transfer	Contains thymine, N-formylmethionine	Does not contain thymine	Contains thymine, N-formylmethionine
Polycistronic mRNA	Present	Present	Absent
Introns of the mRNA	Absent	Absent	Present
Splicing of the mRNA	Absent	Absent	Present
Ribosomes: Size Sensitivity to chloramphenicol and kanamycin	70S Positive	70S Negative	80S Negative
DNA-dependent RNA polymerase Number of enzymes Structure Sensitivity to rifampicin	One Simple subunit (4 subunits) Positive	Few Complicated Subordinate structure (8–12 subunits) Negative	Three Complicated Subordinate structure (12–14 subunits) Negative
Polymerase II promoters	Absent	Present	Present
Metabolism: Similar ATPase Methanogenesis Fixation of nitrogen Photosynthesis with participation of chlorophyll Chemolithotrophs	Absent Absent Present Present Present	Present Present Present Absent Present	Present Absent Absent Present Absent

The comparison of similar properties of organisms, regardless of the presence of the nucleus, is presented in **Fig. 1.1**.

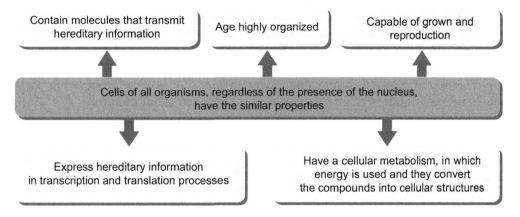

Fig. 1.1. Comparison of similar properties of organisms.

Thus, prokaryotic and eukaryotic cells differ in structural organization, although the cells of all organisms, regardless of the presence of the nucleus, have similar properties. So, bacterial physiology studies the chemical composition of bacterial cells, physical and biological processes occurring in the cells, and transformations of different metabolic compounds.

1.2. Bacteria in the phylogeny of living organisms and diversity of cell shapes

Microbiology is a branch of science that covers the study of viruses, archaea, bacteria, fungi, algae, and protozoa. There are no significant differences in the intracellular structure of bacteria and archaea. There are fundamental biochemical differences between them, which reflects their evolutionary origin. About 10 years ago, most scientists believed that evolution was in two ways: one of them led to the formation of prokaryotic cells (bacteria), and the other to the emergence of eukaryotic cells. The terms "bacteria" and "prokaryotes" were considered synonymous. These views have undergone radical changes in the 80s of the last century, when Carl Richard Woese, a molecular biologist, began analyzing information molecules that directly reflect hereditary cell information.

The analysis of ribosomal RNA showed that there are three principal lines of evolution that form three separate domains of cell evolution:
- bacterial cells;
- archaeal cells;
- eukaryotic cells (mushroom, algae, protozoa, plants, animals).

The conducted studies allowed determining:
- organelles of eukaryotes;
- cells involved in the formation of energy (mitochondria and chloroplasts) come from prokaryotic cells that have lost their ability to live independently;
- organisms evolved in three different ways from one common predecessor, which led to the formation of a large variety of microorganisms, plants, and animals that exist today.

Due to the received information, a new system of classification of living organisms was created. It is based on the analysis of rRNA macromolecules.

It is assumed that there was a certain common ancestor "progenote," which gave rise to the three branches of the evolutionary tree (**Fig. 1.2**). How it was, it is unknown.

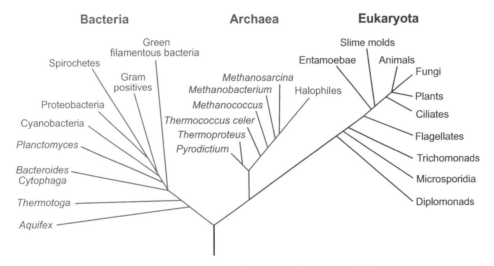

Fig. 1.2. Phylogenetic tree of life by Carl Woese *et al.* (1990).

Archaea are highly specialized prokaryotic organisms, and although they are similar to bacteria in the structural organization, they have a number of fundamental differences. The most important feature of the archaea is the specificity of their ribosomal and transport RNA; their ribosomes differ in shape. Differences were also found in other components of the protein synthesis system. Archaea do not have fatty acids and polyhydric alcohols as part of membrane lipids and usually have from 20 to 40 carbon atoms. The lipid layer of the membrane is formed by a monomolecular layer, which, obviously, gives its strength. Externally, archaea often have surface layers formed in a certain way by structured and regularly packed protein or glycoprotein molecules of the correct and sometimes strange form. The structure of the cell wall of the archaea may include peptides and polysaccharides. Some archaea are characterized by processes that are not intrinsic to other organisms. For example, some representatives of this group of prokaryotes form methane (methanogens) in their process of life. Most archaeas are extremophiles, that is, they develop under extreme conditions, at high temperatures (+90 °C) or in saturated saline solutions. Acidophilic archaea grow in the environment, where the pH is as low as in the concentrated sulfuric acid. There are autotrophic forms of archaea that do not require organic food but are satisfied with the energy obtained through oxidation–reduction reactions, with the involvement of inorganic molecules.

Bacteria that differ in their morphological and physiological properties are an extremely diverse group of prokaryotic microorganisms. They can be spherical, cylindrical, spiral, and pleomorphic (**Fig. 1.3**).

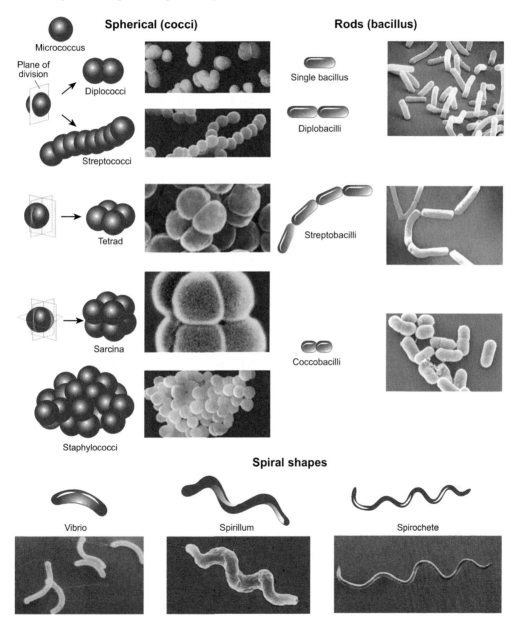

Fig. 1.3. Basic shapes of bacterial cells: spherical (coccus, cocci), rods (bacteria, bacillus), and spirals.

Size, shape, and arrangement of bacterial cells (**Table 1.2**):
- the size, shape, and arrangement of bacterial cells depend on the species;
- even in pure culture, individual cells of the population are not present in the same phase of the cell cycle at the same time;
- therefore, the size (and sometimes the shape) varies;
- for bacterial cells, there is a considerable variation in size due to species affinity and the physiological state of the cell and the external environment.

Table 1.2. The size of some bacterial cells

Organism	Size (μm)	Organism	Size (μm)	Organism	Size (μm)
Chlamydium	About 0.3	Streptococcus	1.0–1.5	Clostridium	0.5–0.8×3.0–8.0
Bdellovibrio	0.3–0.5×0.5–1.4	Sarcina	4.0–4.5	Shigella	0.4–0.6×1.0–3.0
Rickettsia	0.1–0.3×1–4	Azotobacter	2–3×3–6	Spirochaeta	0.2–7×5–500
Staphylococcus	0.8–1.0	Proteus	0.5–1×1–3	Spirillum	0.5×3
Escherichia coli	0.5–1×1–3	Micrococcus	About 1.0	Corynebacterium	0.32–0.8×1.0–8
Bacillus subtilis	0.9–1.1×1.8–4.8	Pseudomonas	0.3–0.5×1.0–1.8	Treponema	0.2×4–14
Mycobacterium	0.3–0.5×1–4	Acetobacter	0.4–0.8×1.0–2.0	Corynebacterium	0.3–0.8×1.0–8.0
Chromatium	3–4×5–6	Lactobacillus	0.7–1.0×3.0–8.0	Brucella	0.3–0.5×0.5–2

Some bacteria are capable of photosynthesis and use the energy of sunlight and CO_2 to accumulate biomass. Among them, there are cyanobacteria, which were called blue-green algae. Another group of bacteria obtains energy by metabolizing such inorganic compounds as ammonium and sulfur. Bacteria can decompose different organic compounds, from glucose to hydrocarbon oil. There are bacteria that can grow only under specific conditions, for example, in human tissues, where they can cause diseases. Purple nonsulfur bacteria can grow both as anaerobes (photoorganoheterotrophs) and aerobes (chemoorganoheterotrophs). They are found in water with a high content of organic matter and low sulfur content. Intracellular parasites (rickettsiae) can use nutrient substances of the host, its ATP, and coenzymes. Representatives of many genera of bacteria, in particular *Rhizobium*, are capable of fixing nitrogen, while *Agrobacterium* causes tumor growth in plants.

Chemolithotrophic bacteria obtain energy by restoring inorganic compounds, and nitrifying bacteria can oxidize ammonia and nitrites. Colorless sulfur bacteria *Thiobacillus* can oxidize sulfur, hydrogen sulfide, and thiosulfates to sulfates. Photolithotrophs (purple sulfur bacteria) can oxidize hydrogen sulfide to sulfur and accumulate it in their cells. Methylotrophic bacteria use methane, methanol, and other monocarbon compounds as the only source of carbon and energy. Representatives of the *Pseudomonas* genus play an important role in the processes of mineralization (and at the same time cause diseases). Various diseases are caused by enterobacteria, which are widely used in experimental studies. The species of myxobacteria, which forms fruiting bodies, moves along the surface in search of food, and excretes enzymes that can cause lysis of bacteria and yeasts, has a complex life cycle.

Representatives of the *Clostridium* genus, which form endospores, cause food poisoning (botulism), gangrene, and tetanus. Staphylococci are important pathogens. At the same time, species of the *Lactobacillus* genus, capable of lactic fermentation, are widely used in the food industry. The *Propionibacteria* genus and streptomyces, capable of degrading many organic compounds and producing antibiotics, are of industrial significance.

Thus, despite the fact that prokaryotes and eukaryotes are very different in structure, the metabolisms of these two groups of organisms are similar. Photosynthetic eukaryotes are algae (unicellular or multicellular), which can be organized into complex structures. Based on the study of the structure and organization, blue-green algae were considered as cyanobacteria, because they do not have a nucleus. Microbial physiology covers the research of thousands of different microorganisms. On examples of the best-investigated bacteria, the basic principles of their life will be discussed.

1.3. Bacterial evolution

About 3.8 billion years ago, the Earth's atmosphere contained hydrogen, methane, ammonia, and water but did not contain CO_2, oxygen, and organic compounds. In the 50s of the last century, S. Miller and G. Urei have shown that under the influence of electrical discharges, organic compounds, in particular, sugars, amino acids, and nucleotides, can be formed in such an atmosphere.

In the early decades of the 20th century, Aleksandr Oparin (in 1924) and John Haldane (in 1929, before Oparin's first book was translated into English) independently suggested that if the primitive atmosphere was reducing (as opposed to oxygen-rich), and if there was an appropriate supply of energy, such as lightning or ultraviolet light, then a wide range of organic compounds might be synthesized (**Fig. 1.4**). These compounds could provide energy, growth, and reproduction necessary for life. An alternative is the hypothesis about the origin of micelles in oceanic thermal waters. Although it is not known how the first organisms evolved, experiments show that in conditions similar to the atmospheric conditions of the Earth at that time, micelles resembling cells are formed in a mixture of organic compounds. The micelles are separated from the environment, and within the micelles, there may be some reactions between organic compounds, which could lead to the formation of cells. These first microorganisms were supposed to be resistant to external factors: high temperature, acid pH, and high salt concentration. They had to be anaerobic, because they had no oxygen; they had to use organic compounds for energy, growth, and development. According to these features, primitive organisms are reminiscent of microorganisms, which are now classified as archaea. They are physiologically highly specialized and can grow under adverse conditions (temperature, acidity). The gradual change in the conditions on Earth led to the emergence of new organisms, called autotrophs (using CO_2 as a carbon source, capable of synthesizing complex organic compounds from CO_2). Primary anaerobic microorganisms developed mechanisms for the capture and use of chemical and solar energy for biosynthetic purposes. In the form in which these processes are observed on Earth now, they received the name of chemosynthesis and photosynthesis. It is believed that the first photosynthetic organisms on Earth were the ancestors of cyanobacteria (blue-green algae). Photoautotrophs are capable of using light as an energy source for the synthesis of organic compounds. The synthesized organic compounds could be a carbon source for other microorganisms called heterotrophs. The first photosynthetic bacteria were obligatory anaerobes. Later, there were new photosynthetic microorganisms that could form oxygen from water. It is believed that it was 2 billion years ago. The formation of oxygen in the process of photosynthesis had a huge impact on evolutionary development. Due to aerobic breathing, the metabolism of organic compounds became possible.

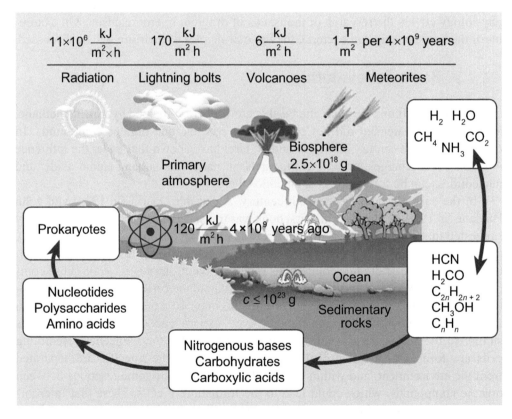

Fig. 1.4. The scheme of life's origin based on Oparin–Haldane hypothesis.

Alberts and Watson *et al.* believe that the emergence of anaerobic prokaryotes in the conditions of the depletion of nutrients from the environment and the ability to eat one another (phagocytosis) was another important stage in the development of life on Earth. According to these authors, while some organisms used phagocytosis, other anaerobic prokaryotes could enter into symbiosis with their aerobic cells absorbed and create a mutually beneficial association. Thus, phagocyted aerobic cells could not be used as a power source but left inside the host cell for a more complete and effective oxidation of organic matter.

This could give significant advantages to host cells in the struggle for existence. During the evolutionary development of new microorganisms, which adapted to new conditions, this process continued.

The idea of symbiosis, that is, the formation of a mutually beneficial association of two or more organisms, arose in the second half of the 19th century. In 1867, A. S. Famincin published that lichens are a mutually beneficial association of mushrooms and algae. The name "symbiosis" was proposed by De Bari. In 1907, based on his research as well as on the work of A. F. Schimper (Germany), who showed the ability of chloroplasts to self-replant in plant cells, Famincin suggested that chloroplasts could be single-celled algae and symbionts of plants. This idea was supported and developed by K. S. Merezhkovsky and B. M. Kozopolyansky in the 20s of the 20th century. At the same time, it was suggested that not only chloroplasts but also mitochondria are symbionts, since both are capable of self-replication in cells of higher plants. However, this hypothesis was not mentioned as too extravagant for a long time. Only in the 50s and 60s of the last century, biochemists received data on the content of plants and mitochondria in chloroplasts. These organelles isolated from different organisms. Mitochondrial DNA that differed from the DNA of nuclei was similar to the DNA of prokaryotes (in particular, it had a circular structure). In addition, the presence of prokaryotic ribosomes in chloroplasts of plants was established by N. M. Sisakyan and other scientific groups. The process of protein biosynthesis in chloroplasts and mitochondria on many grounds resembled the synthesis of protein in bacterial cells.

Chloroplasts and mitochondria appeared to be similar to bacterial cells by other biochemical features, in particular, by the presence of a specifically constructed phospholipid named cardiolipin in their membranes, which is characteristic only of bacterial membranes and absent in mammalian eukaryotes (**Fig. 1.5**). However, there were many arguments against this hypothesis. The main of them is the weak autonomy of life, first of all, the process of biosynthesis of proteins, in chloroplasts and in mitochondria. It turned out that in these organelles, a small amount of proteins is synthesized, first of all, primarily necessary for the formation and functioning of these organelles.

The next step in the development of the hypothesis about the endosymbiotic origin of eukaryotic organisms was the work of Lynn Margulis, an American researcher. It discovered a very large immunological, that is, structural, proximity of the flagella and some elements of the cytoskeleton of eukaryotic cells with spirochetes, which are bacteria that have a spiral form of cells. This enabled L. Margulis to suggest that the cytoskeleton and eukaryotic cell may have originated from spirochete-like prokaryotes.

Recently, it was shown that ATPase (an enzyme that utilizes ATP), isolated from yeast vacuoles, is structurally distinct from ATPases of other organelles of yeast and practically corresponds to ATPase of some *Archaea*. On this basis, it was assumed that the precursors of the vacuoles of lower eukaryotes, for example yeasts, are ancient prokaryotes belonging to the archaea and entered into a symbiotic relationship with the host cell. Recently, it has been shown that another enzyme, polyphosphatase (utilizing inorganic polyphosphates), isolated from various organelles of yeasts (nuclei, mitochondria, cellular membranes, cytosols, and vacuoles) has completely different characteristics and properties. These data also indicate that many intracellular structures of eukaryotic organisms have different origins.

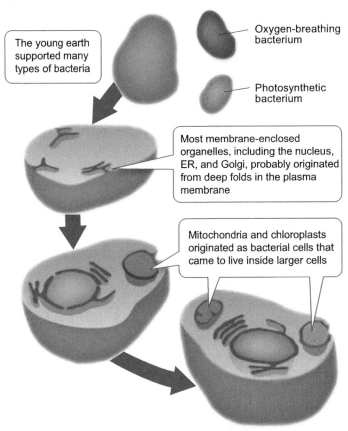

Fig. 1.5. Hypothesis of chloroplast and mitochondria origin from photosynthetic and oxygen-breathing bacteria.

The most significant evidence of the origin of chloroplasts and mitochondria of eukaryotic cells from prokaryotes was obtained in recent years in the study of the structure of one of the most conservative components of living cells — ribosomal RNA. The prominent American molecular biologist Carl Richard Woese and his collaborators have shown that the structure of the 16S RNA from ribosomes of plant chloroplasts is not similar to the structure of the corresponding 18S RNA from the cytoplasmic ribosomes of plants and is almost identical to the structure of similar RNA from some cyanobacteria. At the same time, this group of researchers has found that the ribosomal 16S RNA derived from mitochondria of various eukaryotic organisms, by their structure, are not similar to the ribosomal 16S RNA from the cytosol of the same biological objects but are extremely similar to the 16S RNA of some bacteria, in particular, of the *Paracoccus* genus. All these data clearly indicate in favor of the high probability of the fact that eukaryotic cells have endosymbiotic origin. However, the problem of establishing the nature of the host cell remains unsolved. According to L. Margulis, host cells could be prokaryotes-precursors of modern bacteria, which belong to mycoplasmas (primitive constructed prokaryotes, which practically do not have a cell wall).

In recent years, based on the study of the structure of the ribosomal 16S RNA of many prokaryotes and eukaryotes, T. Oshima, a Japanese researcher, concluded that the host cell was a representative of the kingdom of *Archaea*. Incidentally, in contrast to bacteria, representatives of *Archaea* have many biochemical and molecular genetic properties common with those of eukaryotes.

1.4. Methods of studies of bacterial properties

Different types of microscopy and cultivation of microorganisms are traditional methods that made it possible to investigate their structural and physiological characteristics, in particular, metabolism and growth under different conditions, and to detect the extreme heterogeneity of microorganisms. The use of new types of microscopy (confocal, transmissive, scanning), new methods for obtaining pure cultures of microorganisms, immunological methods, and methods of classical genetics provided much better disclosure of the basic properties of microorganisms and possibilities of the practical use of microbial technologies.

Currently, microbial physiology is undergoing a dramatic revolution. Huge progress in this area is due to the following technological advances:
- personal computers;
- Internet;
- development of DNA sequencing techniques.

In the recent past, the study of the physiology of microorganisms was a long, difficult, and ungrateful process. Even 60 years ago, when the full analysis of *Escherichia coli* was launched, there was still no complete integrated picture of its biochemistry and genetics. Today, we know in detail the biochemistry and genetics of some microorganisms, which are of special interest to humans. Examples are *Rickettsia prowazekii* and *Mycobacterium leprae*, which cannot exist outside the living organism. Currently, we know what special biochemical ways they have, how they obtain energy, which protein factors determine their virulence, and what their place in the evolutionary tree of bacteria is. All this became known thanks to modern sequencing techniques that allowed looking into the genome and decoding it in a relatively short time. An open reading frame (ORF) was identified, which encodes proteins and promoter sequences, and these data were compared with those already known. A comparison of amino acid sequences ORF of unknown function with known sequences (homologous search) made it possible to identify proteins with similar sequences and motifs. Such an analysis provides the prediction of the possible functions of ORF in a cell without conducting biochemical experiments.

Data on the sequence of genes are from EMBL (EU) and GenBank (US). These data can be found on the web pages of the National Institute of Health (http://www.ncbi.nlm.nih.gov) or the European Institute of Bioinformatics (http://www.ebi.ac.uk/Tools/index.html). With programs such as BLAST (basic local alignment search tool) or FASTA, it is possible to compare DNA or amino acid sequences with known ones.

Personal computers and Internet allow you to find this information on websites around the world (**Table 1.3**).

Table 1.3. Genomics and proteomics websites

Website name	Information
http://web.bham.ac.uk/bcm4ght6/genome.html	Gene database of *E. coli*
http://cgsc.biology.yale.edu/	Genetic center of *E. coli*
http://www.ucalgary.ca/%7Ekesander/	Genetic center of *Salmonella*
http://genome.wustl.edu/gsc/bacterial/salmonella.shtml	Genome of *Salmonella*
http://ecocyc.PangeaSystems.com:1555/server.html	Different biochemical paths in *E. coli*
http://www.genome.ad.jp/kegg/	Biochemical pathways of many sequenced organisms
http://bomi.ou.edu/faculty/tconway/global.html	Functional genomics of *E. coli*
http://susi.bio.uni-giessen.de/ecdc/ecdc.html	Database of *E. coli*
http://www2.ebi.ac.uk/fasta33/	Homologues search
http://www.motif.genome.ad.jp/	Search for motifs in proteins
http://www.expasy.ch/cgi-bin/map1	Structures of two-dimensional electrophoresis
http://www2.ebi.ac.uk/clustalw/	Nucleotide and amino acid sequences
http://www.ncbi.nlm.nih.gov/PubMed/	Search for literature in PubMed
http://bmbsgi11.leeds.ac.uk/bmbknd/DNA/ genomic.htm	Genome databases
http://www.biology.ucsd.edu/~msaier/transport/	Information about the transport system

The first complete sequence of the bacterial genome, *Haemophilus influenzae*, was published in 1995. In 2002, more than 50 genomes of microorganisms were screened. Much more of them are in the process of sequencing.

Computer analysis enables to determine the properties of proteins encoded in a certain DNA sequence, their isoelectric points, molecular weights, three-dimensional structures, etc. as well as the potential regulatory regions; promoters can be determined in the DNA sequence.

Particularly complete information can be obtained from KEGG (Kyoto Encyclopedia of Genes and Genomes) at the website: http://www.genome.ad.jp/kegg/. In this encyclopedia, the biochemical systems of some microorganisms are graphically illustrated and genomes of some of them are sequenced. The name of the microorganism and the complete information about the presence of its metabolic pathway, enzymes of this path, the DNA sequence, and proteins of the desired path can be found on "Metabolic paths." This example shows that due to the extensive development of genomics and the Internet, microbial physiology has been completely changed. However, the greatest misconception would be that the homology found guarantees the functioning of this protein or the given regulatory site in the microorganism under investigation. Gene functions must be confirmed by genetic and biochemical methods.

A classic approach to determining the biochemical pathway is to use a selective phenotype or to get a mutant that damages a certain stage of the metabolic pathway using a selective medium. If such a mutant is received, it is possible to map the gene, identify, clone, and sequence. However, it is not always easy to get such a mutant; it is necessary to analyze hundreds of thousands of colonies. If there is a positive selection method (after mutagenization, only mutant clones grow on the medium), it is easier to do so. If selection is negative (the mutant does not grow on the selective medium), then it is necessary to make impressions on an environment that does not contain this growth factor and to select one colony with the desired phenotype from one thousand colonies.

In recent years, the selection of mutants has been greatly facilitated by the development of the method of fluorescence-activated cell sorting (FACS). For this purpose, a gene encoding a green fluorescent protein fused to a gene that is not expressed under certain conditions (e.g., at low pH 4.5) and cannot form a colony is used. After mutagenization of the fusion strain, the mutants are cultivated at high and then changed pH and passed through a device that selects fluorescing and nonfluorescing cells. This approach has greatly improved the work on selecting mutants.

Cloning and DNA analysis. The process of molecular cloning (selective accumulation) of DNA molecules involves several steps:
- fragmentation of DNA by processing endonuclease restriction;
- combining these fragments *in vitro* with a vector DNA molecule (capable of autonomous replication);
- introduction of the vector into the recipient organism, in which recombinant DNA is accumulated.

Genetic engineering was born in 1973, when the first recombinant DNA was created *in vitro* by P. Berg's group in the United States.

Finding out how the cell coordinates biochemical reactions is an important step toward understanding of the concept of life. Unfortunately, a universal method for studying all biochemical, structural, and physiological features of the cell does not exist yet. However, today there are techniques that are most suitable for this purpose. These include gene arrays and two-dimensional division of cellular proteins. These methods bring us closer to understanding how one system in a cell can determine the synthesis of all other systems.

The technology of gene arrays makes it possible to consider the expression of genes at the level of the complete genome. The expression of each gene can be analyzed many times due to the use of microchips in DNA.

The basic idea of this method is as follows: DNA sequences of each gene are placed on a solid surface to allow rapid hybridization with a fluorescently labeled DNA pool (or cDNA). DNA sequences isolated from individual genes are placed on the chip as separate grid spots (**Fig. 1.6**). The square of the area is 1–2 square inches (2.5 cm^2). Further, RNA is rapidly extracted from culture, transferred to complementary DNA (cDNA) using inverse transcriptase (RNA-dependent DNA polymerase). Due to the inclusion of fluorescently labeled nucleotides in the incubation mixture, each cDNA molecule is tagged. It binds to specific loci on the chip, which contains a separate isolated gene.

Laser scanning and fluorescence measurements allow the chip surface to be read out, and computer analysis detects genes that are expressed in original samples taken. By applying such an analysis, one can compare the level of expression of individual genes isolated from cultures grown under different conditions. The cDNA drugs are labeled with different fluorescence labels. If the gene is expressed only under certain conditions, the corresponding spot will be fluoresced in one color; if under other conditions, the spot will have another color. If the gene is expressed in both conditions, the color becomes mixed. Such an analysis can be used to study the effects of conditions of cultivation, stress factors, and mutations on gene expression.

Proteomic analysis. Unlike a genome that has a constant composition, proteins can vary depending on the internal and external environment. Under the influence of stress factors, the level of expression of some proteins is reduced and of other proteins increases. Gene array technology makes it possible to analyze changes in the protein composition caused by the transcription level, and proteomic analysis takes into account post-transcription and post-translational changes.

For proteomic analysis, two-dimensional electrophoresis in polyacrylamide gel is used (**Fig. 1.7**). The cell-free extract containing a mixture of 1000–2000 different proteins is initially subjected to isoelectric focusing; it is a process that allows the proteins to be separated by differences in their isoelectric points. Each protein contains carboxyl groups and amino groups, which are differently protonated and have charge depending on pH. For certain pH values, they are separated. However, identical isoelectric points can have multiple proteins, so electrophoresis should be used in polyacrylamide gel. Proteins move depending on their molecular weight, faster with a lower molecular weight.

Proteins can be visualized by a radioactive label or dye (e.g., Kumasi). Then, computer analysis should be used to compare the proteins. A protein selected for further analysis can be analyzed in detail. For this purpose, genomic and proteomic analysis is used. If a gene of a microorganism is sequenced, the protein is cut and subjected to mass spectrometry. To do this, it is cut by proteases (trypsin) into fragments,

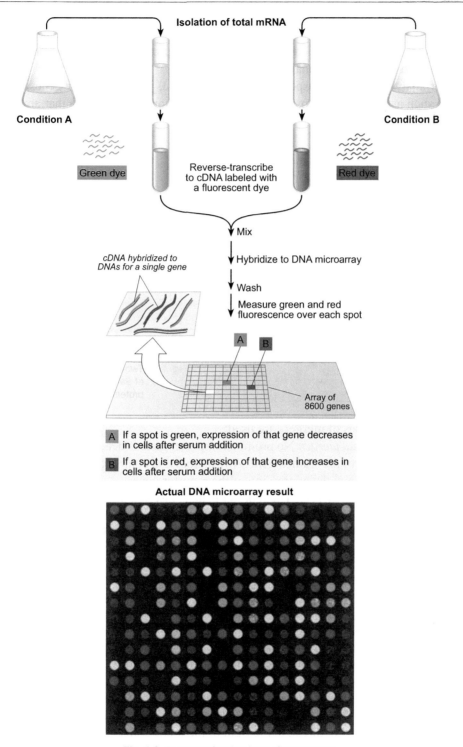

Isolation of total mRNA

Condition A

Condition B

Green dye

Red dye

Reverse-transcribe
to cDNA labeled with
a fluorescent dye

Mix

Hybridize to DNA microarray

Wash

Measure green and red
fluorescence over each spot

*cDNA hybridized to
DNAs for a single gene*

A   B

Array of
8600 genes

A If a spot is green, expression of that gene decreases
in cells after serum addition

B If a spot is red, expression of that gene increases in
cells after serum addition

Actual DNA microarray result

Fig. 1.6. Scheme of technology of gene arrays.

the exact weight is determined, and the fragments are compared with similar ones of the already known protein. If the genome is not sequenced, then the amino acid analysis of the N-terminus is carried out, or it is split by proteases into fragments, which are separated by HPLC and analyzed.

For analysis of various aspects of gene expression, joining reporter genes to promoters such as *lacZ* (β-galactosidase) or *gfp* (green fluorescent protein) that are easily identifiable is very successful.

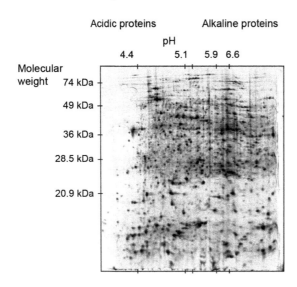

Fig. 1.7. Two-dimensional electrophoresis of *Salmonella enterica* proteins.

Usually, two types of mergers are used:
1 *Transcriptional merger.* No promoter–reporter gene is detected by regulation of the promoter of the studied gene. Any factors that control the expression of this gene will change the expression of the reporter gene.
2. *Gene or protein merger.* The reporter gene is inserted into the target gene, so as both promoters and both ribosome binding sites remain. In this case, all factors that control the target gene at the transcription and translation levels will control the synthesis of β-galactosidase.

Polymerase chain reaction (PCR). DNA obtained in large amount was a daunting problem. In the late 80s of the 20th century, Kary Mullis from California put forward an idea that brought him the Nobel Prize in chemistry (1993). PCR allows obtaining 10^6 DNA amplifications. Using oligonucleotide primers (20–30 bp) to DNA and thermostable DNA polymerase, the target DNA can be replicated (**Fig. 1.8**). The initial DNA molecule is warmed up to destroy the hydrogen bonds that connect the complementary strands. The reaction mixture is then cooled in the presence of two short (12–20 pairs of nucleotides) of DNA fragments, one of which is complementary to the DNA region to the left of the locus to be examined, and the second is the complementary region of another thread to the right of the locus under study. These fragments are called primers that bind to the corresponding DNA regions. The primers linked

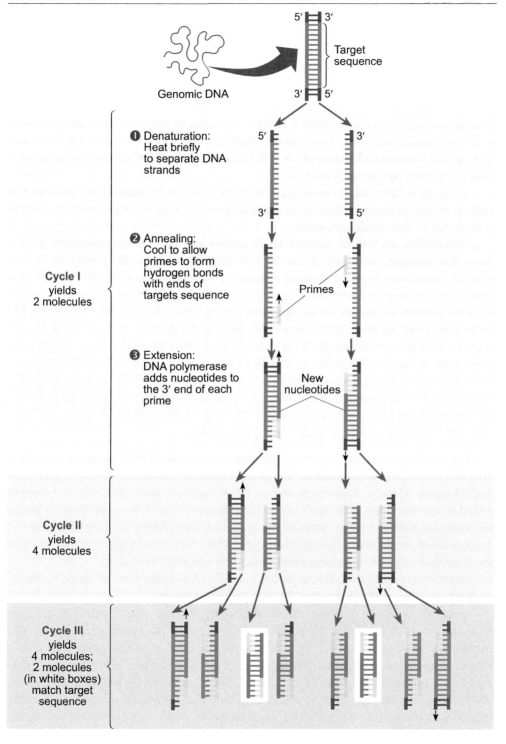

Fig. 1.8. Scheme of polymerase chain reaction.

to the DNA strand define the starting point for the synthesis of a new complementary strand on the DNA matrix. This synthesis is performed by DNA polymerase. In the next cycle, the mixture with the resulting strands of DNA is warmed up again, and the newly synthesized strands of DNA are used as matrices. A new portion of the primers is connected to the corresponding sites, and the synthesis cycle repeats. In the PCR, the synthesis of a fragment is placed between two primers. For 30 cycles, the number of synthesized fragments is about 1 billion. Warming of the mixture is carried out at 95 °C and cooling at 65 °C. These temperatures are too high for the DNA polymerase, so a special enzyme called Taq-DNA polymerase isolated from the thermophilic bacterium *Thermus aquaticus* is used.

These PCR bases can be modified. There can be specially constructed primers that contain certain restriction sites to facilitate cloning, or highly specific primers represented only in one species are used.

PCR techniques can be used to detect certain species of microorganisms in products, for example, pathogen. It can be also used for site-specific mutagenesis. For example, computer analysis showed a potential functional motive (Zn^{2+}-finger protein). The first step in determining if this motive is important for the functioning of the gene product should be the replacement of one amino acid codon by site-specific mutagenesis and the study of the mutant phenotype. To do this, four primers are used. Two external primers are amplified within the gene. Two internal primers overlap the target and are used to obtain the desired mutation. There must be three PCRs. The first two reactions amplify each end of the gene and include a mutation. To reconstruct the complete gene, the products of the initial reactions are mixed and denatured at 95 °C, and the dual desired structure is restored. PCR with external primers restores the complete structure of the gene from site-directed mutation.

Displacement of DNA mobility. An important issue after the regulatory protein allocation is the question whether this protein directly binds to DNA. For this purified protein is added to the corresponding DNA fragment and check whether there is a shift of the electrophoretic mobility of the fragment (**Fig. 1.9**). If the protein binds, the formed complex is larger and will move more slowly. In the mixture of proteins, it is necessary to apply antibodies to this protein. The complex of a specific antibody and the DNA-binding protein is larger and causes the so-called supershift.

Determination of DNA, RNA, proteins, and DNA-binding proteins using southern, northern, western, and southwestern blots is described below.

Genomic *southern blot analysis* is used to map chromosomal rearrangements on a physical map, to determine the position of the gene, and to identify repetitions. The method was discovered in 1975 by E. Southern. The DNA molecule is cut with one or more endonuclease(s), separated by agarose gel electrophoresis, denatured, and transferred to a nitrocellulose filter (**Fig. 1.10**). DNA is captured by high temperature or ultraviolet radiation. The filter is placed in a hybridization buffer containing a probe labeled with a radioactive isotope or by a chemical method. This probe will bind to the complementary DNA region. After filter washing, a signal is detected using autoradiography or measuring the intensity of the fluorescence.

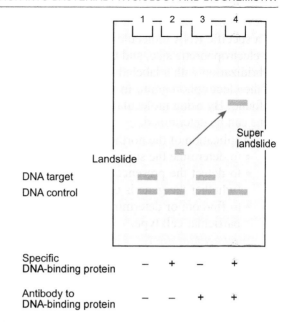

	1	2	3	4
Specific DNA-binding protein	−	+	−	+
Antibody to DNA-binding protein	−	−	+	+

Fig. 1.9. Displacement of electrophoretic mobility of a DNA fragment.

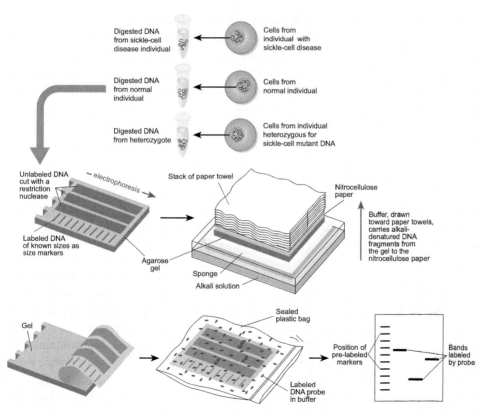

Fig. 1.10. The scheme of southern blot analysis.

Northern blot analysis is conducted to analyze the presence, size, and processing of a specific RNA molecule in a cell extract. RNA is extracted from cells, separated by electrophoretic size, and transferred to nylon or nitrocellulose filter capillary. After hybridization with a labeled single-circuit probe, signals are detected in a certain area of the electrophoregram, in which the homology of the RNA and DNA of the probe is found. By using molecular weight markers, the size of the transcript of a particular gene can be determined.

Application of the northern blot analysis:
- to determine the size of specific mRNAs that are encoded by a particular gene;
- to detect the presence of this type of mRNA cells read from a particular gene (whether the gene is expressed or not);
- to find out or determine the amount and the change in the content of RNA in a particular cell type.

Western blot analyses is used to determine specific proteins in the mixture. Protein extracts are separated electrophoretically, transferred from the gel onto the membrane with antibodies applied to the specific protein. Depending on the source of the antibody, the membrane can be labeled with secondary antibodies that specifically bind to primary antibodies. These secondary antibodies can be labeled, for example, by peroxidase. Then the addition of hydrogen peroxide and luminal will cause a fluorescence reaction where the antibodies are bound to the protein.

Southwestern blot is used to identify proteins that bind to specific DNA molecules. Cell proteins are electrophoretically separated in SDS-polyacrylamide gel and transferred to membranes with labeled oligonucleotides. Using autodiagnostics, it is determined which proteins are associated with the labeled DNA. A protein can be separated and identified in a variety of ways. Mixtures of proteins of different cultures are applied to separate wells of the SDS-polyacrylamide gel. Markers of known molecular weight are applied to the central band. After electrophoresis, the gel is stained with Coomassie brilliant blue or fluorescently labeled. To identify which of the proteins is present in each line, they are transferred to the nitrocellulose membrane. The membrane is treated with the primary antibodies to the desired protein. The binding sites are invisible, so the membrane is then treated with antibodies to the primary antibodies labeled by peroxidase. Peroxidase helps visualize the bound proteins after the treatment with hydrogen peroxide and luminous.

Two-hybrid analysis. The two-hybrid system is important for understanding the protein–protein interactions that occur in the cell. A classic example is the yeast two-hybrid system. The strategy involves the fusion of two genes potentially interacting with the isolated components of the yeast GAL4 transcription activator (**Fig. 1.11**). This is a fragment that contains a DNA binding domain and a fragment that contains a transcription activator. These components should interact with each other to induce expression of the yeast conjugated *GAL1-lacZ* gene (another reporter gene can be used). The *lacZ* gene encodes β-galactosidase, which can easily be determined. Normally, both of these domains are part of the *Gal4* protein. However, when they are separated, they can interact if each domain is merged with one of the pair of interacting proteins.

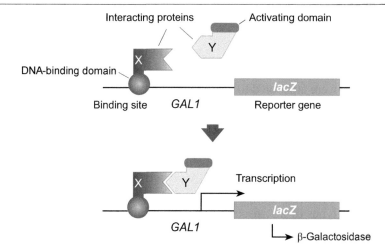

Fig. 1.11. Two-hybrid system for determining protein–protein interactions.

The generated two-hybrid *GAL4* complex binds and activates the transcription of *GAL1-lacZ*. The expression of *GAL1-lacZ* is visualized on agar plates containing the β-galactosidase chromophoric substrate X-gal. In this way, unknown proteins that interact with a known protein can be found. This known protein is merged with one of the *GAL4* domains, using it as a bait. The cells containing the bait are transformed with plasmids containing cloned target genes, fused to another *GAL4* domain. Then the same rendering procedure on the cups follows. The gene encoding the target protein may further be identified by sequencing.

Thus, there are many molecular approaches to the study of physiology of microorganisms that include genomic and proteomic strategies. The number of methods is much larger, but those listed here are sufficient to understand the need for their combination. In order to find out important issues of the physiology of microorganisms, it is necessary to use classical microbiological, biochemical, and genetic methods in combination with modern methods of molecular biology and genetics.

***

Summing up this chapter, it should be noted that the study of the physiology of bacteria is impossible without studying their biochemical properties and understanding the structure and composition of their components. Environmental factors that can influence the organization of bacterial morphology and functional structures are not less important. Cells of all organisms, regardless of the presence of the nucleus, have similar properties, but prokaryotic and eukaryotic cells differ in structural organization. Bacterial physiology studies the chemical composition of bacterial cells, physical and biological processes occurring in the cells, and transformations of their different metabolic compounds.

Bacterial cells are an extremely diverse group of prokaryotic microorganisms, which differ in their morphological and physiological properties. They can be spherical, cylindrical, spiral, or pleomorphic. Although prokaryotes and eukaryotes are very different in structure, the metabolism of these two groups of organisms is similar. The photosynthetic eukaryotes are algae (unicellular or multicellular), which can be organized into complex structures. Based on the study of the structure and organization, blue-green algae were attributed to cyanobacteria, because they do not have a nucleus. Microbial physiology covers the research of thousands of different microorganisms. From the examples of the best investigated of them, the basic principles of their life can be found. The most significant evidence of the origin of chloroplasts and mitochondria of eukaryotic cells from prokaryotes was obtained in the study of the structure of one of the most conservative components of living cells, ribosomal RNA, in recent years.

The traditional methodology in general microbiology, including bacterial physiology, is different types of microscopy and cultivation of microorganisms. This makes it possible to investigate their structural and physiological features, in particular, metabolism and growth under different conditions, and detect extreme heterogeneity of bacterial cells. The use of new types of microscopy (confocal, transmissive, scanning), new methods for obtaining pure cultures of microorganisms, immunological methods, and methods of classical genetics provided much better disclosure of the basic properties of microorganisms and the possibilities of the practical use of microbial technologies. A classic approach to determining the biochemical pathway is to use a selective phenotype or to get a mutant that damages a certain stage of the metabolic pathway using a selective medium. Additional methods applied for the studies of bacterial properties are cloning and DNA analysis, proteomic analysis, polymerase chain reaction, and displacement of DNA mobility. Southern, northern, western, and southwestern blots are used in bacterial physiology for determination of DNA, RNA, proteins, and DNA-binding proteins.

CHAPTER 2

BACTERIAL CHEMICAL COMPOSITION AND FUNCTIONAL CELL STRUCTURES

Bacterial cells are not different from other organisms by their chemical composition. In total, they contain the same elements and chemical compounds as cells of plants and animals. In addition, the proportion of these compounds is also similar, although it may differ between individual bacterial species or depend on the environment in which the species live.

In this chapter, the composition of elements and compounds in the microbial cells, bacterial nucleus (nucleoid), cytoplasm, cytoplasmic membrane, and cell wall is characterized. In addition, different microbial structures, including flagella, pilus, fimbria, capsule, endospores, and pigments, are described.

2.1. Elemental composition

A bacterial cell consists of elements that belong to a group of so-called **biogenic elements**, characteristic also for other living systems. They are usually divided into two groups: **macrobiogenic** and **microbiogenic**.

Macrobiogenic elements are as follows:

Carbon (C) usually makes up to 50% of cell's dry mass and is also the main component of cell's structural, reserve, and other organic compounds.

Nitrogen (N) represents about 8–15% of dry mass, depending on protein content. In organic compounds, it is present mainly in reduced form ($-NH_2$, $=NH$, $=N-$).

Oxygen (O) and **hydrogen (H)** are present mostly as water and in various functional groups of organic compounds. Oxygen level accounts for 20–30%, hydrogen level for 6–8% of dry mass.

Phosphorus (P) is included in the form of phosphate in nucleotides, coenzymes, phospholipids, and energy systems. It accounts for approximately 3% of dry mass.

Sulfur (S) is present especially in proteins and some coenzymes in an amount representing approximately 1% of dry mass. It can also temporarily exist in the elementary form in the cytoplasm of some bacteria (*Beggiatoa*).

Potassium (K), **sodium (Na)**, and **chlorine (Cl)** represent essential inorganic ions of the cell that are used in the activation of several coenzymes and in osmotic regulation. The amount of each of them is approximately 1%.

Calcium (Ca) and **magnesium (Mg)** are present in the cell in an amount of approximately 0.5% of dry mass. Calcium is a cofactor of enzymes, for example, proteinases; magnesium mediates bonding between enzymes and substrates and is a component of bacteriochlorophyll.

Iron (Fe) is a cofactor of multiple enzymes. It is present in the cell in an amount of around 0.2% of dry mass, mainly in the cytochromes.

In bacterial cell, there are also microbiogenic elements, which are also called trace elements: **cobalt (Co)**, **copper (Cu)**, **zinc (Zn)**, **manganese (Mn)**, **boron (B)**, etc. Their function in the cell is being components of metalloenzymes.

The content of some elements can be determined by the location of bacteria. For example, cells of marine halophilic bacteria contain higher amounts of sodium (Na), calcium (Ca), bromine (Br), and iodine (I). The composition of biogenic elements can be almost similar in various bacteria, but the percentage of some compounds containing these elements in bacterial ash differs (**Table 2.1**).

Table 2.1. Composition of ash in some microorganisms

Microorganisms	K_2O	Na_2O	CaO	MgO	P_2O_5	Cl	SO_4	SiO_2	Fe_2O_3	Ash
Serratia marcescens	11.5	29.0	4.1	7.8	37.9	4.9	–	0.5	–	13.5
Mycobacterium tuberculosis	8.2	11.5	8.6	9.8	47.0	1.3	10.8	–	–	9.6
Acetobacter aceti	18.0	2.9	10.7	8.0	47.5	–	–	0.6	10.7	5.9
Azotobacter chroococcum	26.9	–	–	–	55.4	–	–	–	–	8.9
Saccharomyces cerevisiae	57.8	0.9	6.0	2.4	26.1	3.6	8.4	–	1.2	8.8

Thus, the composition of biogenic elements and the percentage of some compounds with these elements in various bacterial species can be different and depends on environmental conditions in which bacteria live.

2.2. Compounds composition

Bacterial cell constitutes of compounds that are of both low- and high-molecular weight (**Table 2.2**). Water represents the greatest fraction of low-molecular-weight substances in the cell. Its amount varies in an interval from 75% to 85% of overall weight. Bacterial spores contain significantly less water — its content accounts for around 15%. Some of the water present in the bacterial cell is in the bound form that is a component of various cell structures, and therefore, cannot serve as a diluent. This function is provided by available *free water* that acts also as a disperse environment for colloid compounds that are present in the cytoplasm. On account of its appropriate physical and chemical characteristics, for example, the ability to dissolve or disperse various compounds, good thermal conductivity, high tension, and the ability to provide hydrogen and hydroxyl ions, water represents an optimal environment for biochemical reactions and vital processes of a cell.

Table 2.2. Compounds in bacterial cells

Compound	% of overall cell weight	Average molecular mass
H_2O	70	18
Inorganic ions	1	40
Saccharides	3	150
Lipids	2	750
Structural components and intermediate products (amino acids, nucleotides a.o.)	2	750
Proteins	15	4×10^4
DNA	1	$3-5 \times 10^9$
RNA	6	$2.5 \times 10^4 - 1 \times 10^6$

Basic molecules found in the bacterial cell:

Small molecules (relative molecular mass 18–50): CO_2, H_2O, NH_4^+, SO_4^{2-}, HPO_4^{3-}, O_2, N_2, etc. are generally adsorbed from the environment as nutrients and can be included in the organic molecules of the following categories.

Intermediates of metabolism (relative molecular mass 50–200) include glycerol, glucose, pyruvate, ribose, glyceraldehyde-3-phosphate, citric acid, etc.

Construction blocks of macromolecules (relative molecular mass 100–350) are represented by amino acids, nucleotides, monosaccharides, and fatty acids.

Macromolecules (relative molecular mass 10^3-10^9) include nucleic acids, proteins, polysaccharides, and lipids. Functional units are formed as combinations of various macromolecules, which may also involve small molecules, for example, ions.

Supramolecular formations (relative molecular mass 10^6-10^9) are enzyme complexes, ribosomes, membranes, respiratory chains, etc.

Nucleic acids make up about 19% of the bacterial cell dry weight (3% DNA and 16% RNA). The nucleic acids present in bacterial cells are composed of polynucleotide chains of different lengths from 80 nucleotides to about 4.2×10^6 nucleotides. Non dividing cells have only one molecule of chromosomal DNA, whereas the copy number of plasmid DNA can be several tens of different types. The base pairing principle has another meaning for the cell.

The presence of adenine and thymine as well as cytosine and guanine must be equal (A = T, G = C) with the molar ratio of purine to pyrimidine bases = 1:1 (always).

However, the ratio of the molar content of A + T / G + C bases in DNA varies in different species and depends on the frequency of occurrence of individual bases. It is usually expressed in molar percentages of G + C = % GC.

Bacterial proteins rare macromolecules of considerably different masses. Relative molecular mass is in a wide range of $5–2,000 \times 10^3$. They represent about 50% of the dry matter of the bacterial cell and play an integral role in its construction and activity. The bacterial cell is thought to contain more than 3,000 different types of proteins.

The function of the protein molecule is given by the specific arrangement in the space, the shape, surface structure, size, and conformation of the molecule.

At each level of the structure, protein has a certain meaning:

- The primary structure determines the nature of biological activity.
- The secondary and tertiary structures guarantee specificity, because the active center is only available to molecules with appropriate structure.
- The quaternary structure allows intramolecular regulation of biological activity.

Proteins have many functions in the cell. The most important of them are the following:

- Building (structural) function: Proteins are parts of the cytoplasmic membrane, cell wall, flagella, fimbria, S-layer, etc.
- Transport function: carrying substances by protein molecules across membranes. Transport proteins can be specific (certain substances are carried by only certain proteins) or "universal," which carry smaller or larger groups of substances.
- Signaling function: Proteins fulfill the function of receptors in the signaling system. The receptors may be surface or intracellular: protein kinases, protein phosphatases, G proteins, etc. A target cell reacts to the signal molecule via receptors only.
- Catalytic function: The basic principle is to reduce the activation energy of the reaction; the equilibrium state of the reaction is not altered (the composition of the resulting equilibrium mixture is not changed), but the equilibrium relationship is speeded up. Catalytic action does not produce byproducts.

Carbohydrates (a synonym is *saccharides*) represent an important structural component of cellular structures. The carbohydrate content of the cell is very variable and depends not only on the bacterium type but also on metabolic activity, the age of the population, and the composition of the medium. The percentage of carbohydrates in the dry matter of bacteria is 12–28%.

Monosaccharides are oxidation products of polyhydric alcohols and are simple carbohydrates made up of only one polyhydroxy aldehyde or ketone given by the empirical formula: $(CH_2O)_n$. *Oligosaccharides* are synthesized from 2 to 12 molecules of monosaccharides linked by glycosidic bonding. They are formed by the condensation of two (*disaccharides*) or three (*trisaccharides*) monosaccharide molecules. *Polysaccharides* are copolymers of monosaccharide repeat units (the macromolecule consists of only one type of monosaccharide). *Surface polysaccharides* mainly contain glucosamine (streptococci, staphylococci, etc.) and glucuronic acid.

Lipids have primarily a structural function in the bacterial cell (being part of membranes) and are a significant reservoir. For the bacterial cell it is characteristic that the saturated fatty acids, hydroxylated, methylated, or containing the cyclopropane ring, predominate. The *structure of the fatty acid* (carbon number, position of the double bonds, and their number) present in the cell is determined by the type of bacteria and the external environment. The nature of the lipid can change significantly in a given bacterium depending on the culture conditions.

Simple fats:
- Glycerides (acylglycerides, acylglycerols) are esters of fatty acids and glycerol. Depending on the number of esterified glycerol hydroxyl groups, they are divided into monoglycerides (monoacylglycerols), diglycerides (diacylglycerols) and triglycerides (triacylglycerols).
- Fatty acids have long linear chains differing in carbon number as well as the number and position of unsaturated bonds.
- Mono-, di-, and triglycerides occur mainly in plants and animals, while in bacteria, they are present in relatively low concentrations.
- Linear saturated and unsaturated fatty acids predominate in *E. coli*, while saturated branching is more common in *Bacillus subtilis*.

Complex fats:
Phosphoglycerides (phospholipids, phosphatides) have one –OH group of glycerol esterified with phosphoric acid, on which the polar alcohol is linked by an ester bond (**Fig. 2.1**). Two other hydroxyl groups are esterified with fatty acids. The phospholipid molecule is amphipathic.

Phospholipids can be arranged in the classical membrane and/or (less often) bend the membrane. Most phospholipids are found in Gram-negative bacteria, especially in enterobacteria. **Cardiolipin** (diphosphatidylglycerol) is found in *Haemophilus influenza*. Examples of glycerophospholipids are as follows:
- **Plasmalogens** are a type of phosphoglyceride. The first carbon of glycerol has a hydrocarbon chain attached via an ether linkage. This linkage is more resistant to chemical attack than an ester linkage. The second (central) carbon atom has a fatty acid linked by an ester. The third carbon links to an ethanolamine or choline by means of a phosphate ester. These compounds were found in only strictly anaerobic bacteria of the genera *Bacteroides*, *Clostridium*, *Desulfovibrio*, *Peptostreptococcus*, *Propionibacterium*, *Ruminococcus*, *Selenomonas*, *Treponema*, and *Veillonella*.
- **Phosphatidates** are lipids, in which the first two carbon atoms of the glycerol are fatty acid esters, and the third is a phosphate ester. The presence of charges gives a "head" with an overall charge. The phosphate ester portion ("head") is hydrophilic, while the remainder of the molecule, the fatty acid "tail," is hydrophobic. These are important components for the formation of lipid bilayers.

Phosphatidylethanolamines, phosphatidylcholines, and other phospholipids are examples of phosphatidates.

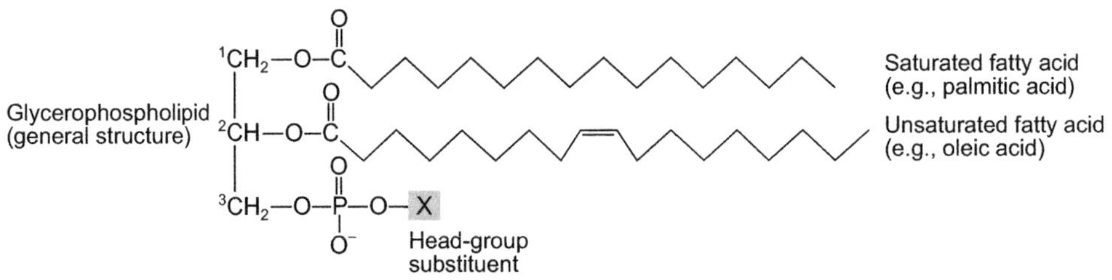

Name of glycerophospholipid	Name of X	Formula of X	Net charge (at pH 7)
Phosphatidic acid	—	$-H$	–1
Phosphatidylethanolamine	Ethanolamine	$-CH_2-CH_2-\overset{+}{N}H_3$	0
Phosphatidylcholine	Choline	$-CH_2-CH_2-\overset{+}{N}(CH_3)_3$	0
Phosphatidylserine	Serine	$-CH_2-CH(COO^-)-\overset{+}{N}H_3$	–1
Phosphatidylglycerol	Glycerol	$-CH_2-CH(OH)-CH_2-OH$	–1
Phosphatidylinositol 4,5-diphosphate	*myo*-Inositol 4,5-diphosphate	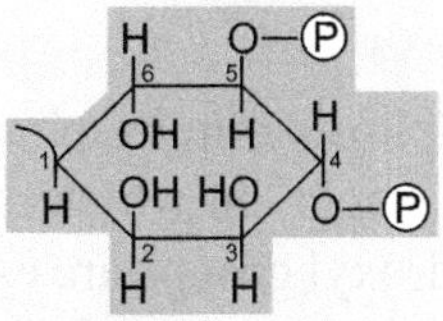	–4
Cardiolipin	Phosphatidyl-glycerol	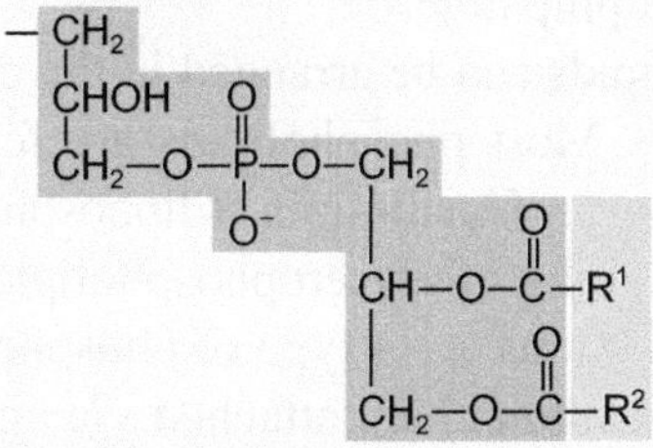	–2

Fig. 2.1. Structure of phosphoglycerides.

Glycolipids (glycosyl diglycerides) are detected mainly in Gram-positive bacteria. The sugar component may be glucose, galactose, or mannose (**Fig. 2.2**). Glycolipids are readily extracted from bacterial cells, together with phospholipids, by stirring with chloroform–methanol mixtures. They may be preferentially removed to some extent by extraction with acetone, but the rate of extraction is comparatively slow, and subsequent fractionation is usually required to remove traces of phospholipids. It is assumed that glycolipids replace sterol in the structure or in some membrane functions:

- Ramnolipid is found in *Pseudomonas aeruginosa.*
- Cord factor is found in the cell wall of *Mycobacterium tuberculosis* and associated with strain virulence.

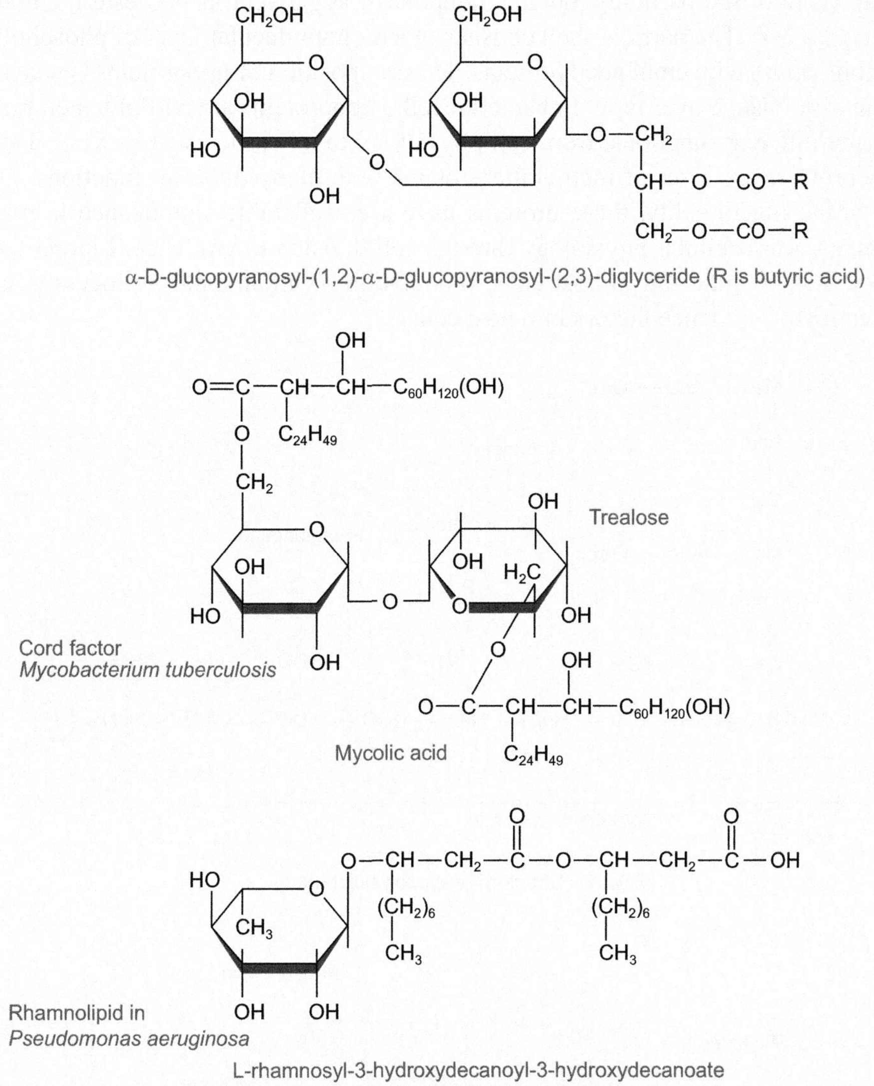

α-D-glucopyranosyl-(1,2)-α-D-glucopyranosyl-(2,3)-diglyceride (R is butyric acid)

Cord factor
Mycobacterium tuberculosis

Trealose

Mycolic acid

Rhamnolipid in
Pseudomonas aeruginosa

L-rhamnosyl-3-hydroxydecanoyl-3-hydroxydecanoate

Fig. 2.2. Structure of glycolipids.

Lipopolysaccharides are complex molecules involved in the structure of the outer membrane of G⁻ bacteria, whose composition is species-dependent (**Fig. 2.3**). The molecule is composed of three components: *lipid A, basic polysaccharide*, and *specific polysaccharide*. Lipid A is "general," that is, it is the same for all species. The lipopolysaccharide molecule is a hydrophobic moiety. It is also sometimes referred to as endotoxin, which can be released in the environment by lysis of bacterial cells.

Lipoproteins have a specific structure that is more or less determined by the degree of water solubility of fats. Typically, it creates a spherical structure. The center of the structure (core of the particle) contains nonpolar triacylglycerols and esterified cholesterol (**Fig. 2.4**). The particle shell consists of a monomolecular layer of phospholipids (amphipathic) with embedded proteins. The composition of lipoproteins is not constant but may change over time. In bacterial cells, lipoproteins are part of a membrane with many different functions, from cell physiology to its virulence (**Fig. 2.5**). Bacterial lipoproteins are a set of membrane proteins with many different functions. Due to this broad functionality, these proteins have a considerable significance in many phenomena, from cellular physiology through cell division to virulence. Lipoproteins play key roles in adhesion to host cells, modulation of inflammatory processes, and translocation of virulence factors into host cells.

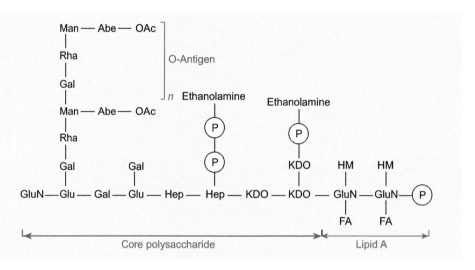

Fig. 2.3. Structure of lipopolysaccharides.

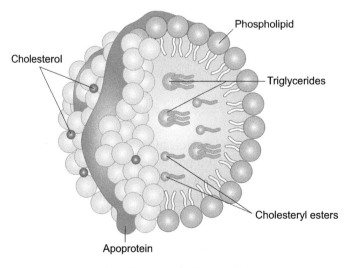

Fig. 2.4. Scheme of lipoproteins.

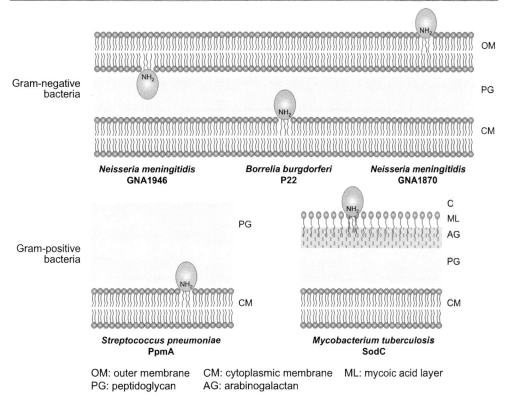

Gram-negative bacteria

OM

PG

CM

Neisseria meningitidis **GNA1946**   *Borrelia burgdorferi* **P22**   *Neisseria meningitidis* **GNA1870**

Gram-positive bacteria

PG

C
ML
AG

PG

CM

CM

Streptococcus pneumoniae **PpmA**   *Mycobacterium tuberculosis* **SodC**

OM: outer membrane   CM: cytoplasmic membrane   ML: mycoic acid layer
PG: peptidoglycan   AG: arabinogalactan

Fig. 2.5. Lipoproteins and their localization in a bacterial cell.

The presence of other compounds of both low- and high-molecular weight, often labeled as biomolecules, is closely related to the characteristics and function of individual cell structures and organelles, namely:
- bacterial nucleus (nucleoid);
- cytoplasm;
- cytoplasmic membrane;
- cell wall.

These structures are present in all bacteria. However, the structure of a bacterial cell of classical Gram-negative bacteria and phototrophic cyanobacteria is somewhat different (**Fig. 2.6**).

Besides, the following structures might be present in some bacteria:
- flagella;
- fimbriae and pili;
- capsule.

While bacterial capsule and cytoplasmic membrane represent inner cell structures, the rest of the structures are external or superficial. Endospores of sporulating bacteria have a different function and composition. Finally, various chromatic substances — pigments — also belong to chemical components of bacterial cell.

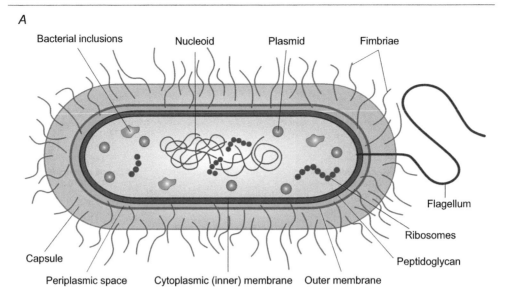

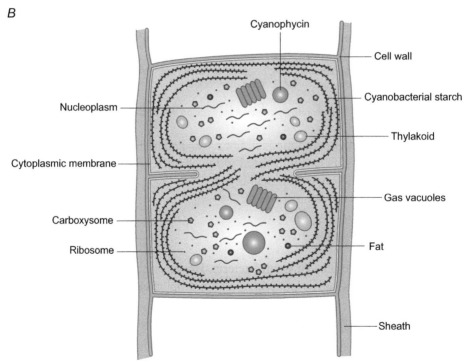

Fig. 2.6. Structure of a bacterial cell: Gram negative bacteria (*A*); cyanobacteria (*B*).

Consequently, despite the fact that the chemical composition and most of functional cell structures are similar in bacteria, there are differences in between phototrophic and Gram-negative bacteria.

2.3. Bacterial nucleus

Bacterial nucleus (nucleoid) accounts for approximately 10% of overall cytoplasm volume. As opposed to eukaryotic nucleus, bacterial nucleus has no membrane; it means that it is not separated from the basic cytoplasm by a membrane. The basic element is deoxyribonucleic acid (DNA), a carrier of genetic information (**Fig. 2.7**). Helical DNA is circular double-stranded. It does not contain histones (about four types of proteins are bound to the DNA molecule) and forms a compact complex consisting of more than 50 loops (in *E. coli*). In addition, it forms a higher order complex with RNA and protein molecules.

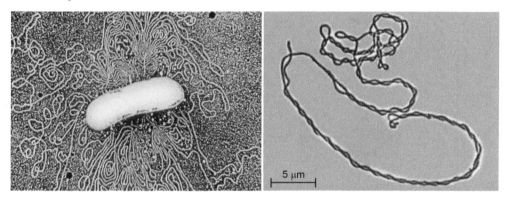

Fig. 2.7. DNA of *Bacillus subtilis*.

The nucleotide of the bacteria is a compact body because it contains a number of copies of small *multifunctional proteins* (**HU** protein is important for chromosome replication and for regulation of gene expression; **H–NS** protein plays a major role in the "thickening" of DNA in the formation of a compact nucleotide). The shape of the nucleotide in most bacteria is spherical or elongated. An exception is the toroidal nucleotide of *Deinococcus radiodurans*. A prokaryotic cell contains one copy of the chromosome. Bacteria *Azotobacter vinelandii* or *Deinococcus radiodurans* have more than one copy of the chromosome in the cell (from "several to many"). Bacterium *Vibrio cholerae* has two types of chromosomes in the cell: chromosome I and chromosome II, the former being larger than the latter.

The molecular weight of the DNA is 10^9–10^{10}. A typical bacterium contains 4×10^6 base pairs. The size of the chromosome and the number of bases in the chain varies in each species. The smallest chromosome, containing about 640×10^3 nucleotides, is in bacterium *Buchnera aphidicola* (endosymbiont aphids), and the largest known chromosome (about 4.4×10^6 nucleotides) is detected for *Mycobacterium tuberculosis*. Fast-growing bacteria may contain two to eight copies of the chromosomal DNA. The *E. coli* chromosome contains 4,629,221 base pairs and encodes 4,397 genes, including 108 genes for RNA, 4,289 genes for proteins, and 952 genes for enzyme synthesis (including 703 mapped).

The smallest prokaryotic genome containing only 490,885 nucleotides is detected for *Nanoarchaeum equitans* (*Nanoarcheota* phylum, *Archaea* domain).

Its molecule consists of two complementary, antiparallel polynucleotide chains coiled to a shape of a helix. Nucleotide bases of both polynucleotide chains are connected by hydrogen bonds and oriented to the inner side of the helix. Molecules of deoxyribose, linked by phosphodiester bonds in 3′C and 5′C positions, are oriented to the external side (**Fig. 2.8**).

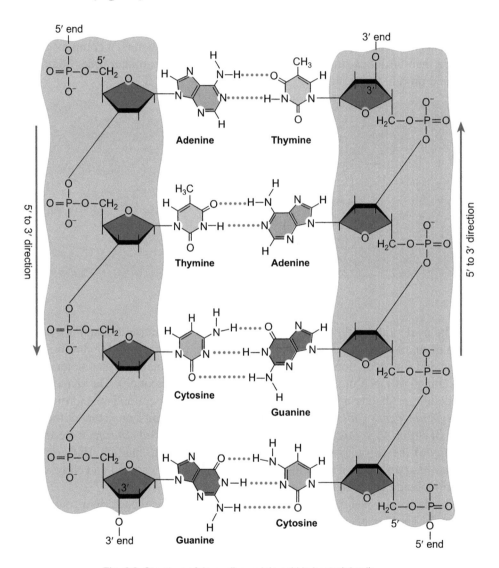

Fig. 2.8. Structure of deoxyribonucleic acid in bacterial cells.

Pairing of the bases is caused by binding adenine to thymine by two hydrogen bonds (AT) and binding guanine to cytosine by three hydrogen bonds (GC). The value of AT/GC ratio, usually provided as GC percentage, depends on the amount of individual bases and is genus specific. This value varies from 25% to 75% GC in bacteria.

The bacterial nucleus contains one molecule of DNA (molecular weight approx. $3–5\times10^9$, length around 1 mm). The whole molecule representing bacterial chromosomes consists of $3–5\times10^6$ nucleotide pairs and is organized to a circular shape (**Fig. 2.9**).

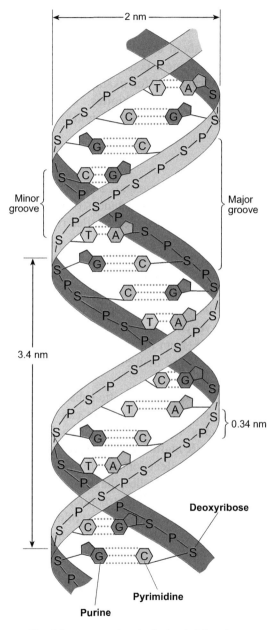

Fig. 2.9. DNA structure in the bacterial nucleus.

Negative charges of phosphate and one ionized hydroxyl group present on each of the base pairs must be balanced out by cation groups. In bacteria, this neutralizing function is provided by polyamines spermine and spermidine that are present in the cell in high concentrations. Mg^{2+} partially contributes to neutralization as well.

$NH_2–(CH_2)_2–NH(CH_2)_4–NH–(CH_2)_3–NH_2$ Spermine

$NH_2–(CH_2)_3–NH(CH_2)_4–NH_2$ Spermidine

A small amount of DNA, about 0.1–0.2%, may be present in an extrachromosomal form as plasmids or episomes, connected to cytoplasmic membrane. They usually comprise genetic information, for example, for the synthesis of enzymes determining resistance against certain antibiotics. They are capable of autonomic replication or merging. Unlike chromosomes, plasmids are not essential for the cell.

2.4. Cytoplasm

The cytoplasm (cytosol) is a heterogenous solution containing several membranous structures, dissolved and dispersed substances of both low- and high-molecular weight. Apart from that, there are solid particles called ribosomes and, in some bacteria, various inclusions and gas vesicles.

The cytoplasm fills the entire interior space. It is a water based fluid containing nucleoid, ribosomes, inclusions, nutrient molecules, metabolites, intermediates of metabolism, amino acids, vitamins, nucleic acids (DNA: chromosomal and extrachromosomal, RNA: all types), proteins, storage compounds, and salts of organic and inorganic acids. The cytoplasm is **very viscous** and resembles a gel where all the reactions necessary for cell life take place (from the search for the corresponding molecules to the actual reaction). It contains more than **50% of all cell proteins**, most of which have an enzymatic function.

Besides these "common" substances, specific substances (inclusions) may be present in some bacterial species. For example, there can be a parasporal inclusion in *Bacillus thuringiensis* ("remnants" after spore formation, first bioinsecticides), magnetosomes in *Magnetospirillum*, or **R bodies** (proteins) present in some *Pseudomonas* and *Caedibacter* species with negative effects on certain species of protozoa. Because the bacterial cell does not have specific organelles, the cytoplasm assumes a number of their functions. Although it does not have a visible structure, it is very unlikely that it would be without any organization, taking into account most of the metabolic reactions. Above all, it is about making an effective collision between the substrate molecule and the enzyme. In the simplest case, the cytoplasm must flow, so that the collision occurs. In addition, the flow must be "somehow" regulated. The organization of these functions could be carried out in a simplified, even in some primitive form by the cytoskeleton.

Soluble substances like glycides, organic salts, fatty acids, amino acids, and coenzymes, including precursors, are present as well. A fraction of these compounds is transported into the cell from external environment, but majority of them are generated as intermediate products of metabolism. Compounds of high-molecular weight present in the cytoplasm are mostly proteins and ribonucleic acids, majority of them is present in special bodies — ribosomes.

Proteins consist of amino acids linked together in a certain sequence into peptide chains. These chains have a characteristic spatial arrangement (conformation). The structure of proteins is stabilized by both covalent and noncovalent bonds. According to the role of chemical bonds in molecule stabilization, primary, secondary, tertiary, and quaternary molecular structures are distinguished.

The primary structure is given by the sequence and number of covalently bound amino acids. Hydrogen bonds between atoms of nitrogen in amino acids result in the formation of the secondary structure of a polypeptide chain. The tertiary structure is spatial arrangement of the molecule determined by bonds between different lateral substituents of the polypeptide chain. The preservation of this structure is maintained by hydrogen, nonpolar, and ionic bonds. Transversal links of the chain's lateral substituents may be provided also by covalent bonds, for example, disulfide bonds emerging during–SH group oxidation. Aggregations of polypeptide chains by intermolecular interactions allow the spatial arrangement of the molecule consisting of several subunits. This is called the quaternary structure.

The overall protein content of a bacterial cell is 50–80% of the cell's dry mass. Bacteria capable of utilizing molecular nitrogen (*Azotobacter*) or bacteria that create thick slime capsules (*Acetobacter, Leuconostoc*) contain less proteins, around 15%.

A majority of bacterial proteins is globular, including mostly albumins and globulins. Besides single proteins, there is an outbalancing amount of physiologically active complex proteins. Proteins of a special significance are those with catalytic function — enzymes, determining the metabolic activity of bacteria. Overall number of enzymes in the bacterial cell is estimated to be around two or three thousand.

Ribonucleic acids (RNA) are polynucleotide chains that, in contrast to DNA, contain ribose instead of deoxyribose and uracil instead of thymine. Moreover, they usually exist in a single-stranded form, although, some regions of the molecule can exist in a double-stranded form. In the cell, there are three types of RNA:

Messenger (mRNA) comprises 2–4% of total RNA. It originates in the bacterial nucleus in the process of transcription of complementary chromosomal DNA strand. After the transcription, it transfers to the ribosomes where it is used as a matrix that determines the sequence of amino acids in polypeptide synthesis. Molecular weight of mRNA varies from 2.5×10^4 to 10^6 Da according to the type of encoded protein.

Transfer (tRNA) is of a significantly lower molecular weight (23–30 kDa). The unique feature of tRNA is the presence of uncommon bases such as dihydrouracil, dihydrocytidine, and various methylated nucleotides. Its function is the transfer of individual amino acids to ribosomes during protein synthesis. For each of 20 amino acids, there is at least one tRNA. Of the total cell's RNA content, tRNA comprises a fraction of 16%.

Ribosomal (rRNA) is a component of cytoplasmic bodies — ribosomes, in which synthesis of polypeptide protein chains takes place. According to their sedimentation rate, they are usually distinguished as 5S rRNA (molecular weight of 35×10^3 kDa), 16S rRNA (molecular weight of 44×10^4 kDa), 23S rRNA (molecular weight of 10^6 kDa). rRNA comprises a significant part of the overall RNA in a cell, about 80%. The function of this ribonucleic acid has not been completely clarified so far.

Ribosomes are particles with dimensions of 16×18 nm. Their composition includes approximately 30% proteins and 70% rRNA. Polyamines spermine and spermidine, which have a stabilizing function, are present as well. The number of ribosomes in bacterial cell's cytoplasm varies from 5,000 to 50,000, depending on cell's growth stage and other conditions. The highest amounts are detected in fast growing cells in the exponential phase. According to the sedimentation rate, bacterial ribosomes are known as 70S. Their stability is determined by concentration of Mg^{2+} ions in the surrounding environment. In the concentration lower than 10^{-3} M Mg^{2+}, 70S particles dissociate into subunits, niosomes, 50S and 30S. 30S subunits contain one molecule of 16S rRNA (molecular mass $5×10^5$ kDa) and 21 protein molecules. 50S subunits are composed of one 23S rRNA molecule (molecular mass $1.2×10^6$ kDa), one molecule of 5S rRNA (molecular mass $4×10^4$ kDa), and about 35 protein molecules (**Fig. 2.10**).

Ribosomal RNA molecules are single-stranded and have a different GC content. Most ribosomal proteins are structurally different. Each protein is present in one copy only, except for L7/L12, which is present in two copies. Binding between RNA and proteins is accomplished by ion and hydrogen bonds.

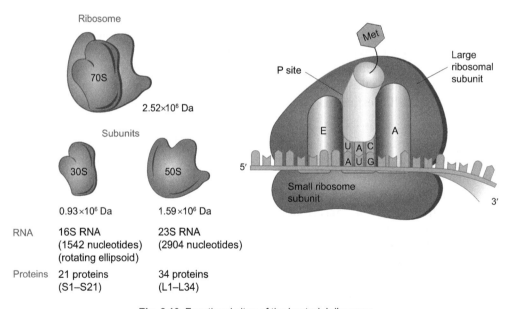

Fig. 2.10. Functional sites of the bacterial ribosome.

Ribosome proteins have specific functions and specific binding sites, for example:
- **A site**: proteins L1, L5, L7, L12, L30, L33;
- **P site**: proteins L7, L12, L14, L18, L24, L33.
  - mRNA binding: S1;
  - binding of fMet-tRNA: S2, S3, S10, S14, S19, S21;
  - codon recognition: S3, S4, S5, S11, S12;
  - binding of aminoacyl-tRNA: S9, S11, S18.

It is a supramolecular complex that is rapidly created (without energy supply) and quickly dissociated (without energy generation). To produce a ribosome from subunits and mRNA, the presence of Mg^{2+} is required. The number of the ribosomes is dependent on physiological activity. In an average fast growing cell, there are about 15,000 (sometimes up to 30,000) ribosomes.

The ribosome is a "body" of about 6,000 nm and a weight of about 4.4×10^{-18} g. The distribution of ribosomes in a cell depends on its need. Ribosomes are located either loosely in the basal cytoplasm (about 20%), nucleotide proximity (about 60%, protein synthesized according to mRNA immediately after leaving the nucleotide), or bound to the cytoplasmic membrane (synthesis of cytoplasmic membrane proteins, outer membranes, cell walls, or extracellular proteins).

The rate of ribosome activity is not regulated; it is almost constant. The chain is extended by about 15 amino acids per second. In *E. coli*, the translation rate is about 170,000 amino acids per second on the genome, that is, 500,000 amino acids per second per cell. Higher protease synthesis rates are achieved by a greater number of ribosomes.

Polyribosome. The system works so that once the ribosome moves, it releases the binding site to which the next ribosome will be attached (**Fig. 2.11**). The distance between the ribosomes is about 80 bases. Binding of other ribosomes and protein synthesis continues as long as no mRNA hydrolysis occurs. This can happen for a few minutes or, in exceptional cases, hours.

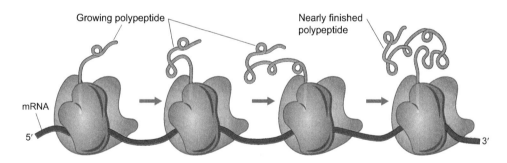

Fig. 2.11. Synthesized peptide chain.

Ribosomal proteins are of molecular weight $1-6 \times 10^4$ kDa. In total, 80% of ribosomes in cytoplasm are present in the form of aggregates containing 70S particles linked together by mRNA, resulting in the formation of **polyribosomes**. At lower Mg^{2+} concentrations, these aggregates disintegrate into individual 70S ribosomes.

Inclusions are usually present in the cytoplasm in the form of granules or liquid drops, mainly in older cells. They are often made of supplementary substances like polysaccharides, β-hydroxybutyric acid, polyphosphates, or sulfur (**Fig. 2.12**). The formation and amount of these substances depend on the surrounding environment characteristics and cell's metabolic activity. In original state, they are osmotically inert and not soluble in water. In case of change of conditions, they can be engaged into energetic metabolism and processes of biosynthesis.

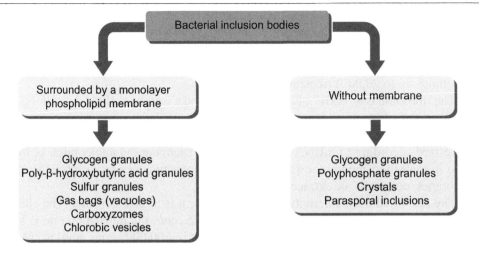

Fig. 2.12. Scheme of bacterial inclusion bodies.

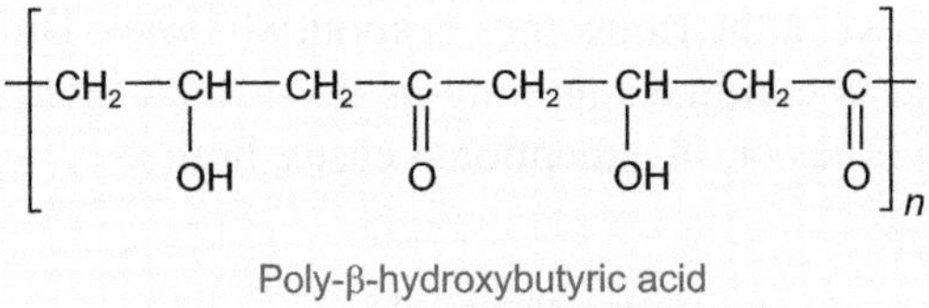

In cells of aerobic genera belonging to *Bacillus* and *Pseudomonas* species and in anaerobic phototrophic bacteria, a polymer whose chains contain around 60 subunits of β-hydroxybutyric acid (its molecular mass varies from 60 to 250,000 daltons) is often present (**Fig. 2.13**). This

Poly-β-hydroxybutyric acid

Fig. 2.13. Summary formula: $(C_4H_6O_2)_n$.

acid can be accumulated in a notable amount, up to 80% of cell's dry mass. It is of neutral character because its carboxyl groups are linked by ester bounds. It represents an adequate supplementary material, especially in terms of energy.

Sulfur is temporarily stored in the form of light refracting drops mostly in cells of sulfur bacteria. Its amount depends on the environment's H_2S content. In its absence, sulfur is oxidized to sulfate. Anaerobic H_2S-oxidizing sulfur bacteria (*Beggiatoa, Thiothrix, Achromatium*) can use it as a source of energy in conditions with lack of H_2S. Phototrophic purple sulfur bacteria use it as an electron donor.

Solid cellular inclusions occur as specific crystals in *Bacillus thuringiensis* and related genera. This parasporal bodies are composed of a protein that is markedly toxic to caterpillars. Therefore, the crystal bodies are used as an effective preparation for eliminating these infestants.

Gas vacuoles (aerosomes) are characteristic for bacteria living in water, typically purple or green sulfur bacteria (*Lamprocystis, Pelodictyon, Thiodictyon*) and fibrillar bacteria (*Pelonema, Peloploca*). The presence of gas vacuoles apparently makes it possible for bacteria to regulate their movement in water according to nutrition conditions, oxygen content, and light radiation intensity.

Glycogen is an insoluble glucose polymer, α-1,4-glucan with numerous branching α-1,6-bonds (more than 100,000 molecules). Branching occurs on every 8th to

10th glucose molecule (**Fig. 2.14**). It is randomly distributed in the cytoplasm in the form of body visible in a light microscope after staining. Glycogen is a reserve and accumulates in cells after culturing in an environment with excess carbon or nitrogen deficiency. Glycogen and starch are polysaccharides comprised in granules. They accumulate in environment abundant in carbon. Glycogen is mostly present in cells of *Bacillus* and *Enterobateriaceae* (*Salmonella*, *E. coli*, and other genera). Amount of glycogen can comprise up to 50% of cell's dry mass in some genera. Starch grains were found in *Clostridium* sp., *Acetobacter*, and *Neisseria*.

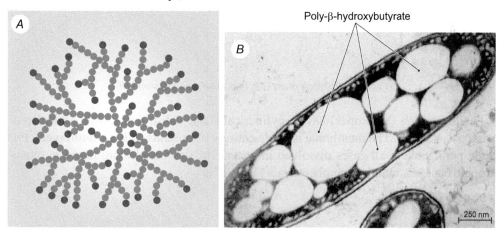

Fig. 2.14. Glycogen structure (*A*). Poly-β-hydroxybutyric acid granule (*B*).

A poly-β-hydroxybutyric acid forms droplets in the cytoplasm, visible in light microscopy as light-colored bodies. It can make up to 60% of the bacteria dry matter. Granules contain up to 98% poly-β-hydroxybutyric acid and 2% protein (sometimes fat grafts). The granules are mainly found in aerobic bacterial species of *Bacillus*, *Pseudomonas*, and *Azotobacter* (**Fig. 2.15**) as well as in phototrophic bacteria.

Sulfur is accumulated as a reserve compound for those bacteria that use it as a source of energy (chemolithotrophic sulfur bacteria) or deposited in the cytoplasm (**Fig. 2.16**) as S^0 after the use of sulfur compounds as electron donors in phototrophic bacteria (purple

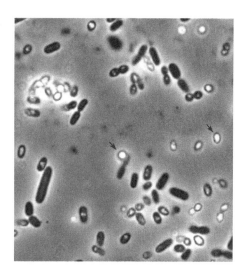

Fig. 2.15. *Azotobacter* sp.

sulfur or green sulfur). Some bacteria may also store sulfur in cells resulting from the induction of the cytoplasmic membrane. In this case, however, this is not about inclusion.

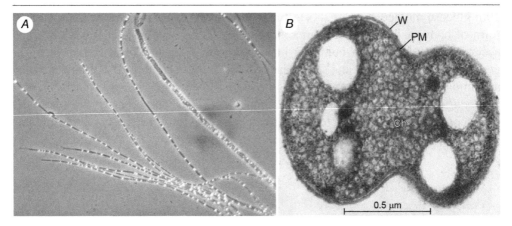

Fig. 2.16. *Thiothrix nivea (A); Thiocapsa roseopersicina (B).*

Gas vesicles (aerotopes) have a cylindrical shape substructure with conical ends (as a polyhedron). The membrane is made entirely of proteins (**it is a monolayer**). It is fully permeable to all gases dissolved in water. The gas vesicles are found in phototrophic bacteria, mainly cyanobacteria (**Fig. 2.17**).

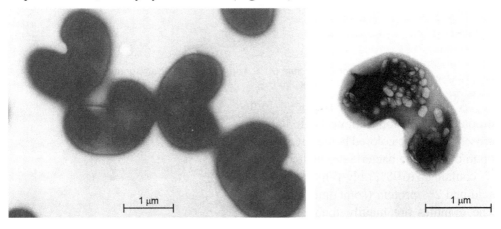

Fig. 2.17. Gas vesicles of *Ancylobacter aquaticus.*

Carboxysomes are present in bacteria using CO_2 as a carbon source (**Fig. 2.18**). Carboxysomes contain the enzyme ribulose-1,5-diphosphate carboxylase. They are usually located near the nucleotide. The most common is the presence of carboxysomes in nitrifying bacteria, cyanobacteria, and thiobacilli. Carboxysomes are polygonal-shaped structures of a size of 50–500 nm that are enfolded by a single-layer membrane that is 3.5 nm thick. They can be found mainly in photosynthetic purple

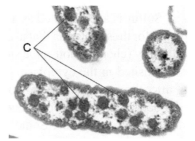

Fig. 2.18. *Thiobacillus neapolitanus* (C — carboxysomes)

bacteria. They are probably the place of autotrophic CO_2 reduction since they contain ribulosodiphosphate carboxylase (carboxydismutase), which is the key enzyme of this reduction in Calvin cycle.

Polyphosphates are present mostly as volutin or metachromatic granules, first described in *Spirillum volutans.* They form long chains of Na-poly-metaphosphate occurring mostly with lack of some nutrient. With change of conditions, they can be used as a source of phosphorus. The energetic value of polyphosphates is of secondary importance.

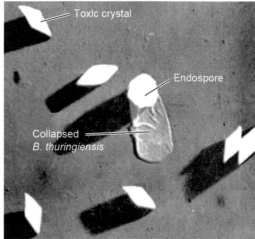

Polyphosphoric acid

Polyphosphate granules (metachromatic granule, volutine). Volutine is a polyphosphate that accumulates in the cell in the form of spherical bodies of 45 nm – 1 μm (**Fig. 2.19**). It is a phosphate molecule reservoir consisting of chains of up to 500 orthophosphate molecules and being, therefore, water insoluble. The bond between the molecules is energy-rich and requires the supply of energy in the form of ATP:

$$ATP + (HPO_3)_n \rightarrow ADP + (HPO_3)_{n+1}$$

The reaction is catalyzed by polyphosphate kinase.

Inclusion without membrane (parasporal inclusion). Parasporal inclusions are created by some *Bacillus* genus. They are bipyramidally octahedral-symmetric and consist of $4 \cdot 12$ nm rod-shaped polypeptide subunits (**Fig. 2.20**). The molecular weight of these subunits is about 230,000. They arise as a result of the overproduction of the proteins forming the endospore packaging. The parasporal inclusion of *Bacillus thuringiensis* was used as the first bioinsecticide against the flour worm.

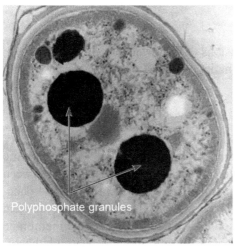

Fig. 2.19. Polyphosphate granules accumulated in a bacterial cell.

Fig. 2.20. *Bacillus thuringiensis.*

The comparison of all described cytoplasmic inclusions and their functions is presented in **Table 2.3**.

Table 2.3. Common inclusion bodies in microorganisms

Cytoplasmic inclusions	Where found	Composition	Function
Gas vesicles	Aquatic bacteria especially cyanobacteria	Protein hulls or shells inflated with gases	Buoyancy (floatation) in the vertical water column
Parasporal crystals	Endospore-forming bacilli (genus *Bacillus*)	Protein	Unknown but toxic to certain insects
Magnetosomes	Certain aquatic bacteria	Magnetite (iron oxide) Fe_3O_4	Orienting and migrating along geomagnetic field lines
Carboxysomes	Many autotrophic bacteria	Enzymes for autotrophic CO_2 fixation	Site of CO_2 fixation
Phycobilisomes	Cyanobacteria	Phycobiliproteins	Light-harvesting pigments
Chlorosomes	Green bacteria	Lipid, protein, and bacteriochlorophyll	Light-harvesting pigments and antennae
Glycogen	Many bacteria, e.g., *E. coli*	Polyglucose	Reserve carbon and energy source
Poly-β-hydroxybutyric acid	Many bacteria, e.g., *Pseudomonas*	Polymerized hydroxybutyrate	Reserve carbon and energy source
Polyphosphate (volutin granules)	Many bacteria, e.g., *Corynebacterium*	Linear or cyclical polymers of PO_4	Reserve phosphate, possibly a reserve of high energy phosphate
Sulfur globules	Phototrophic purple and green sulfur bacteria and lithotrophic colorless sulfur bacteria	Elemental sulfur	Reserve of electrons (reducing source) in phototrophs, reserve energy source in lithotrophs

Thus, summing up the cytoplasmic inclusions of bacterial cells described above, it should be noted that their diversity and functions depend on the bacterial species or genus.

2.5. Plasma membrane

The cytoplasm is surrounded by the plasma membrane (also called the cytoplasmic or cell membrane). It is a complex of lipoproteins consisting of three layers; between two monomolecular protein layers, there is one bimolecular layer of lipids and phospholipids (**Fig. 2.21**). The thickness of each layer is approximately 2–3 nm, protein content is about 60%, the amount of lipids is 40%. Apart from these, the membrane also comprises a small quantity of saccharides. Glycolipids and phosphatidylethanolamines are major components of the lipids present.

The structure of the membrane is created by hydrophobic ends of the molecules being oriented toward each other, that is, inside, while hydrophilic ends are oriented toward protein layers.

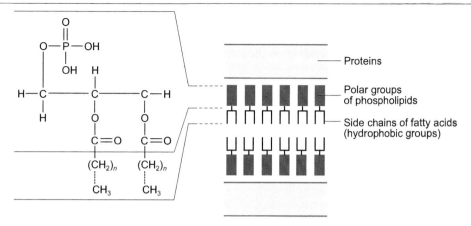

Fig. 2.21. Davidson–Danielli tri-layer (protein–lipid–protein) model of membrane structure.

The nature of the lipid molecule, a derivative of **glycerol-3-phosphate**, is evident from its formula, in which C_1 and C_2 carboxyl groups are esterified by long-chain fatty acids (R_1 and R_2), while the phosphate group on C_3 is esterified by various substituents including glycerol, its aminoacyl derivatives, and ethanolamine.

$$CH_2-O-R_1$$
$$CH-O-R_2$$
$$CH_2-O-\overset{\overset{\displaystyle O}{\|}}{P}-O-R$$
$$OH$$

Glycerol-3-phosphate

This membrane structure is stabilized by the mutual electrostatic influence of lipid polar groups and side chains of protein layer amino acids. According to more recent models, the membrane structure is of a more asymmetric nature. As globular protein molecules cross the lipid layer, their nonpolar hydrocarbon chains are located between convoluted polypeptide chains (**Fig. 2.22**).

The plasma membrane plays an important role in compound exchange, especially as an osmotic barrier regulating transport of compounds between the cell and the environment. This function is determined by its semipermeable character and the presence of substrate-specific enzymes, the so-called permeases, localized in the membrane, mediating active transport. It is presumed that hydrogen bonds cross the lipid layer of the membrane. These bonds operate as pores, through which a regulated substance flow moves. Membrane lipids also facilitate maintaining the suitable Mg^{2+} ion concentration needed for the activation of enzymes bound to the membrane. Some of these lipids, for example, undecaprenyl phosphate, act as glucose transporters in cell-wall biosynthesis. In some bacteria, there are special membrane-derived bodies called mesosomes, chromatophores, and carboxysomes.

The membrane separates the basic cytoplasm from the external environment. Its structure corresponds to the structure of the biological membrane (the same in all prokaryotes). The plasma membrane makes up from 10% to 26% of bacterium's dry weight and is about 8 nm thick. It is based on the liquid continuum of two-layer phospholipids with embedded proteins. Proteins are scattered from inside or outside or pass through it and represent 10–20% of all cell proteins. Degraded proteins have a structural or enzymatic function.

Plasma membrane proteins and lipids are not statically located but can "move." The cytoplasmic membrane is asymmetric. The ratio between saturated and unsaturated fatty acids varies depending on the culture temperature. The proteins are binding to the phospholipid bilayer.

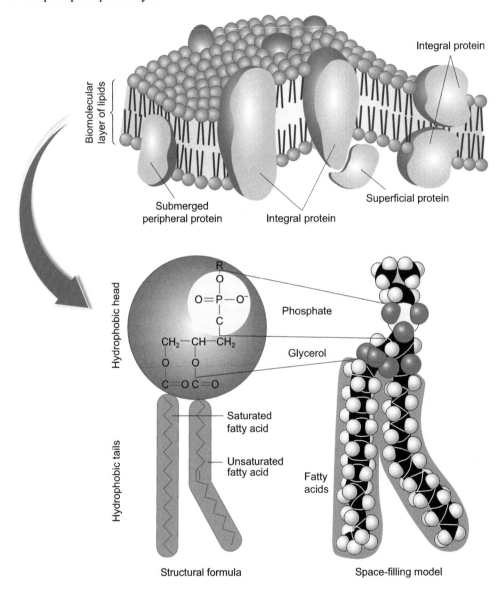

Fig. 2.22. Model of the plasma membrane.

The proteins pass through the layer, and the transmembrane proteins protrude on both sides. They have a hydrophilic region directed to the aqueous environment on both sides of the membrane, are hydrophobic in the nonpolar region of the membrane, and interact with nonpolar chains of phospholipids. The proteins bound to the outer surface are attached by C-terminal amino acid by one or more covalent bonds to lipid groups (glycolipid anchor). Peripheral proteins bind mediated on both sides of the membrane by interacting with other membrane proteins.

The characteristics of plasma membrane are as follows:

- It is a place of energy transformation (respiratory chain, photosynthesis apparatus, ATPase, etc.).
- It is a site of bacterial chromosomal and extrachromosomal DNA replication.
- The membrane is a site of lipid synthesis and hydrolysis.
- The last phase of cell wall synthesis takes place on it, or some of its components are synthesized.
- The membrane is semipermeable and responsible for the passive and active transport of substances into the cell and the cell. For polar components, it is impermeable.
- It can contain flagellum inside.
- The cytoplasmic membrane cannot be "made" *de novo*.
- Invaginations are very common to increase the active area.

Invagination of the plasma membrane

Mesosomes are membrane structures located in the cytoplasm (**Fig. 2.23**). They are vesicle-shaped and occur by invagination of the membrane into cytoplasm. During this process, vesicles are curled to a ball shape that is internally divided into tubes or vesicles. Mesosomes are particularly often found near transversal septa where they reach the bacterial nucleus or the cell-division initiation site.

Mesosomes have the shape of a bladder; they are divided into tubular formations or pouches. There are two or more mesosomes, depending on the metabolic activity of the cell. Mesosomes are responsible for replicating the bacterial chromosome and dividing the cell. It is usually in contact with the nucleotide. Chemolithotrophic bacteria have additional invaginations near mesosomes, allowing the oxidation of inorganic substances.

With quick drop of temperature in dividing cells or with sudden change of osmotic pressure, the mesosomes can be pushed out

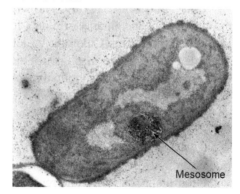

Mesosome

Fig. 2.23. Mesosome of *Bacillus subtilis*.

into the space between cytoplasmic membrane and cell wall. If bacterial cells are turned into protoplasts, their mesosomes are lost completely. In the process, vesicular structures are transferred to the external environment.

Mesosomes are involved in processes of aerobic respiration related to electron transport and oxidative phosphorylation. These processes are available because of the presence of ubiquinone, ferredoxin, and cytochrome enzymatic complexes. Apart from that, they obviously affect the initiation and development of cell division and formation dividing septum. Some enzymes of Gram-negative bacteria are localized in between the cell wall and the membrane in the so-called periplasmic space. Most of them are phosphatases that cleave phosphorus from phosphorylated compounds and thus facilitate their transfer through the membrane. Besides phosphatases, penicillinase is also present.

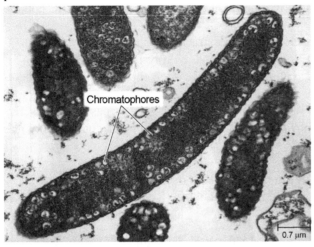

Fig. 2.24. Chromatophores of *Rhodospirillum rubrum*.

Chromatophores, similarly to mesosomes, exist in the form of vesicles and are present mainly in phototrophic bacteria (**Fig. 2.24**). A complete separation of chromatophores from the membrane occurs only with the disruption of the cell and homogenization. Chromatophores are functionally parallel to chloroplasts in green cells. Inside, there are pigments that absorb radiation, the so-called bacteriochlorophylls and carotenoids. Chromatophores have the shape of a bell divided into segments. They are the site of ATP synthesis in phototrophic bacteria.

Chlorobium vesicles (also known as **chlorosomes**) are created by invagination of the cytoplasmic membrane forming a follicular body that is eventually divided by septa into multiple segments. Chlorobic vesicles contain bacteriochlorophyll *a*, *c*, *d*, and carotenoid pigments and are in phototrophic bacteria. Their width is 30–40 nm. They are bounded by a simple phospholipid membrane and located in close proximity to the cytoplasmic membrane.

Chlorosomes are a photosynthetic antenna complex found in green sulfur bacteria (GSB) and some green filamentous anoxygenic phototrophs (species of *Chloroflexaceae*, *Chlorobiaceae*, and *Oscillochloridaceae* family). Chlorosomes contain bacteriochlorophyll, carotenoids, quinones, and lipids (**Fig. 2.25**). A special antenna system without proteins is created and represents the most efficient system of energy recovery in nature. The structure of each chlorosome is unique.

Thylakoids are probably formed by the removal of the cytoplasmic membrane and have a lamellar structure (**Fig. 2.26**). Thylakoids consist of a thylakoid membrane surrounding a thylakoid lumen. A thylakoid is a membrane-bound compartment inside cyanobacteria. It contains pigments for phototrophic synthesis: chlorophyll *a*, α- and β-carotene, and xanthophylls (echinenone, myxoxanthophyl, zeaxanthin). The phycobilisomes are located on the surface of thylakoid sacs. They contain specific dyes: phycobilins (usually three or two are blue (*c-phycocyanin* and *allophycocyanin*) and one is red (*c-phycoerythrin*)). They are responsible for chromatic adaptation.

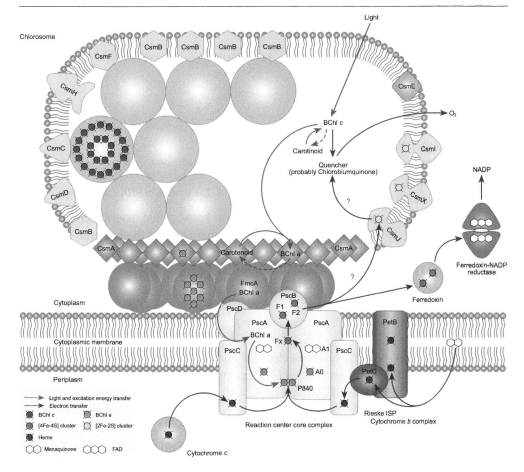

Fig. 2.25. Structure of a chlorosome.

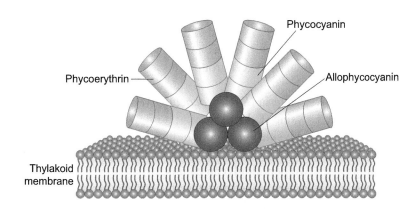

Fig. 2.26. Structure of a thylakoid.

The RNA degradosome is a multienzyme complex responsible for mRNA degradation (**Fig. 2.27**). It consists of endoribonuclease (***RNAse E***), helicase (***RhIB***), enolase, exoribonuclease (PNPase). It is strictly bound to cytoplasmic membrane and *RNase E* is plated into the cytoplasmic membrane.

Consequently, the structure of the cytoplasmic membrane in

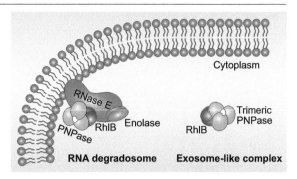

Fig. 2.27. The scheme of RNA degradosome.

the bacterial cell plays an important barrier function between the external environment and the cytoplasm. It is a multifunctional structure that protects the bacterial cell from many factors of the external environment. Invaginations of cytoplasmic membrane can form the following structures: mesosomes, chromatophores, chlorosomes (or chlorobic vesicles), RNA degradosome, and carboxysomes.

2.6. Cell wall

The cell wall of bacteria has a mechanical and protective function, since it determines the shape of the cell and protects it against osmotic pressure changes of the environment. It is compact and elastic. Unlike cell membrane, it is permeable for salts and numerous other low-molecular-weight compounds. In comparison to eukaryotic cell walls, it comprises neither cellulose, as higher plants do, nor chitin that is present in fungi and arthropods. Strains of *Sarcina ventriculi* are joined by cellulose to form larger bundles, and cells of *Acetobacter xylinum* and *A. antigenum* secrete cellulose to surrounding environment in a form of thin fibrils.

The main component and supporting material of cell wall is a relatively simple heteropolymer, peptidoglycan murein. It is a macromolecule composed of regularly alternating subunits of *N*-acetylglucosamine (NAG) bound by 1,4-glycosidic bond with *N*-acetylmuramic acid (NAM) — an ester of lactic acid and *N*-acetylglucosamine. Peptides of amino acids, in Gram-negative and Gram-positive mostly L-alanine (L-Ala), D-glutamic acid (D-Glu), D-alanine (D-Ala), and L-diaminopimelic acid (DAP), are bound to muramic acid by their carboxyl group. Structural units of peptidoglycan are illustrated in **Fig. 2.28**.

In some Gram-positive bacteria (*Staphylococcus, Micrococcus*, etc.), diaminopimelic acid is replaced by L-lysine (L-Lys).

Diaminopimelic acid and lysine are important in the process of murein net formation in a way that they facilitate the creation of another peptide bond and thus serve as a linking element between two peptidoglycan chains.

Heteropolymerous chains are linked together by peptide bonds, creating murein macromolecule that is characterized by the content of elements NAM:NAG:DAP (Lys):D-Glu:D-Ala in a ratio of 1:1:1:1:2.

Bacteria significantly differ from higher organisms, since the bacterial cell wall contains compounds that are not present in cells of animals and plants (*N*-acetylglucosamine, muramic acid, DAP, D-forms of alanine and glutamic acid). This fact also determines effective application of medical preparations (e.g., penicillin) that specifically affect bacterial cell wall biosynthesis while being harmless to other organisms.

The murein layer has a function of cell wall's supportive skeleton and determines its rigidity. Various compounds, different in Gram-positive and Gram-negative bacteria, can be layered upon this skeleton.

The cell wall of Gram-positive bacteria is relatively thick (15–80 nm) and accounts for 10–25% of cell's dry mass. The murein net has more layers and is often saturated by teichoic acid — a polymer of ribitol joined into chains by phosphodiester bonds that is covalently linked to glycopeptide.

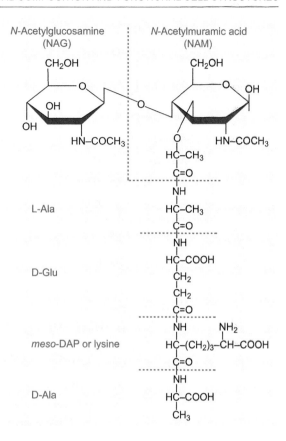

Fig. 2.28. Structural unit of peptidoglycan.

Teichoic acids are linear polymers of ribitol phosphate or glycerophosphate and represent 15–40% of the weight of the wall. Their chemical structure is generic or species-specific (**Fig. 2.29**).

This acid is joined to the cell membrane by its acyl groups. In many Gram-positive bacteria, teichoic acid reaches the cell surface where it exists as a part of specific receptor spots for bacteriophages. The alcohol group of ribitol can be substituted with D-alanine, glucose, or *N*-acetylglucosamine. In *Staphylococcus aureus*, the peptide chain is joined to *N*-acetylmuramic acid of another mucopeptide by a pentaglycine bond (**Fig. 2.30**).

A special cell wall composition was observed in *Micrococcus lysodeikticus*, where about a half of NAM units do not contain peptide chains, and thus, the transversal links are relatively weak. This explains why the cell wall of this organism can be easily degraded by lysozyme (muramidase): to lower the rigidity of the murein net, it is sufficient to disrupt only one glycosidic bridge. The cell wall of some streptococci contains also protein antigen (*N*-antigen).

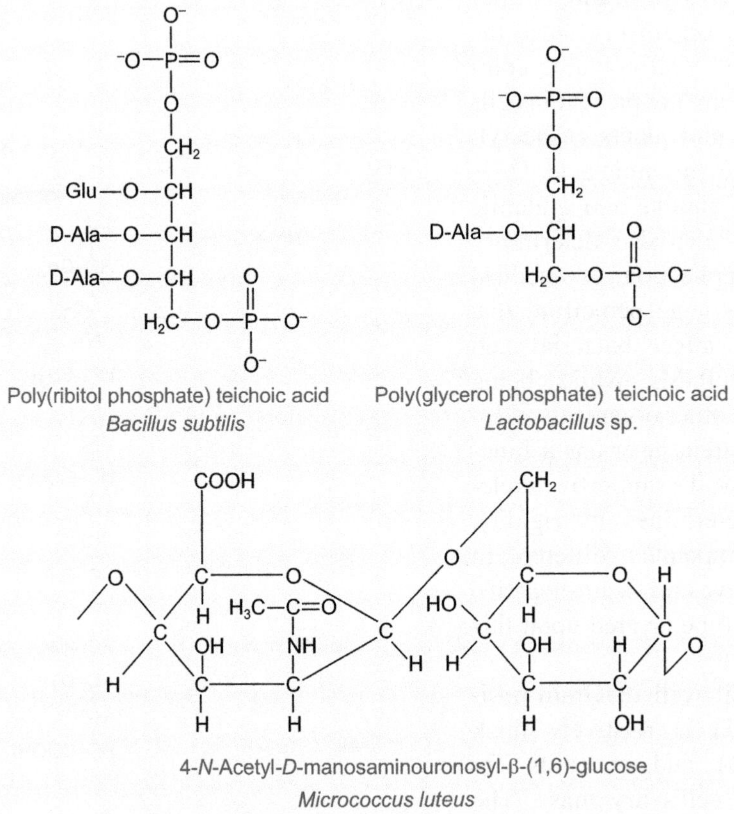

Poly(ribitol phosphate) teichoic acid
Bacillus subtilis

Poly(glycerol phosphate) teichoic acid
Lactobacillus sp.

4-*N*-Acetyl-*D*-manosaminouronosyl-β-(1,6)-glucose
Micrococcus luteus

Fig. 2.29. Structural units of peptidoglycan in bacterial cells.

Besides peptidoglycan, the cell wall of Gram-positive acidoresistant bacteria, mainly mycobacteria, contains paraffins, commonly called waxes. These are often made up of various groups of substances. The most prominent are ***mycolic acids*** and ***lipopolysaccharides***.

Mycolic acids are often long β-hydroxy acids, substituted in α-position by an aliphatic side chain. For example, the mycolic acid isolated from the cell wall of *Mycobacterium smegmatis* has this structure:

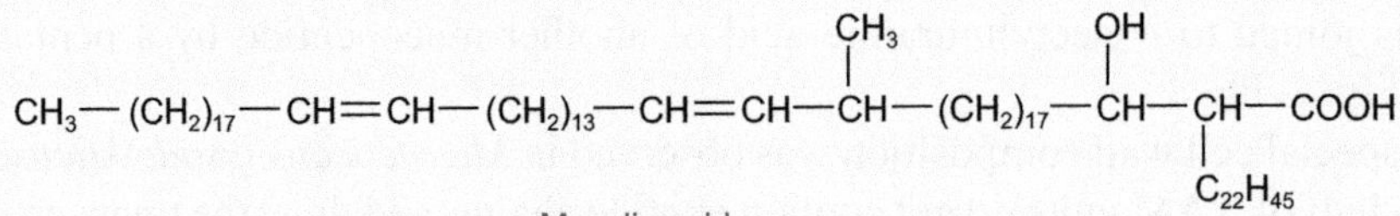

$$CH_3-(CH_2)_{17}-CH=CH-(CH_2)_{13}-CH=CH-\underset{CH_3}{CH}-(CH_2)_{17}-\underset{OH}{CH}-\underset{C_{22}H_{45}}{CH}-COOH$$

Mycolic acid

Extracted mycolic acids are resistant to the effect of acids or ethanol during staining procedures (such as Gram staining), evading decolorization; hence, the cells containing these compounds are referred to as acid-fast microorganisms.

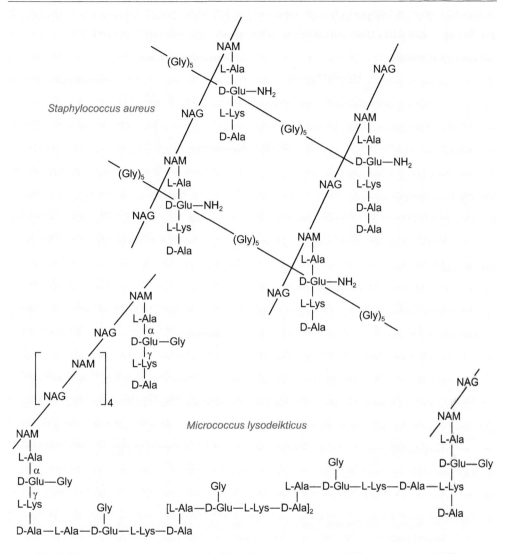

Fig. 2.30. Structure of peptidoglycan in *Staphylococcus aureus* and *Micrococcus lysodeikticus.*

It is widely accepted that the main cause of this effect is a reaction between mycolic acids and basic fuchsine. This reaction results in the creation of a complex, which acts as a permeation barrier; thus it prevents the penetration of the mineral acid into the stained cell wall.

As was mentioned above, there are differences between cell walls of Gram-negative and Gram-positive bacteria, mycobacteria, and fungi. These differences are presented in **Fig. 2.31** and will be described below.

The differences between Gram-negative and Gram-positive bacteria

Outer membrane: The spectrum of proteins in the outer membrane is highly variable and their diversity is dependent on both external and internal conditions.

A number of proteins are inducible; they are only created if their presence is necessary for the cell, and after the "completion of the task," they are hydrolyzed.

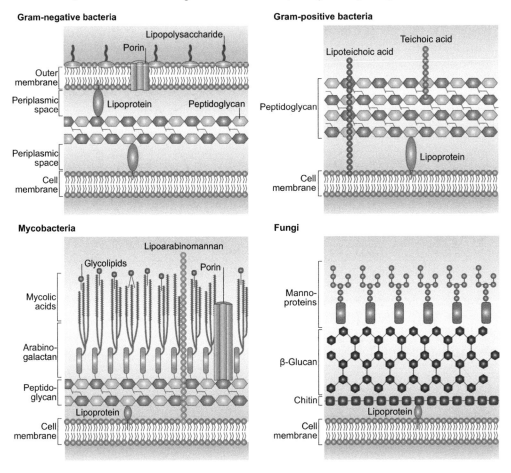

Fig. 2.31. Differences in cell wall structure among Gram-negative and Gram-positive bacteria, mycobacteria, and fungi.

Porins: Integral proteins (oligomers, usually trimers) forming pores, allowing "free" movement of small polar molecules through the outer membrane and the periplasmic space. All pore molecules are noncovalently attached to the peptidoglycan but are tightly bound. Other functions are as follows:

- *OmpC* produces nonspecific pores (for hydrophilic molecules), phage TuIb, T4.
- *OmpA* binds the recipient cell to F pilus (conjugation), phage TuII receptor, specific interactions with LPS.
- *OmpF* proteins diffusion channel for small molecules, phage receptor TuIa, T2;
- *OmpB* proteins diffusion channel for maltose and other metabolites;
- *LamB* specific maltose and maltodextrin porin, phage receptor λ;
- *PhoE* anion selective channel, induced by phosphorus deficiency.

Lipopolysaccharides. The cell wall of Gram-negative bacteria is much thinner (8–18 nm) but its composition is more complex (**Fig. 2.32**). The murein microstructure has just one layer and it often makes up less than 12% of the dry weight of the cell wall. However, it serves as a site of connection for lipopolysaccharides and phospholipids

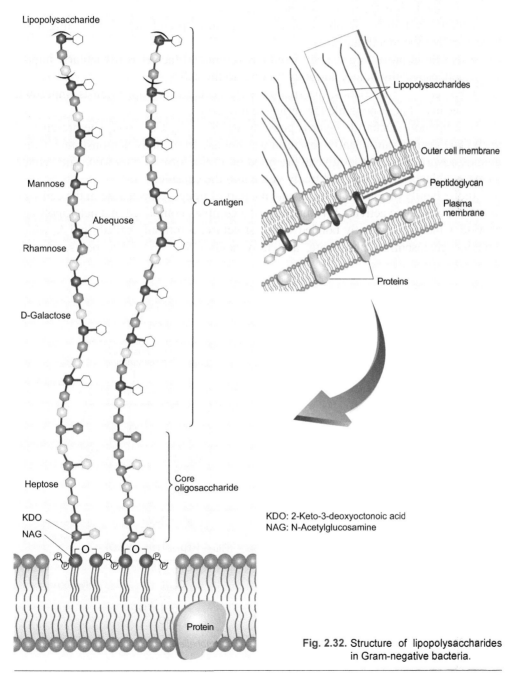

Lipopolysaccharide

Mannose

Abequose

Rhamnose

D-Galactose

Heptose

KDO

NAG

O-antigen

Core oligosaccharide

Protein

Lipopolysaccharides

Outer cell membrane

Peptidoglycan

Plasma membrane

Proteins

KDO: 2-Keto-3-deoxyoctonoic acid
NAG: N-Acetylglucosamine

Fig. 2.32. Structure of lipopolysaccharides in Gram-negative bacteria.

and partly for lipoproteins that are connected by a noncovalent bond. This additional layer comprises the remaining 80% of the dry weight of the cell wall.

The lipopolysaccharide layer of the outer membrane of Gram-negative bacteria is characterized as follows:

- The lipid part is a "component" of the phospholipid bilayer, and the polysaccharide is directed to the external environment.
- Under certain conditions, bacteria may form a "pseudolayer," including up to 45% of the membrane area.
- Its role in protecting the cell from environmental factors is particularly important, especially before antibody binding to the cell.
- For cell characteristics, the specific polysaccharide composition is important, because it is responsible for antigenic properties.

The lipopolysaccharide layer is possibly stabilized by Ca^{2+} ions. The presence of teichoic acids in Gram-negative bacteria was not experimentally demonstrated. Compounds covering the peptidoglycan are piled up in the form of a mosaic. They create an external layer of the cell and freely verge into the capsule.

In the family of *Enterobacteriaceae*, the part of lipopolysaccharide in the cell wall may have endotoxic activity. It is made up of a complex comprised of a heteropolysaccharide, which is covalently bonded with a structure, commonly called lipid A, which contains glucosamine, phosphate, and fatty acids (**Fig. 2.33**). The polysaccharide

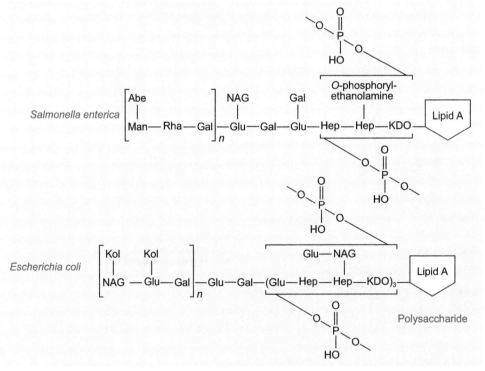

Fig. 2.33. Structure of lipopolysaccharide in cell walls of *Salmonella enterica* serovar Typhimurium and *Escherichia coli*.

fraction then serves as an antigen. The core of the polysaccharide contains glucose, galactose, N-acetylglucosamine, heptose, and 2-keto-3-deoxyoctonic acid (KDO). The core is then connected with various side chains, which consist of serologically specific saccharides including dideoxyhexoses (mannose, fucose, 6-deoxytalose, colitose, galactosamine, etc.).

The existence of the lipopolysaccharide layer, which overlaps the murein, may be testified by the fact that lysozyme activity, which splits the bond between C_1 N-acetylmuramic acid (MurNAc) and C_4 N-acetylglucosamine (GlcNAc or NAG), is only observable after the destruction of this layer. For example, it can be provided by the application of ethylenediaminetetraacetic acid (EDTA). This compound with chelating activity "sequesters" the Ca^{2+} ions into the EDTAmetal complex and dissociates the layer; thus, lysozyme may disintegrate murein.

Lipid A is a lipid component of an endotoxin held responsible for the toxicity of Gram-negative bacteria. It is the innermost of the three regions of the lipopolysaccharide (LPS), also called endotoxin molecule, and its hydrophobic nature allows it to anchor the LPS to the outer membrane. Lipid A consists of two glucosamine (carbohydrate/sugar) units, in a $\beta(1 \rightarrow 6)$ linkage, with attached acyl chains ("fatty acids"), and normally containing one phosphate group on each carbohydrate. The chemical composition of lipid A and extractable lipids replacing or complementing LPS in bacterial cells are presented in **Fig. 2.34** and **Fig. 2.35**.

A polysaccharide or protein surface layer serves as specific adhesins. To achieve adherence to host surfaces effectively, many bacteria produce multiple adherence factors called adhesins. It should be noted that polysaccharides play an important role in attaching a bacterial cell to the host cell. Adhesive sites are located between the outer and the cytoplasmic membranes (e.g., in *Escherichia coli*). They provide a stable position of the cytoplasmic membrane and the cell wall (**Fig. 2.36**).

When lysozyme is used in a 0.1–0.2 M solution of sucrose, cells within such solution form round **protoplasts**, which are stable only in a hypertonic environment. These cells have their walls removed resulting in an intact state of the cells; they do not adsorb phages and are completely lacking traces of muramic and diaminopimelic (DAPa) acids. Another type is **spheroplasts**, cells whose cell walls were not completely removed, usually as an effect of penicillin on a growing cell culture in isotonic or hypertonic environment. Another way of creating spheroplasts is the usage of D-amino acids during cultivation or letting the cells undergo the so-called "anaerobic lysis." The difference with protoplasts is that spheroplasts are not completely lacking a cell wall and are usually bearing its fragments; thus, we are able to detect muramic acid. They are also able to undergo binary fission.

In addition to lysozyme, there are also other enzymes that possess the ability to deconstruct the murein skeleton of the cell wall. These are bacterial muropeptidases that split peptide bonds forming transverse connection within the murein microstructure (**Fig. 2.37**). For example, endopeptidase isolated from the cells of *Escherichia coli* cleaves the bond between D-alanine and meso-DAP acid. Similar effect has amidase isolated from actinomycete of genus *Streptomyces*, which splits the bond between the carboxyl group of lactic acid and the amino group of L-alanine.

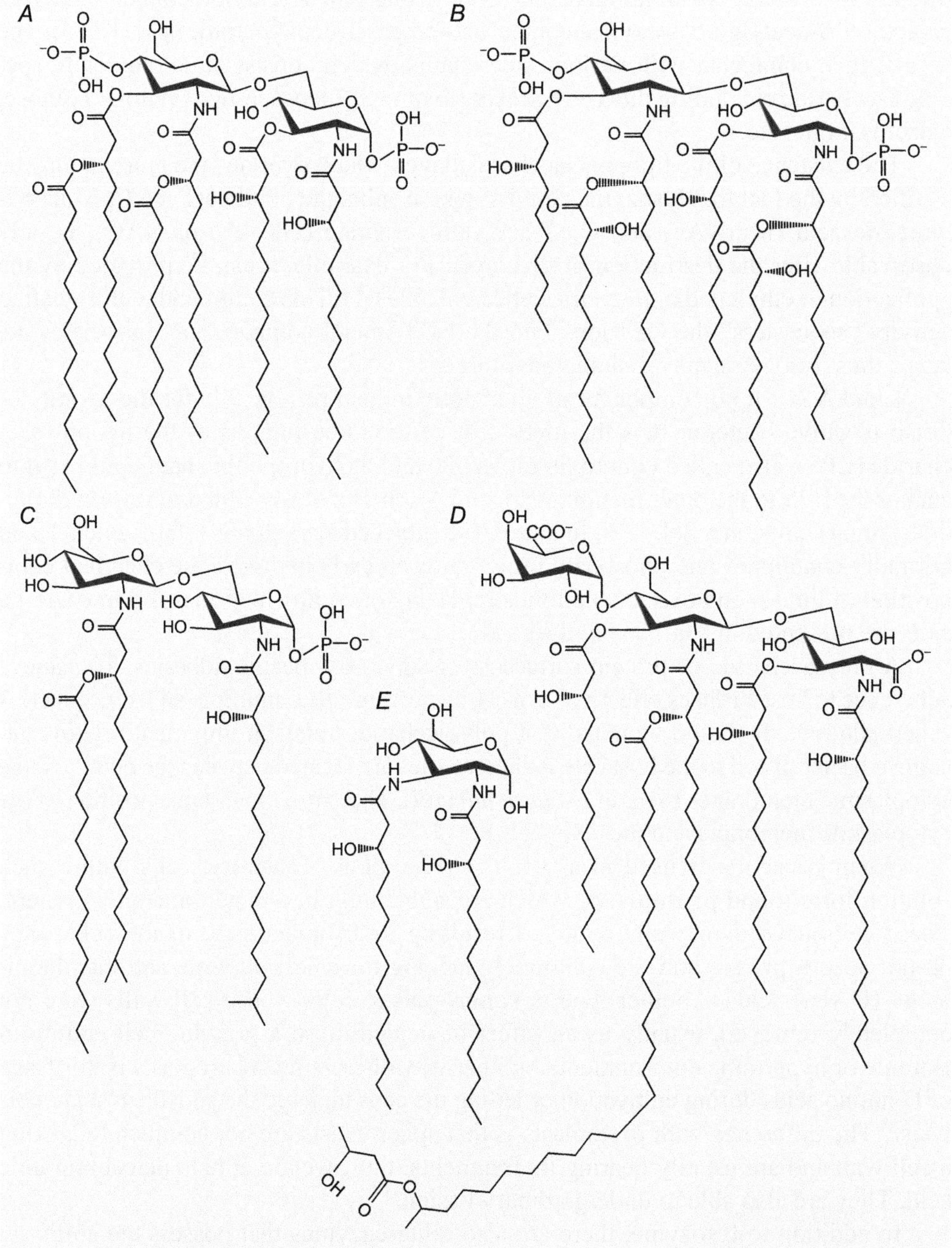

Fig. 2.34. Lipid *A* with nonclassical structure: lipid *A* in *E. coli* (*A*), lipid *A* from *Pseudomonas aeruginosa* (fatty acids are shorter than in *E. coli* and are distributed symmetrically over two glucosamine residues) (*B*), lipid *A* from *Porphyromonas gingivalis* (*R* is **Hydrogen** in one strain and ***Acyl group*** in another) (*C*), lipid *A* from *Rhizobium leguminosarum* (the reducing glucosamine residue is replaced by 2-amino-2-deoxygluconic acid; the usual phosphate group in the 4′-position is replaced by a galacturonic acid residue) (*D*), lipid *A* from *Rhodopseudomonas viridis* is based on a 2,3-diaminoglucose monomer and completely lost phosphate substituents.

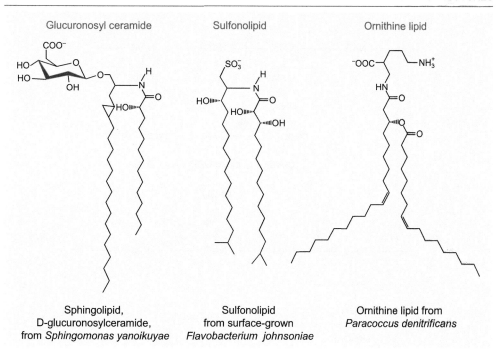

Glucuronosyl ceramide

Sulfonolipid

Ornithine lipid

Sphingolipid,
D-glucuronosylceramide,
from *Sphingomonas yanoikuyae*

Sulfonolipid
from surface-grown
Flavobacterium johnsoniae

Ornithine lipid from
Paracoccus denitrificans

Fig. 2.35. Extractable lipids replacing or complementing LPS in different bacteria.

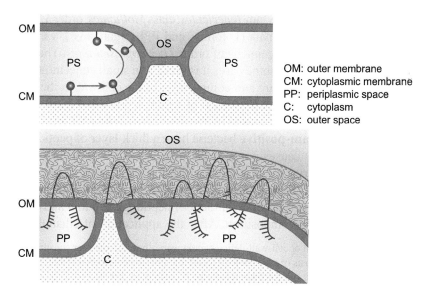

OM: outer membrane
CM: cytoplasmic membrane
PP: periplasmic space
C: cytoplasm
OS: outer space

Fig. 2.36. Adhesive sites are located between the outer and the cytoplasmic membrane of *Escherichia coli*.

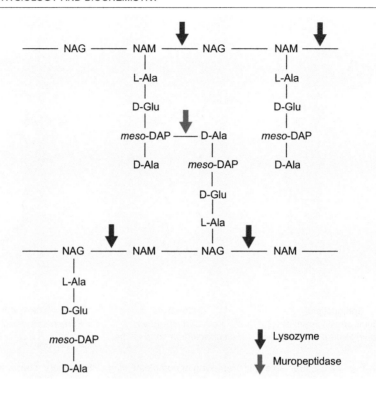

Fig. 2.37. Putative structure of murein in *Escherichia coli* (arrows are pointing to the cleavage sites → red one = muropeptidase, brown ones = lysozyme).

A rigid cell wall accompanied with the murein layer is not present in a rather small group of microorganisms from phylum *Tenericutes*, class *Mollicutes*, order *Mycoplasmatales*, commonly called mycoplasmata (L forms[1] and PPL organisms[2]), various archaeal cells like genus *Halobium*, or specimens of order *Thermoplasmatales*. These organisms mostly lack muramic acid, or in some cases it is present in small concentrations.

The variability in the structure and composition of the cell wall is the quintessence of various staining techniques, the most used of which is possibly **Gram staining**. The difference in the stain is possibly caused by the thickness of the wall itself and its effect on the elution process. **Gram-positive** bacteria have a thick layer of peptidoglycan. The usage of ethanol probably causes pores in the microstructure to shrink, hindering the way for the complex of crystal violet-iodine, disallowing it to wash out from the cell wall. As a result, the cells remain stained purple. However, in **Gram-negative** bacteria, the cell wall is much thinner with a less amount of peptidoglycan in it and even the microstructure is simpler in its composition. After the application of ethanol on the fixed

[1] Also known as L-phase variant or cell-wall-deficient organisms, these are microorganisms lacking the cell wall temporarily — spheroplasts or are unable to revert and thus are considered stable. Parasitic bacteria such as genus *Mycoplasma* are not considered L-forms (Leaver *et al.*, 2009).

[2] PPL is a common abbreviation for pleuropneumonia-like organisms, which are usually parasitic lifeforms colonizing respiratory system. It is a usual tag for mycoplasma *sensu lato*, an organism lacking a cell wall.

cells, the pores remain wider, and the complex of crystal violet-iodine is extracted out; thus, the cell is stained by safranin. A comparison of individual cell wall structures of Gram-positive and Gram-negative bacteria is presented in **Table 2.4**.

Table 2.4. Some cell wall features of G$^+$ and G$^-$ bacteria

Features	G$^+$	G$^-$
Peptidoglycan	Thick (multilayer)	Thin (single layer)
Teichoic acids	Many	Absent
External membrane	Absent	Present
Lipopolysaccharide content	Practically no	Large (outer membrane)
Lipids and lipoproteins	Little	Much
Production of toxins	Predominantly exotoxins	Mainly endotoxins
Mechanical damage	Small	Big
Effect of lysozyme	Strong	Weak
Sensitivity to penicillin	Strong	Weak
Sensitivity to streptomycin	Weak	Strong
Sensitivity to tetracyclines	Weak	Strong
Inhibition of basic dyes	Strong	Weak
Resistance to drying	Strong	Weak

The cell wall of mycobacteria is extremely resistant to environmental factors and the main determinant of virulence. It may contain more than 60% of lipids. The structure of the cell wall consists of three main components (**Fig. 2.38**):
- characteristic long chains of mycolic acids;
- branched arabinogalactan;
- peptidoglycan layer (**Fig. 2.39**).

Membrane segments contain extractable lipids, waxes, and glycolipids. A capsule consisting of polysaccharides and proteins is usually present over this layer. The main feature is the presence of **mycolic acids** (branched fatty acids with 26–60 to sometimes 90 carbon atoms). Mycolic acids are **covalently** attached to arabinogalactan. There is a layer of glycolipids outside the mycotic acids. Under these layers, a layer of peptidoglycan is located above the plasma membrane. For muramic acid, peptidoglycan replacing *N*-acetyl is *N*-glycol. The cell wall structure is structurally similar to the cell wall of Gram-negative bacteria (although they are G$^+$, including the peptidoglycan structure). The surface is **highly** hydrophobic and has low permeability. These properties are considered the cause of mycobacterial resistance to a large number of antibiotics, disinfectants, and drying. They are acid-resistant.

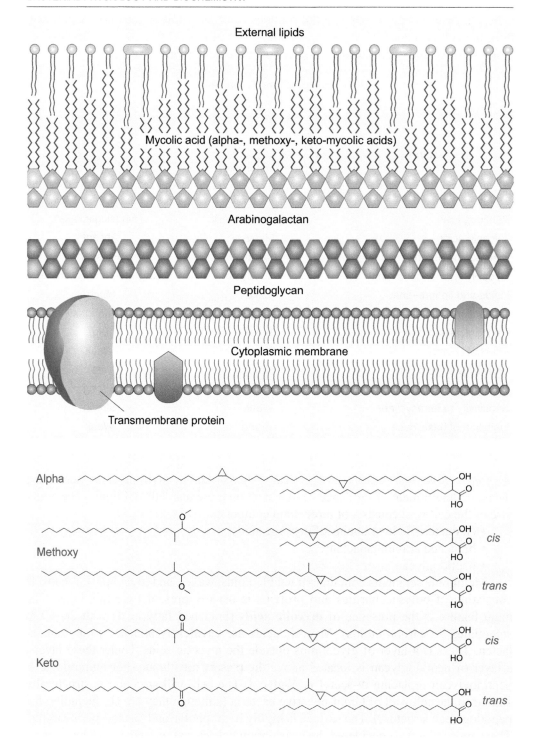

Fig. 2.38. Structure of the cell wall of mycobacteria.

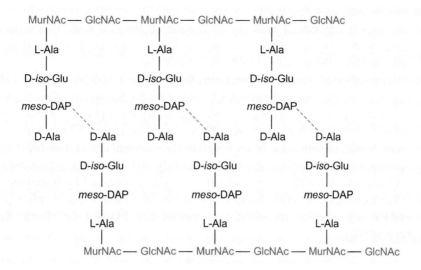

Fig. 2.39. Structure of the peptidoglycan layer in *Mycobacterium* sp.

- α-Mycotic acids make up 70% of the mycotic acids present (contain several cyclopropane rings).
- Methoxymic acid forms 10–15% of the mycolic acid present (contains several methoxy groups).
- The remainder is formed by keto-mycolic acids (containing several keto groups).

In mycobacteria, the structural modification of peptidoglycan is significant; *N*-acetylmuramic acid is oxidized to *N*-glycolylmuramic acid. This modification results in an increased peptidoglycan strength due to hydrogen bonding and potentially decreases susceptibility to lysozyme.

Cell wall of cyanobacteria. Cyanobacteria usually have a similar structure as Gram-negative bacteria. Some of the specific features of the cyanobacterial cell wall differ from other Gram-negative bacteria, and cyanobacteria appear to be a mixture of Gram-negative and Gram-positive cells. The conventional Gram-negative bacteria have the cell wall with a thickness of only about 2–6 nm. In single-cell cyanobacteria, it is about 10 nm thick; in fibrous cyanobacteria, it is 15–35 nm thick, and rarely, the thickness can be up to 700 nm (e.g., in *Oscilalatoria princeps*). The peptides forming cross-linking are rather characteristic for Gram-negative bacteria.

The functions of the cell wall:
- The bacterial cell wall plays the role of an outer cell skeleton.
- It provides a shape to the cell and protects it from the external environment.
- Peptidoglycan is responsible for strength and resistance (it is specific for prokaryotes).
- The synthesis of peptidoglycan is catalyzed by enzymes of the periplasmic space and the external cytoplasmic membrane.
- There are two types of the cell wall: Gram-positive and Gram-negative.

Bacteria without a cell wall:

- Genetically embedded *inability to synthesize the cell wall*. It is especially in intracellular parasites (*Mycoplasma, Acholeplasma,* etc.). The cell is usually bounded by a three-layer cytoplasmic membrane.
- *Protoplasts* are formed in G^+ bacteria after treatment with lysozyme (by cleavage of the glycosidic bond of peptidoglycan) or by the action of antibiotics blocking cell wall synthesis (penicillins, cephalosporins). Protoplasts can metabolize and grow, but they do not reproduce. They must stabilize in a hypertonic environment. Under certain conditions, reversion to a normal cell is possible (**Fig. 2.40**).
- *Spheroplasts* metabolize, grow, and multiply. In G^- bacteria, protoplasts are not formed, because only the peptidoglycan layer is removed by lysozyme, but the outer membrane remains. Subtraction of the factor that induces spheroplast formation is a possible reversion to a normal cell. Fixed L-cell forms have lost reversibility.

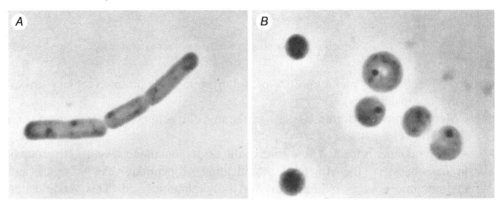

Fig. 2.40. Normal cell of *Bacillus megaterium* (*A*) and protoplasts formed after lysozyme stabilized in sucrose (*B*).

Thus, there are three forms of organisms that are devoid of the cell wall, namely protoplasts, spheroplasts, and L forms. The first two forms can reverse and acquire a cell wall upon termination of a factor that inhibits or cleaves its synthesis. L forms lost their ability to reverse after termination of the factor. There is also a separate group of microorganisms that are genetically incapable of synthesizing the cell wall.

2.7. Flagella, pilus, and fimbria

The motility and ability to adhere on surface are discriminatory factors in bacterial characteristics. The purpose of motility is not only an option to mindlessly swim through the environment, but rather an ability to orientate the cell and navigate to nutrient rich and stress absent spaces.

Some bacteria have gentle protein fibers on outside cells (over the cell wall), which are commonly referred to as *flagella* and *fimbriae* (pili). However, it is more appropriate to divide them into three groups: flagella, fimbria, and pili, based on their different composition of protein subunits. In their characterization, it is necessary to take into account not only the chemical structure, the architecture, but also different functions of individual structures.

The ability of bacterial motile to some factor is called taxis. The motile cells are logically capable to direct their movement as a reaction to the presence of various positive or negative factors. Based on such factors, we can distinguish *chemotaxis*, induced by the presence or absence of a chemical component, *aerotaxis*, a reaction to an oxygen gradient, which depends on the physiological properties, and *phototaxis*, a reaction to light, which is the most important for photosynthetic organisms. The reaction is commonly focused on avoiding the factors, which are too high or low, balancing the organism's access to essential resources in its environment.

Flagella are rigid and rotate by power of the *M-ring* (rotor) and the *S-ring* (stator). The energy source is a proton gradient. One revolution of flagella consumes about 250 protons. The movement of the bacteria is rotary, locomotive, and Brown's. They move forward along the curved track, stop, wobble, and turn (**Fig. 2.41**). The new direction of movement is entirely random. The cell turns counterclockwise. Motion speed is about 50 μm/s. The movement of flagella also facilitates chemotaxis and phototaxis.

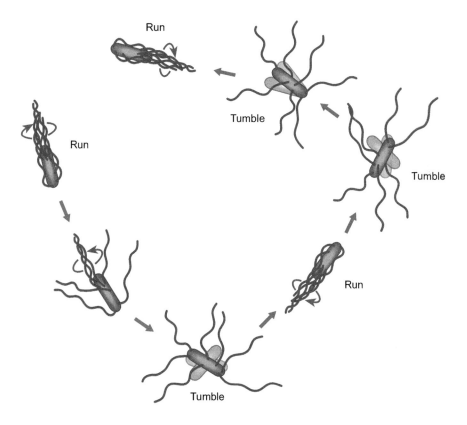

Fig. 2.41. Movement of bacterial cells.

The location of flagellum is mostly varied and depends on bacterial species. It can usually be found at the poles of the cells (***monotrichous*** — having just one flagellum, ***amphitrichous*** — having single or multiple flagella on both poles of the cell, ***lophotrichous*** — having a tuft of flagella on just one pole of the cell). A less common location is ***peritrichous*** when flagella cover the whole cell (**Fig. 2.42**). The diameter of such flagellum may range from 10 to 20 nm, with the moderate length of 20 μm. The number of flagella and their structure are also an important discriminative factor with impact on prokaryotic taxonomy.

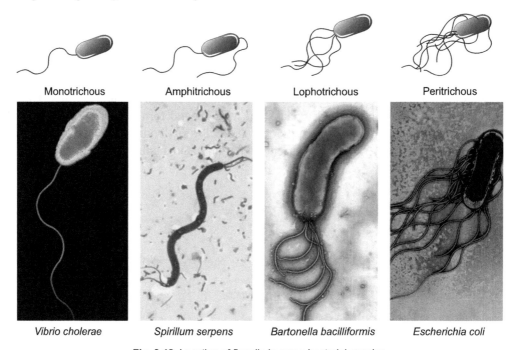

Fig. 2.42. Location of flagella in some bacterial species.

The main parts of flagella (**Fig. 2.43**) are as follows:

- ***The filament of the flagellum*** formed of a globular protein, flagellin (molecular mass 30,000–60,000 Da), connected to several filaments (5–10) spirally twisted. The filament is long up to 20 μm with a diameter of 10–30 nm. The flagellin does not contain cysteine and tryptophan. The filament is terminated by a "closing" protein. Some bacteria have also a structure wrapping the flagellum (the lipopolysaccharide layer in *Vibrio cholerae*). Grow in the nutrition medium is about 10–20 min.
- ***The hook*** is composed of identical globular proteins (other than flagellin). It is probably a connection between the filament and the basal part of the flagellum (**Fig. 2.44**).
- ***The basal part*** anchors the flagella into the cell wall and the cytoplasmic membrane. It is responsible for the movement of the flagellum and composed of discs made up of nine different proteins. Their number and location differ in G^+ and G^- bacteria.

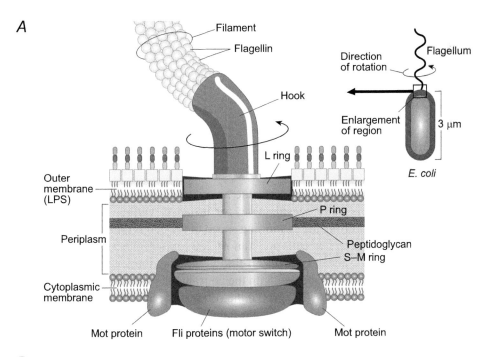

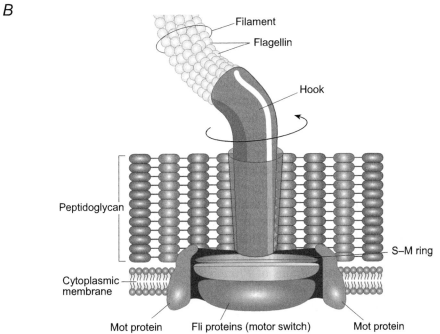

Fig. 2.43. Comparison of the flagella structure in Gram-negative (A) and Gram-positive (B) bacteria.

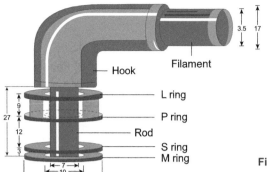

Fig. 2.44. Structure of the flagella attachment in Gram-negative bacteria.

The movement is somewhat similar to the function of a ship propeller. The rotation of such cellular motor is quite considerable; for instance, various flagellated spirilla achieve 40–60 rps. In addition, the swimming speed of the cell is quite remarkable, with 27 μm/s in *Bacillus megaterium* cells and almost 200 μm/s in *Vibrio cholerae*. In fact, such bacteria can travel a distance that is 50-fold larger than the length of their cells.

The movement itself is connected with the continuous production of proton, creating a gradient which is then used to propel the rotor of the flagellum, causing the cell to move. On the other hand, the eukaryotic flagella work on a different basis, using ATP-dehydrogenase molecular motors to generate movement, which is not based on rotation, but rather on a wriggling-like motion.

The building blocks of the flagellum are flagellins, proteins similar to myosin found in muscle cells. Flagellins are globular proteins with a molecular weight around 20–40 kDa. Fagellin composition includes the low amount of amino acids like cysteine, histidine, proline, tyrosine, and tryptophan. Apart of proteins, flagella of various bacteria contain a lesser number of saccharides. The microstructure of the flagellum is built of molecules of flagellin with a diameter around 5 nm. Those form fibers of 3 – 11 molecules, which are shaped into a helical structure around the common axis. The helical structure is not the sole one; besides it, a form of vertical rods exists. The synthesis of the flagellum is carried out by the polymerization of flagellins, which come out through the basal disks, located under the plasma membrane (**Fig. 2.45**).

Flagellated strains of *Proteus vulgaris* are able to spread around the surface of a medium in a fine film and called H-forms (from the German word Hauch for film, literally breath or mist). Other strains are lacking flagella and do not form the film; hence, they are called O-forms (ohne Hauch—without film). This nomenclature rose from the observation of Edmund Weil (1879–1922) and Arthur Felix (1887–1956). They noticed that flagellated strains of *Proteus* sp. were forming a thin film, resembling the mist produced by breath on a glass (E. Weil and A. Felix, 1917). This denomination was later accepted and is commonly used in bacterial diagnostics, referring to somatic bacterial antigens — *O*-antigens and flagellar antigens — *H*-antigens.

Bacterial flagella synthesis is a complex process involving the expression of at least 30 genes. More than 10 genes control the synthesis of hook and basal body proteins. Other genes control flagellin synthesis, its construction, or function. Flagellin is synthesized inside the cell, and the type III secretion system is transferred through the outer membrane (**Fig. 2.45**).

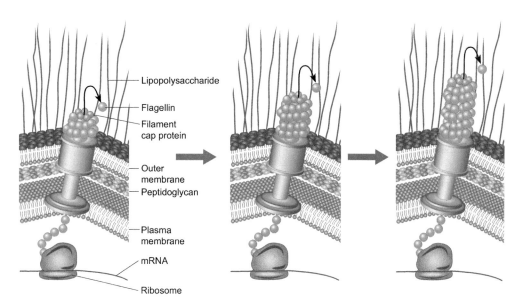

Fig. 2.45. Synthesis of the flagellum filament.

As was described above, most bacterial species have flagella, which are located outside the cells; however, spirochetes have a completely different type of its placement, *axially placed flagella* (also called *endoflagella* or *periplasmic flagella*) (**Fig. 2.46**). This type of flagella is located between the cytoplasmic (inner) and the outer membranes in the periplasmic space above the cell wall (**Fig. 2.47**).

Beside the flagella, bacterial cells (especially in Gram-negative bacteria) may be accompanied with long filamentous structures called *fimbriae* (**Fig. 2.48**). Unlike the flagella, fimbriae are thinner and shorter. Their diameter ranges from 7 to 10 nm, and length from 1 to 4 µm only. Their chemical composition resembles a polymer, composed of protein subunits creating a helical structure. The primary function of fimbriae is adherence; they help their bearer to adhere to specific surfaces. Such may be plant or animal cells or inert compounds like cellulose or glass.

Fimbriae are numerous relatively short rigid straight protein filaments. They are very fragile and easy to mislead. They occur only in Gram-negative bacteria. There may be several hundred different types on one cell. The fimbria has the nature of a tube 2–8 nm wide and a length from 0.1 to several microns. It can include more components. The main structural unit is a protein subunit called pilin or fibrillin. The pilin has a linear sequence of identical protein subunits that can be supplemented with different specific subunits. The type, number, and composition of fimbria depend on a number of physical and chemical factors. The structure is encoded on the chromosome or on plasmids. A switching mechanism is applied to the transcription, which determines whether the given protein is a subunit of a given type of fimbria.

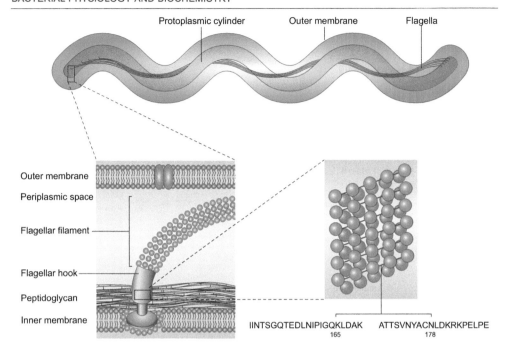

Fig. 2.46. Periplasmic flagella in the spirochetes.

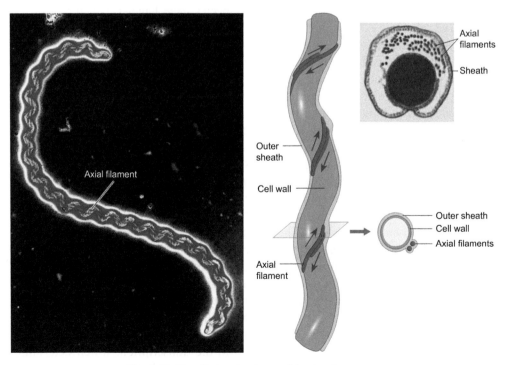

Fig. 2.47. Flagella in spirochetes of *Leptospira* genera.

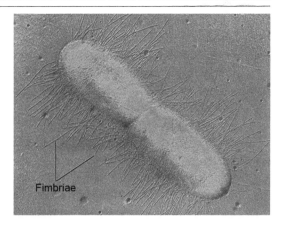

Fimbriae

Fig. 2.48. Bacterial fimbriae.

The classification of the fimbriae in Gram-negative bacteria is according to **Table 2.5**:
- homology of the amino acid sequence of the major subunit;
- physical, adhesive, and antigenic properties;
- mechanism of secretion and assembly.

Table 2.5. Some types of fimbriae

Fimbriae	Species	Comments
Type I (a) Basic (b) P fimbriae	*E. coli* strains causing inflammation of the urinary bladder, kidneys	Encoded on the chromosome, *pap* operon; expression is regulated by heat
(c) K88	*E. coli* pathogenic strains for piglets	Encoded on plasmids; the main unit is adhesin
(d) K99	*E. coli* pathogenic strains for calves and lambs	Encoded on plasmids; the main unit have adhesion properties
Type II	Enterobacteria, primarily *Salmonella* species	Similar to type I, but adhesive properties have been lost
Type III	Enterobacteria	Mannose-insensitive adhesion
Type IV	Various G⁻ pathogenic strains, including *Neisseria gonorrhoeae*, *Neisseria meningitidis*, and *Vibrio cholerae*	Mostly encoded on the chromosome, some of them on plasmids

Another type of structure, which may be similar to fimbriae, are F-pili. These are protein tubes with a diameter of 5–10 nm and a length of 0.5–10 μm. They can be found among donor cells of *E. coli* K12 strain and strains containing a specific plasmid (F⁺ and HFr strains), which is a vital condition for conjugation. Pili are made of the protein called pilin with a molecular weight around 16 kDa. The composition of a molecule of pilin is characterized by a number of amino acids containing side chains.

Pili are protein structures on the surface of a cell occurring specifically in Gram-negative bacteria, mostly in organisms capable of ***DNA conjugation***. One to several pili can be located on the cell (usually one and a maximum of 5 – 6). They are longer than fimbria, and their diameter is around 10 µm. They are determined on *conjugative plasmids* (sex factor). There are several types of pili that differ in size and shape. Some are thin and flexible, while others are short and rigid. For a given organism, the pilus type correlates with the physical conditions of the conjugation process (e.g., in the fluid environment or on the surface of the agar layer).

Pili encoded on plasmid F (F pili in *E. coli*) are flexible, one to several microns long, and have an average diameter of 8–9 nm. Their structure is tubular with an axial channel of about 2.5 nm in diameter. It is composed of protein subunits of one type — *pilin*, which contains 70 amino acid residues, one D-glucose, and 2 phosphate residues. Assembling models include five pilin subunits, which are joined in a circle, and the next circle is rotated by 29° with respect to the previous. The subunits are arranged helically.

2.8. Bacterial capsule

In many genera of bacteria, the cell wall itself is accompanied by a hydrated layer called the capsule. The presence of the capsule is an important diagnostic sign, and it usually serves as a key to bacterial resistance, giving rise to the virulence of certain strains. The main cause is often an elevated resistance to an effect of lysozyme, or ability to evade bacteriophage infection as well as a phagocytosis. Another major benefit is the ability to endure drought owing to the significant amount of water contained in this layer. Furthermore, the components of these polymeric compounds are usually high in carboxyl groups, enabling the layer to absorb cations from the environment (**Fig. 2.49**).

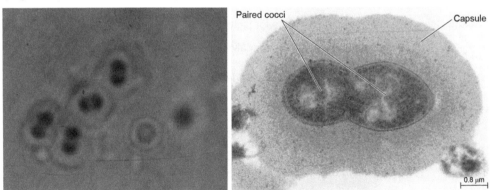

Fig. 2.49. Bacterial capsules.

Strains that are known to produce the capsule generally grow on agar in the form of so-called **S-forms** (smooth), referring to the appearance of their colonies. The S-forms are considered to be virulent strains; however, they can spontaneously mutate to **R-forms** (rough), which do not produce the capsules. This process often goes along with the change of the biological features of such cells.

Capsule production is strongly dependent on the nutrient spectrum and the physical conditions of cultivation. Some bacteria can produce both a capsule and a slimy layer. *S. mutans* produces soluble and insoluble forms of dextrans. The insoluble form is referred to as the ***mutan***, being a branched glucose polymer (a linear chain having an α-1,6 and an α-1,3 bonds). Mutan is responsible for cell binding on the tooth enamel and the formation of dental plaque.

Microcapsule (0.2 nm thick) can be mistakenly considered as a part of the cell wall under certain conditions. It is a complex containing protein, polysaccharide, and sometimes lipid traces in the species of *Enterobacteriaceae* family. This complex is behind the effect of *O*-antigen, produced by the cells in *S*-form. Microcapsules of certain species of genus *Streptococcus* may contain the *M*-, or *T*-protein, which has antigenic features.

Macrocapsule is significantly stronger and consists mainly of polysaccharides and proteins. Its composition is different depending on the type of bacteria. The macrocapsule is layers with thickness above 0.2 μm, composed of polysaccharides (*Pneumococcus* sp.) or peptides (*Bacillus* sp.). The capsules of pneumococci (probably strains of *Streptococcus pneumoniae*) contain both homo- and heteropolysaccharides with antigenic properties. In genus *Bacillus* (in particular, *B. subtilis, B. anthracis,* etc.), the main components of the capsule are *D*- and *L*-residues of glutamic acid in the form of glutamyl polypeptide (**Fig. 2.50**). The serological groups A and C of genus *Streptococcus* have the envelope made of hyaluronic acid.

Hyaluronic acid

The structural components of hyaluronic acid are glucuronic acid and *N*-acetyl-glucosamine.

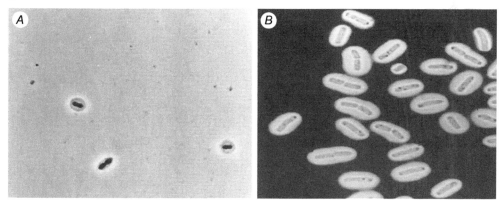

Fig. 2.50. *Streptococcus pneumoniae* (A); *Bacillus* sp. (B).

The chemical compositions of dextran (in *Leuconostoc mesenteroides*), polyglucose gluconate (in *Streptococcus pneumoniae* III), hyaluronic acid (in *Streptococcus species*), and poly-γ-D-glutamic acid (in *Bacillus anthracis*) are presented in **Table 2.6.**

Table 2.6. Chemical compositions of some bacterial capsules

Bacterium	Name of capsular material	Structure of repeating unit
Leuconostoc mesenteroides	Dextran	α-1,6-Poly-D-glucose
Streptococcus pneumoniae III	Polyglucose glucuronate	Glucuronic acid — Glucose
Streptococcus spp.	Hyaluronic acid	Glucuronic acid — N-Acetyl glucosamine
Bacillus anthracis	Poly-γ-D-glutamic acid	γ-D-glutamic acid

As mentioned above, the strains producing a capsule are referred to as *S*-forms and non-producing as *R*-forms. R-forms may arise from S-forms by mutations. The presence of the capsule is an indication of virulence, and its absence shows avirulence. The capsule has antigenic properties (*capsular antigen*). It is mainly made up of water and organic substances, usually homopolysaccharides (cellulose, dextran), heteropolysaccharides (alginate, hyaluronic acid), or proteins or a combination of a protein and a polysaccharide.

Thus, according to the thickness and the consistency of the bacterial layer around the cell, various types of polymeric sheaths may be distinguished: **macrocapsules**, **microcapsules**, **slime layers**, and **slime itself**.

Capsule significance. The capsule protects the cell from drying out, represents an effective barrier against the penetration of toxic substances, prevents adsorption of phage particles, can serve as a reservoir of nutrients, promotes cell adhesion to the mat (e.g. when creating dental plaque), and helps the cell to avoid phagocytosis. A metal can accumulate in the alkaline environment in the capsule. Capsule formation is largely related to pathogenicity (encapsulated strains, *S*-forms, are usually pathogenic). If cells lose their ability of capsule formation (*R*-form mutation), they also lose the ability to cause disease. The capsule may mask the adhesives and has antigenic properties. Some exopolysaccharides are produced on an industrial scale (xanthan from *Xanthomonas campestris* and dextran from *Leuconostoc mesenteroides*).

Some building blocks of the bacterial envelope may be excreted from the cell in the form of a slime. Quite rich slime production can be observed in strains of *Leuconostoc mesenteroides* cultivated in a medium containing sucrose, which is transformed into the polysaccharide **dextran**. Dextran is a complex branched glucan comprising subunits of D-glucose, which serves as a blood substitute in surgery (features as antithrombotic, reduces blood viscosity, and tackles hypovolemia) (A. Gronwall and B. Ingelman, 1945) and is used in the preparation process of dextran gel — sephadex. Strains of *Leuconostoc mesenteroides* are used for the mass production of dextran, which is created extracellularly and catalyzed by transglycosidase.

Other bacteria, such as *Streptococcus salivarius* and some species of genus *Bacillus*, are capable of creating **levan polysaccharide** (group of fructans) from sucrose. As the name suggests, levans are polymers made up from fructose subunits. Many strains of *Enterobacteriaceae* family (e.g., *Escherichia coli*, *Enterobacter cloacae*, and *Salmonella enterica* subsp. *enterica* serovar *Typhimurium*) are able to excrete slime in certain circumstances. However, this capsular polysaccharide, called *colanic acid*, is different in its composition. It comprised glucose, glucosamine, galactose, fucose, and pyruvic acid.

In order *Burkholderiales*, representatives of *Incertae Sedis III. Leptothrix* — *Leptothrix ochracea* and *Incertae Sedis VI. Sphaerotilus* — *Sphaerotilus natans* are capable of producing cellular envelopes called *sheaths*. These sheaths are composed of heteropolysaccharides containing glucose, glucuronic acid, galactose, and fucose and can serve as an attachment or a means of movement. Some strains of bacterium *Gallionella ferruginea* as well as some *Myxobacteria* are capable of limited movement.

Sometimes, the slime serves as a cementing factor, which bonds a colony of cells together into more stable units. For example, some strains of *Acetobacter xylinum*, which produce cellulose from glucose, connect cell by a leathery membrane. This membrane was formerly called ***Mycoderma aceti***. *A. xylinum* is a soil bacterium that synthesizes and secretes cellulose as part of its metabolism of glucose. The cellulose produced by *A. xylinum* is structurally and chemically identical to cellulose found in higher plants, and in addition, not contaminated by lignins or other cellulosic derivatives. For this reason, *Acetobacter xylinum* serves as a potential model organism for the study of cellulose biosynthesis. Cells of *Sarcina ventriculi* are connected in a similar manner, forming typical aggregates. However, cellulose in these bacteria is different from the compounds, which make up the cell wall and cellular envelopes; thus, the mutations that stop the production of cellulose do not have effect on the growth of the culture.

Glycocalyx is made up of long polysaccharide fibers (**Fig. 2.51**) responsible for the adherence of bacteria to surfaces (strains in the creek, tooth enamel, plant cell, epithelial cell, etc.). Fibers have the function of adhesins. Binding to surfaces is not often (tooth enamel) or very specific (urethra).

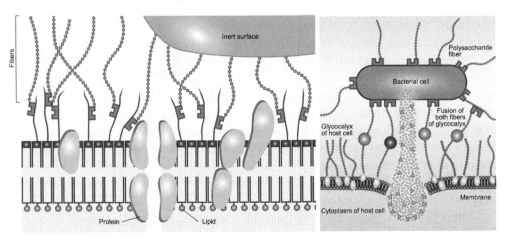

Fig. 2.51. Scheme of bacterial interaction and adhesion with host cell: adhesion of bacterial cells and host cells (specific adhesion — lectins, nonspecific — divalent ions), fusion of both fibers of glycocalyx, and creating a structure to prevent the loss of mutually exchangeable molecules to the environment.

S-layer is a two-dimensional array of proteins or glycoproteins. Generally, it is made up of one protein that is species-specific. The layer consists of symmetrical sub units that have the shape of a square, triangle, or hexagon of 3–30 nm (**Fig. 2.52**). In the subunits, pores with a diameter of 2–8 nm are created. The presence of regularly arranged functional groups (–OH, –COOH, and –NH$_2$ on the surface), which form specific motifs for certain ions or small molecules, is significant for the layer. This creates specific sites for ions and small molecules. It is mainly used to protect against extracellular enzymes, pH change, and phage adsorption.

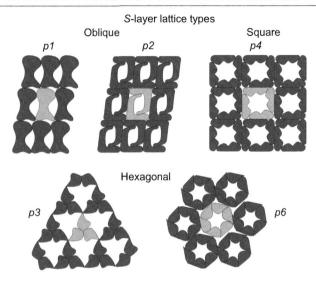

Fig. 2.52. Types of S-layers.

The number of protein units in this "crystalline grid" is 1–6. The monomolecular layers form a regular shape on solid substrates. It is also possible to use them in biotechnology or molecular nanotechnology.

2.9. Endospores

Some Gram-positive rods, especially species of *Bacillus* and *Clostridium* genera, produce endospores under certain conditions, which helps to overcome a hostile and harsh environment, and are resistant to many external factors. The creation of endospores is preceded by DNA multiplication and a division of the bacterial chromosome; one genome is then staying in the mother cell, and the second serves as a genome of the spore. The division of the nucleoid is accompanied by the creation of the *spore septum*, which encloses the mother cell and the emerging spore. The spore septum is formed by invagination of the plasma membrane, which engulfs the protoplast of the oncoming spore. This results in the creation of two membranes, the inner one and the outer one made by the protoplast of the original cellular body. The formation of the forespore is succeeded by the maturation of the endospore, during which the additional layers and spore walls are created.

The endospores in bacteria are generated by the process of *sporulation*. An endospore is not a reproductive but a *resting stage* to allow for an unfavorable external condition. About 30 operons involving more than 200 genes participate in the sporulation process. Whether the cell is going to be sporulated is decided in the *G1* phase of the cell cycle and is a random process. It is essentially programmed cell death. Spores are created in a nutrient-rich environment (usually at the end of the exponential growth phase), but the lack of some of the nutrients is a significant signal for sporulation.

Endospore formation (sporulation process) and the structure of the spore (**Fig. 2.53**):
- inner membrane (*intine* originates from the cell membrane);
- thin peptidoglycan layer (the basis of the "future" cell wall of the peptidoglycan);
- cortex (peptidoglycan) is present only in the bacterial spore and consists of concentrated layers of a specific peptidoglycan. It is responsible for the mechanical protection of the spore;
- outer membrane (extine originates from the cell membrane);
- cover protein layers (*coat spore* representing about 30–60% of spore dry matter), proteins are rich in cysteine;
- exosporium is a fine layer that is formed in some bacteria and responsible for the outer structure of spores (protein, polysaccharide, lipid, phosphate).

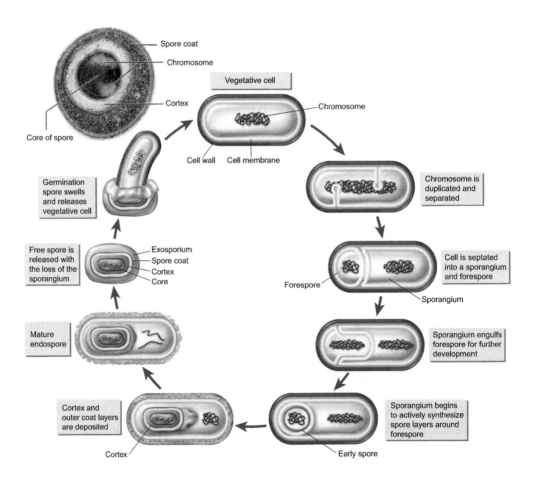

Fig. 2.53. Schema of sporulation process and structures of bacterial endospore.

The presence of dipicolinic acid indicates the thermoresistance of the spores (**Fig. 2.54**). The spore contains a minimum amount of water and is very light refracted.

It contains calcium dipicolinate (10 – 20% of the weight of the spore) with the molar ratio of 1:1. Spores have high thermal resistance and resistance to extreme environmental factors (radiation, acids, solvents, high hydrostatic and osmotic pressure, etc.), immeasurable metabolism, and long survival time (several dozen years).

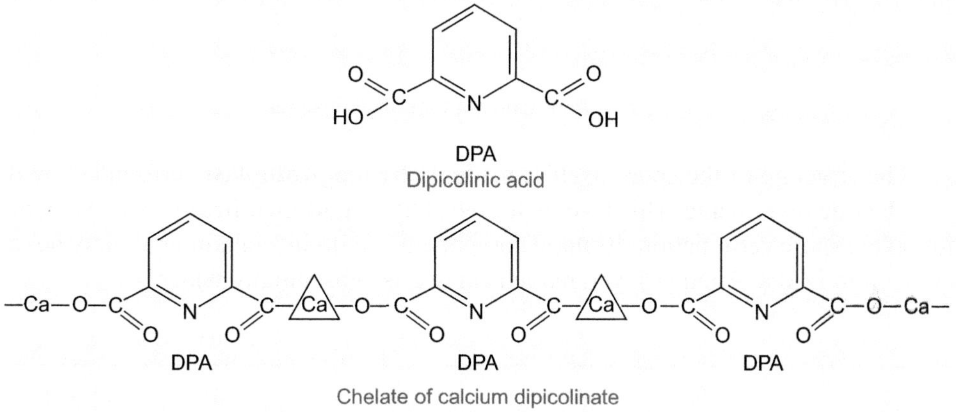

HO
OH
DPA
Dipicolinic acid
DPA
DPA
DPA
Chelate of calcium dipicolinate

Fig. 2.54. Structure of dipicolinic acid and calcium dipicolinate chelate.

The location of the endospore in the cell can differ and depends on bacterial species (**Fig. 2.55 and Fig. 2.56**).

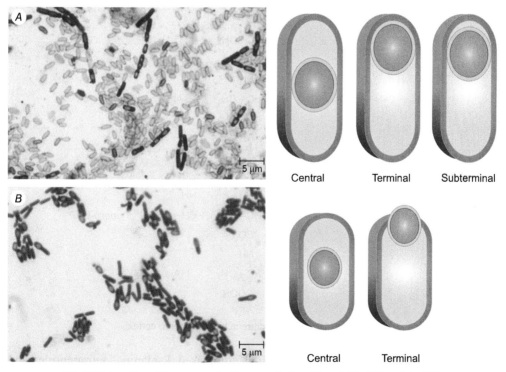

A
5 µm
Central
Terminal
Subterminal
B
5 µm
Central
Terminal

Fig. 2.55. Location of the spore: *Bacillus megaterium* (*A*) and *Clostridium tetani* (*B*).

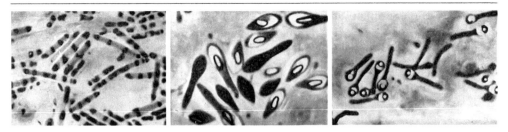

Fig. 2.56. Endospores in *Clostridium* species.

The structure of the spore itself is made up by the **protoplast** surrounded by the **cytoplasmic membrane**. This layer is followed by a **rigid membrane**, which perishes during the spore germination. Its main component is murein, containing mainly hexosamines and DAPa. Another layer, named **cortex**, is substantially wider (**Fig. 2.57**).

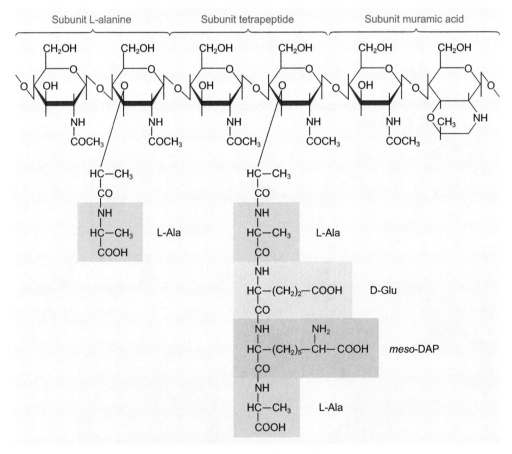

Fig. 2.57. Chemical structure of bacterial spore cortex.

It is made up of muropeptide, which comprised L-alanine, tetrapeptide, and muramic acid subunits. In addition, there is a huge amount of calcium dipicolinate (calcium salt of dipicolinic acid), which stabilizes the spore and helps it to maintain dormancy. Cortex is accompanied by a thin membrane, which originates from the cytoplasm of mother cell and has identical composition. The following layer is called **spore coat** which makes up almost 50% of the total volume. It consists primarily of proteins with a high amount of certain amino acids, especially cystine. Apart from various proteins, spore coats contain phosphomuramyl polymer in the form of peptide or protein. In some bacteria, an additional layer/sheath called exosporium (**Fig. 2.58**), containing proteins, lipids, saccharides, and phosphates, may be present. The most diverse are spores in different species of the bacteria of the *Clostridium* genus (**Fig. 2.59**).

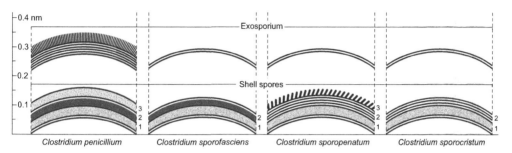

Fig. 2.58. Scheme of the construction of spores.

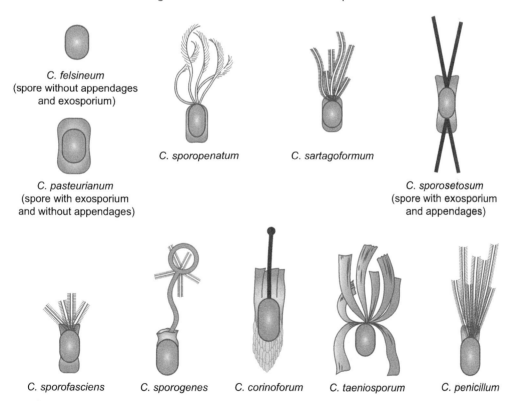

Fig. 2.59. Surface structures of spores of *Clostridium* genus.

Bacterial endospores are considerably resistant to the effects of toxic chemical agents, which is possible thanks to high impermeability of external layers. Thermoresistance is given by the extremely low content of water, which never exceeds 15% and lowers the possibility of denaturation of essential proteins by heat. Apart from the low water content, the presence of calcium dipicolinate in a high concentration also helps overcome the heat damage, since the endospores from environment with absent calcium ions have minimal concentrations of calcium dipicolinate and thus are sensitive to denaturation by means of high temperature. During sporulation process, the morphological, physiological, and biochemical changes are observed (**Fig. 2.60**).

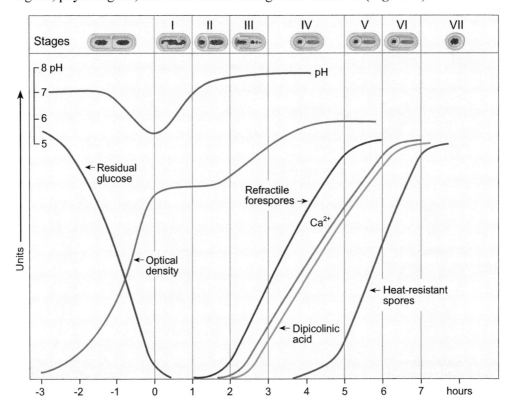

Fig. 2.60. Morphological and physiological changes in spore production in aerobic sporulating bacteria. The upper part of the graph represents types of spores (I–VII).

Sporulation in streptomycetes. A radically different type of sporulation is characteristic for streptomycetes. However, sporulation in streptomycetes is not a way of survival during adverse conditions. It is a way of reproduction of these microorganisms (**Fig. 2.61**). In this case, not endospores, but **exospores** are formed.

Aerial hyphae grow at their tips and have apical assemblies of ***DivIVA*** proteins in the extension (**Fig. 2.62**). The formation of apical sporogenic cells leads to growth inhibition. The ***FtsZ*** protein is assembled into helical fibers that convert into regularly spaced **Z-rings** that control spore septum formation. Upon the completion of the septa, preservatives create strong walls of spores that require the presence of the bacterial actin *MreB*. It is located first at the septum closure site and then appears in the vicinity of the spores. *MreB7* assures the correct construction of the wall of the spore.

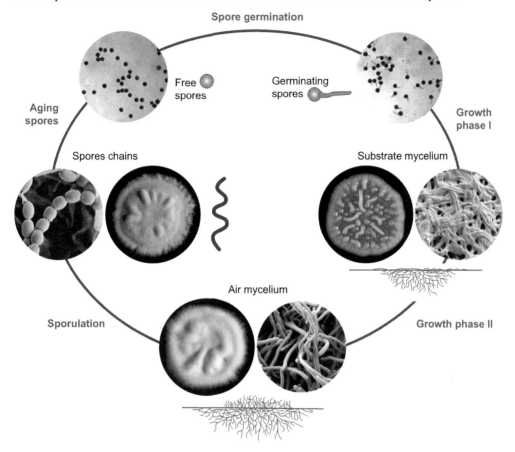

Fig. 2.61. Schematic of the location of hydrophobic surface proteins in the formation of airborne mycelium.

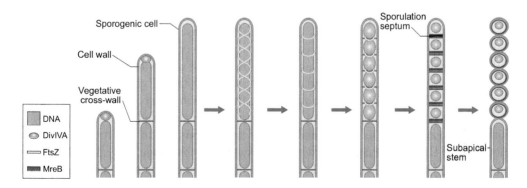

Fig. 2.62. Scheme of the implementation of processes leading to the differentiation of air aerial hyphae into a sporulation.

Germination of bacterial endospores. The transformation of an endospore into a vegetative cell is called **germination** (Fig. 2.63). Most of the bacterial endospores germinate spontaneously; however, some species require the presence of special activation factors. Activation is mostly done by heating the spore up to 65 °C for a short period. The mechanism behind this effect has not been completely understood yet. Moreover, the process must be reversible by cooling it down. Besides, certain amino acids, mostly alanine, are promoting the percentage of germinating spores in a population.

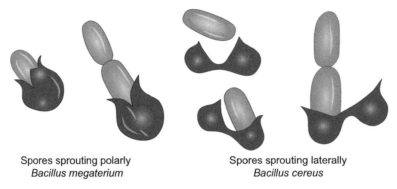

Spores sprouting polarly
Bacillus megaterium

Spores sprouting laterally
Bacillus cereus

Fig. 2.63. Germination of endospores of various species of genus *Bacillus*.

The germination process includes three phases:

Spore activation consists in disturbing the spore (mechanical abrasion, elevated temperature, low pH, presence of small molecules, especially amino acids and vitamins, etc.).

Self-sprouting spores require the presence of water, triggering chemical impulses from the environment, hydrolyzing the cortex, releasing the spore protoplast. Ca-dipicolinate is released into the environment; water, K^+, Mg^{2+}, and other ions and larger molecules are introduced into the cell. More than 30% of the weight of the spill is released into the media. During the germination, thermoresistance and brightness disappear. Undergoing processes are degradative in nature and result from the activity of enzymes present in the spores. New proteins do not arise. The entire process of germination in an individual spore lasts about 1 minute.

The differential phase, in which the spore protoplast is transformed qualitatively into a vegetative cell. Individual specific genes are transcribed and individual types of mRNA appear in a certain order, and the formation of appropriate proteins and the replication of DNA occur. Germination ends with the first division of the cell; the daughter cells are of the same size as the cell, from which the spore originated.

Besides alanine, the process of germination of the activated spore requires ribosides, glucose, and various inorganic cations and anions. In an environment rich in nutrients, the spores first absorb water and swell. At this stage, the spores undergo profound physiological changes. The respiration is increased greatly, and the activity of many enzymes present in the protoplast of the spore rises. At the same time, the spore is excreting calcium salts of dipicolinic acid and soluble proteins, associated with a loss of 25–30% of the spore's dry mass. From this moment, the spores lose their thermoresistance and ability to appear as a bright object in the optical microscope. The cortex is degraded, and only the spore coat remains intact. A new vegetative cell pierces the spore coat on the ends — polarly or laterally. It leaves the coat and prepares itself for the first cell fission. The spore coat remains in the form of an empty sack or slowly dissolves.

2.10. Pigments

Bacterial cells can be accompanied with various chemical substances, which do not take part in its composition directly and enhance the physiological and biochemical properties of the cell. One group of these compounds is called pigments, which are basically colorful substances. These pigments are being produced by various bacteria and especially those from the order *Actinomycetales*. These compounds may be contained inside the cells and then are called endopigments, or they can be excreted from the cells into the environment and are called exopigments. The production of pigments is considered an important diagnostic feature of microorganisms; however, in certain cases, it may depend on environmental factors.

The chemical structures of pigments are diverse, but it is considered that main groups are formed by derivatives of **pyrrole**, **carotene**, **phenazine**, **indole**, and **naphthoquinone**; some less occurring compounds, which are also produced by microorganisms, are **anthocyanins** and **melanins**.

Pyrrole-like rings are foundations of many complicated biomolecules, such as **bacteriochlorophyll** and **prodigiosin**.

The function of bacterial pigments is determined mostly by the physiological properties of the cell. They are mostly a part of **photosynthetic processes** (chlorophylls and carotenoids) or serve as a proton transporter in **respiratory reactions** (pyocyanine and phthiocol). The occurrence of pigments in aerial microbiota is explained by **defensive functions** of pigments against light and **UV rays**. Some pigments also bear **antibiotic properties** (phenazines and naphthoquinones). The overall role and functions of pigments are yet not fully understood.

Bacteriochlorophyll is a green pigment, which occurs in photosynthesizing bacteria and has the same function as chlorophyll in green plants. According to the structure of its molecule, we can divide it into **five groups**: *a*, *b*, *c*, *d*, and *e* (**Fig. 2.64**).

In purple bacteria[3], the main bacteriochlorophylls are types *a* and *b*, accompanied by type *d* with lesser occurrence. In contrast with green plants, bacteriochlorophylls have the first pyrrole ring substituted with acetyl group instead of the vinyl group. In types *c* and *d*, the isocyclic ring is not substituted and phytol on the third pyrrole is replaced by farnesyl. Light is then absorbed in two spectrum intervals: **purple** with a wavelength around 400 nm and **red** (or infrared) with a wavelength ranging from 600 to 800 nm.

---

[3] The term "purple bacteria" is comprising various groups of photosynthesizing bacteria from the phylum Proteobacteria. They are split between two main groups, **nonsulfur** (divided into classes α — orders *Rhodospirillales* and *Rhizobiales* and β — order *Rhodocyclales* and family *Comamonadaceae*) and **sulfur** that contains species of class γ, especially the order *Chromatiales.*

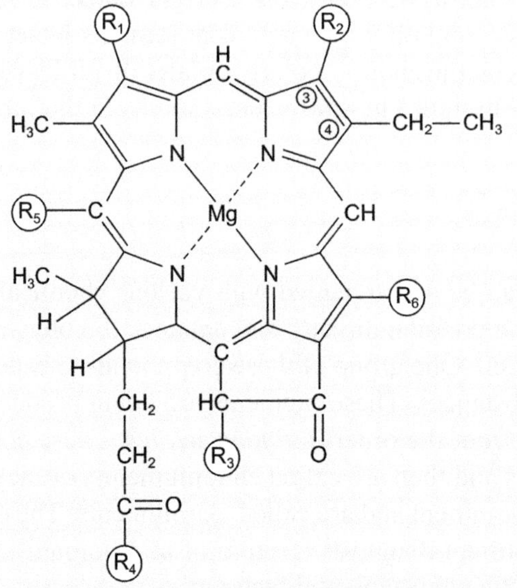

Fig. 2.64. Bacteriochlorophyll and its derivatives.

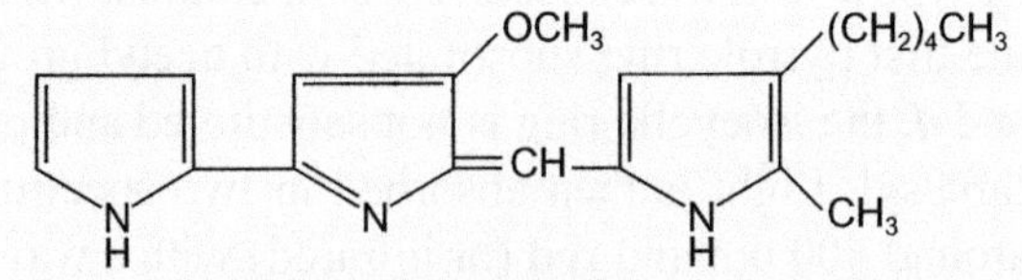

Chlorophyll	R_1	R_2	R_3	R_4	R_5	R_6
Chlorophyll a	$-CH=CH_2$	$-CH_3$	$-\underset{O}{\overset{\parallel}{C}}-OCH_3$	Phytyl	$-H$	$-CH_3$
Bacteriochlorophyll a	$-\underset{O}{\overset{\parallel}{C}}-CH_3$	$-CH_3$	$-\underset{O}{\overset{\parallel}{C}}-OCH_3$	Phytyl	$-H$	$-CH_3$
Bacteriochlorophyll b	$-\underset{O}{\overset{\parallel}{C}}-CH_3$	$-CH_3$	$-\underset{O}{\overset{\parallel}{C}}-OCH_3$	Phytyl	$-H$	$-CH_3$
Bacteriochlorophyll c	$-\underset{OH}{\overset{\mid}{C}H}-CH_3$	$-CH_3$	$-H$	Farnesyl	$-CH_3$	$-C_2H_5$
Bacteriochlorophyll d	$-\underset{OH}{\overset{\mid}{C}H}-CH_3$	$-CH_3$	$-H$	Farnesyl	$-H$	$-C_2H_5$
Bacteriochlorophyll e	$-\underset{OH}{\overset{\mid}{C}H}-CH_3$	$-\underset{H}{\overset{\mid}{C}}=O$	$-H$	Farnesyl	$-CH_3$	$-C_2H_5$

Prodigiosin is a red pigment which is produced by strains of the species *Serratia marcescens* (invalid and obsolete name is *Bacterium prodigiosum*) and some species from the order *Actinomycetales*. Under elevated temperature (above 37 °C) the production of prodigiosin is halted.

Prodigiosin

Derivatives of this compound are key for a big number of pigments, ranging in color from yellow through orange to red. Many of these compounds are found in cells of photosynthesizing bacteria. The structure of carotene significantly varies in different clades of bacteria. For example, in the family *Chlorobiaceae*, carotenoids have an aromatic ring on one end of its chain, while in purple bacteria, carotenoids with acyclic residues or aliphatic in their nature can be mostly found.

Carotene

The main carotene derivatives are the following: β-carotene, lycopene, spirilloxanthin, hydroxysteroid, chlorobactene, β-isorenieratene, isorenieratene, and staphyloxanthin (**Fig. 2.65**). Carotenes contribute to photosynthesis by transmitting the light energy they absorb to chlorophyll. They also protect oxygenic phototrophic bacteria by helping to absorb the energy from the singlet oxygen, an excited form of the oxygen molecule O_2 which is formed during photosynthesis. Chemically, carotenes are polyunsaturated hydrocarbons containing 40 carbon atoms per molecule, variable number of hydrogen atoms, and no other elements. Some carotenes are terminated by hydrocarbon rings, on one or both ends of the molecule. All are colored to the human eye, due to extensive systems of conjugated double bonds. Structurally, carotenes are tetraterpenes, which means they are synthesized biochemically from four 10-carbon terpene units, which, in turn, are formed from eight 5-carbon isoprene units.

The absorption spectrum of carotenoids in green bacteria corresponds to a range from 735 to 755 nm, and in purple bacteria, it is in a range from 850 to 910 nm. The presence of carotene-based pigments is typical for many more types of bacteria. It is commonly found in all phyla, but as an example, we give a list of important species bearing carotenoids in their cells: *Micrococcus* sp., *Staphylococcus* sp., *Sarcina* sp., *Flavobacterium* sp., *Erwinia* sp., *Cellulomonas* sp., *Mycobacterium* sp., and *Nocardia* sp.

These compounds are mainly produced by species of the genus *Pseudomonas*. This group includes dark-blue **pyocyanin**, which is produced by *Pseudomonas aeruginosa*, green **phenazine** produced by *P. chlororaphis*, purple **iodinin**, and orange **phenazine-1-carboxylic acid**. Pyocyanin is one of the many toxins produced and secreted by the Gram-negative bacterium *P. aeruginosa* (**Fig. 2.66**). Pyocyanin is a secondary metabolite with the ability to oxidize and reduce other molecules and, therefore, kill microbes competing with *P. aeruginosa* as well as mammalian cells of the lungs, which are infected by *P. aeruginosa* during cystic fibrosis. There are three different states, in which pyocyanin can exist: oxidized, monovalently reduced, and divalently reduced. In order for pyocyanin to be synthesized by *P. aeruginosa*, two specific genes must be functional. *P. aeruginosa* strains, which are unable to synthesize pyocyanin, can still benefit from its effects if such a strain co-infects the lung together with wild-type strains that can produce pyocyanin.

Derivatives of indole are represented by purple pigment **violacein** produced by *Chromobacterium violaceum* by tryptophan oxidation.

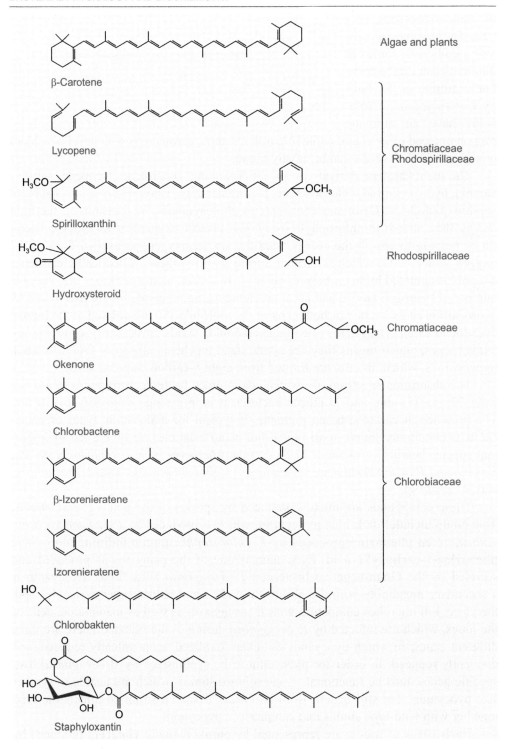

β-Carotene — Algae and plants

Lycopene

Spirilloxanthin — Chromatiaceae Rhodospirillaceae

Hydroxysteroid — Rhodospirillaceae

Okenone — Chromatiaceae

Chlorobactene

β-Izorenieratene — Chlorobiaceae

Izorenieratene

Chlorobakten

Staphyloxantin

Fig. 2.65. Derivatives of carotene.

From **naphthoquinone**, yellow **phthiocol** isolated from strains of *Mycobacterium tuberculosis* is derived.

Anthocyans represent red and blue pigments with properties of natural indicators reacting on the properties of a given environment, especially its pH. By their structure, we can regard them as **glycosides**, which can be hydrolyzed to saccharide and **chromone** derivatives. Mostly species of genus *Streptomyces* are producers of these pigments.

Melanines are compounds comprising brown, black, orange, and red pigments. They are complex in structure with usually high-molecular weight and produced during the degradation of amino acids, especially tyrosine. Besides micromycetes, they are produced by various species of the order *Actinomycetales* and bacterial strains of *Azotobacter chroococcum* and *Bacillus subtilis* var. *niger*.

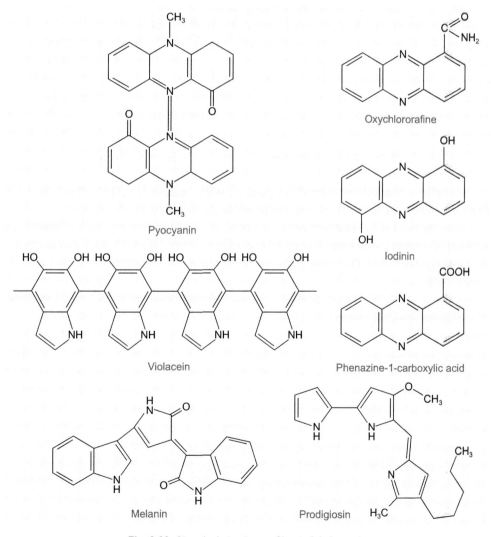

Fig. 2.66. Chemical structures of bacterial pigments.

Violacein is a naturally occurring diindole pigment with antibacterial, antiviral, antifungal, and antitumor properties. Violacein occurs in several species of bacteria and accounts for their striking purple hues. Violacein shows an increase in industrial applications, especially in cosmetics, medicines, and fabrics (**Fig. 2.67**).

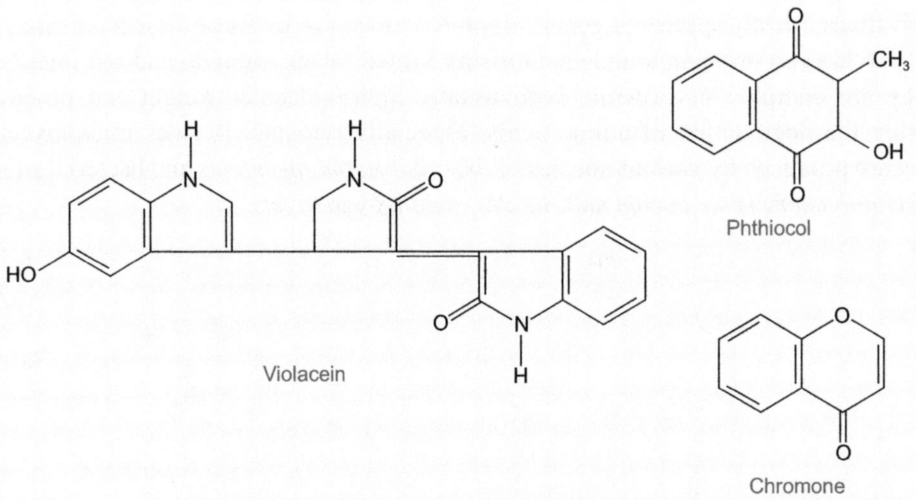

Fig. 2.67. Chemical structures of violacein, phthiocol, and chromone.

Phthiocol is a yellow crystalline quinone with vitamin K activity that is isolated from the human *Mycobacterium tuberculosis* and also made synthetically.

Chromone (or 1,4-benzopyrone) is a derivative of benzopyran with a substituted keto group on the pyran ring. It is an isomer of coumarin. Derivatives of chromone are collectively known as chromones. Most, though not all, chromones are also phenyl-propanoids.

Thus, all pigments play a protective role, and some of them are involved in bacterial photosynthesis process.

CHAPTER 3

BACTERIAL NUTRITION AND GROWTH

All bacteria require nutrition compounds for their growth and necessary for ensuring their vital functions. These substances serve as material for construction of the cellular components as well as for the formation of energy systems, which are essential for the cell life. Bacteria take these substances either in their original state or after exoenzymatic treatment.

3.1. Basic sources of nutrition

For bacterial life, growth, and multiplication, the following nutrients in the environment are needed:
- **Carbon source** is necessary for the synthesis of small organic molecules serving as a macromolecular skeleton.
- **Nitrogen source** is used for the formation of amino and imino groups as part of organic matter. An imino acid is any molecule that contains both imino (–C=NH) and carboxyl (–C(=O)–OH) functional groups. Imino acids are related to amino acids, which contain both amino (–NH$_2$) and carboxyl (–COOH) functional groups, differing in the bonding to the nitrogen.
- **Energy source** for the biosynthesis of low-molecular substances and biological macromolecules, forming the cell structure.
- **Mineral elements**: O$_2$, H$_2$, S, P, Na, K, Mg, Fe, Ca, Mn, Zn, Co, Cu, and Mo.
- **Growth factors**: Some microorganisms require these factors if they cannot synthesize them. The scheme showing nutrient conversion in a bacterial cell is presented in **Fig. 3.1.**

Bacterial Physiology and Biochemistry
http://doi.org/10.1016/B978-0-443-18738-4.50003-6,

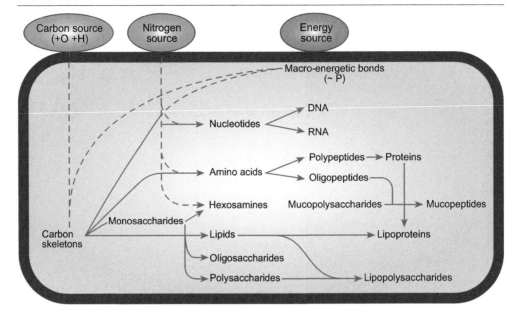

Fig. 3.1. Scheme of nutrient conversion in a bacterial cell.

The nutritional value of nutrition sources depends on their physical state, chemical structure, and the physiological properties of the cell. Water solubility, molecule size, spatial configuration, and degree of oxidation are the most important physical and chemical properties. Macromolecules serving as a source of nutrition must first be broken down into smaller fragments. For insoluble substances (fat, cellulose, etc.), decomposition takes place at the interface between the compound and water.

3.2. Sources of carbon

Carbon sources are referred to as nutrients that contain basic substrates for the synthesis of monosaccharides, amino acids, lipids, and carbon skeletons of organic bases as well as macromolecular compounds (**Fig. 3.2**). The amount of substances that can be used as carbon sources is unusually large, and their nature is various. Except for pure mineral forms (such as graphite), bacteria can use all natural carbon compounds, including very stable substances (e.g., naphtha, paraffins, and gaseous hydrocarbons) and even those that are toxic to other organisms.

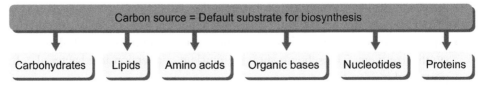

Fig. 3.2. Carbon sources are involved in the biosynthesis process.

Depending on the ability to use one or more compounds as a carbon source, bacteria can be divided into monophagous and polyphagous. Monophagous bacteria, for example, *Cytophaga*, require only cellulose as a carbon source. The polyphagous bacteria are predominantly represented by the genus *Pseudomonas* that has the ability to utilize many organic compounds as a carbon source.

The nutritional value of carbon sources depends on their physical state, chemical structure, and the physiological properties of the bacteria. Molecular size, solubility, degree of oxidation, and spatial configuration are the main physicochemical properties that determine the availability of carbonaceous compounds for bacteria.

High-molecular compounds cannot pass into the cell. First, they need to be degraded into low-molecular substances, which are processed by appropriate exoenzymes. Most bacteria also use water-soluble compounds as a carbon source.

The availability of carbon sources for bacteria is limited (especially in the case of highly oxidized compounds), in particular, by the degree of oxygenation of carbon atoms. Therefore, the nutritional value of, for example, carboxylic acids is relatively low and only a few bacteria can use substances like formic acid and oxalic acid as carbon sources. Carbon dioxide, which is a fully oxidized compound, has an even higher degree of oxidation.

Carbon dioxide can serve as a carbon source only in the presence of a suitable source of energy, especially in autotrophic bacteria. However, current evidence indicates that even under these energy-intensive conditions, CO_2 is used by most bacteria, including heterotrophic forms, for which it is not a basic but a complementary source of carbon nutrition. The inclusion of carbon sources into oxidation reduction processes is also related to the degree of oxygenation of carbon atoms. Semioxygenated carbon atoms occurring in groups $-CH_2OH$, $-CHOH$, and $=COH-$ are the most easily involved in this process. Because of this, the compounds containing these groups (and mainly including simple carbohydrates, alcohols, and organic acid salts) generally have a high nutritional value. In contrast, substances with multiple methyl and methylene groups that do not instantly participate in oxidoreduction processes (higher fatty acids, lipids, hydrocarbons, etc.) are used only by certain bacteria with appropriate enzymatic equipment (*Corynebacterium*, *Mycobacterium*, *Pseudomonas*).

The usability of organic carbon sources by bacteria is often conditioned by the spatial arrangement of these substances. For example, glycides that are incorporated into bacterial cell metabolism belong to *D*-isomers, while amino acids belong to *L*-isomers (except for some *D*-amino acids in the bacterial cell wall).

Categorization of the bacteria based on carbon nutrition. Depending on whether the carbon source is CO_2 or organic substance, bacteria are divided into the following groups:

- *Autotrophic bacteria* use solely carbon dioxide (CO_2) as the carbon source, from which they synthesize all carbon compounds. Other sources of nutrition have inorganic character as well.

- *Heterotrophic bacteria*: Most bacteria are heterotrophic and use various organic substances as carbon sources. For many of them, CO_2 can only serve as a supplemental source of carbonaceous nutrition. Other sources of nutrition may include both organic and inorganic substances. Such bacteria include saprophytic forms (using carbonaceous nutrition from dead organic matter) and parasitic forms (their way of life is bound to host cells and tissues).

For bacteria, autotrophy is not restricted to photosynthesis. In addition to two ways to get carbon, *autotrophy* and *heterotrophy*, there are two ways of obtaining energy: *phototrophy* and *chemotrophy*, and all of them are mutually combinable. Thus, according to nutrition, these microorganisms can be divided into photoautotrophic, photoheterotrophic, chemoautotrophic, and chemoheterotrophic.

Photoautotrophic bacteria (green bacteria, cyanobacteria) use water as a source of hydrogen for CO_2 reduction. Green sulfur bacteria (*Anoxyphotobacteria*) use H_2S or H_2 as the source of hydrogen. The species of *Chlorobium* genus can use also acetate (*mixotrophy*) as a carbon source in addition to CO_2.

Photoheterotrophic bacteria (purple sulfur bacteria — *Chromatium* genus, purple nonsulfur bacteria — *Rhodospirillum* genus) use a light quantum as a source of energy (cyclic phosphorylation) and an organic substance as a carbon source. Anaerobic purple sulfur bacteria can be capable of aerobic respiration in the dark in the presence of oxygen.

Chemoautotrophy (*chemolithotrophy*) is known only for bacteria and archaea. Energy is obtained by the oxidation of reduced inorganic compounds (sulfur, nitrogen, iron, etc.), and the carbon source is CO_2.

Chemoheterotrophy (most organisms) uses virtually any oxidizable organic substance as a source of carbon and energy.

Methylotrophic bacteria use mostly monohydrate compounds containing a methyl group $–CH_3$ (methane, dimethyl ether, methanol) as sources of energy and carbon. Methylotrophy can be obligatory (*Methylomonas* genus) or facultative (*Hyphomicrobium* genus).

Carbon sources used by heterotrophic organisms are:
- salts of organic acids (monocarboxylic with saturated and unsaturated bonds, monocarboxylic hydroxy and keto acids, dicarboxylic acids with saturated bonds, tricarboxylic hydroxy acids);
- carbohydrates (*monosaccharides*: pentoses, hexoses; *disaccharides*: sucrose, trehalose, maltose, cellobiose, lactose, melibiose; *trisaccharides*: raffinose; *polysaccharides*);
- lipids;
- amino acids, peptones, proteins.

3.3. Sources of nitrogen

The role of nitrogen sources is primarily the formation of amino ($–NH_2$) and imino ($–NH$) groups occurring in molecules of amino acids, nucleotides, heterocyclic bases, and other compounds that are components of the cytoplasm. Unlike other organisms, bacteria can use different compounds as a source of nitrogen.

Molecular atmospheric nitrogen. Bacteria that use this form of nitrogen are called free nitrogen-fixing bacteria. They include primarily the genera *Azotobacter*, *Rhizobium*, and *Clostridium pasteurianum*. The same properties of nitrogen fixation were also observed in some phototrophic, methane, and desulfurizing microorganisms and bacteria of the *Achromobacter* genus.

Ammonium salts and ammonia (NH_4^+, NH_3). These substances penetrate easily into cells where they are converted to amino and imino groups. That is why most bacteria use them as a source of nitrogen. In addition to mineral ammonium salts, ammonium salts of organic acids can be also used as suitable nitrogen sources.

Nitrates are a source of nitrogen for actinomycetes, while for other bacteria, their use is rarer.

Urea is a physiologically neutral source of nitrogen. When it is used, carbon dioxide is released into the environment. As a source of nitrogen, it is mainly used by bacteria that have the ability to form urease enzyme (***urobacteria***).

Amino acids, peptones, and proteins. These compounds are a source of nitrogen, especially for heterotrophic bacteria. They can also serve as a source of energy for many of them. Proteins and peptones that do not penetrate across the cell wall can only be used by those bacteria that produce enzymes that break down these substances to amino acids. The presence of enzyme deaminases in bacteria is a precondition for the use of amino acids as a source of nitrogen, except for bacteria that require amino acids without deamination. These include mainly pathogenic species, lactic acid bacteria, and particularly bacteria that are in contact with amino acids in their natural environment. Whole molecules of amino acids are incorporated as building blocks directly into high-molecular substances during the metabolism of these bacteria.

3.4. Mineral nutrition

Besides carbon and nitrogen, also other elements incorporated either directly into cell compounds or acting as catalysts for biochemical reactions are important components of bacterial nutrition. Oxygen and hydrogen significantly affect the overall metabolism of the cell. Some organisms require molecular oxygen, and some can use molecular hydrogen.

Depending on their relationship to oxygen, bacteria are divided into:
- obligate (strict) aerobes: aerobic respiratory metabolism;
- obligate (strict) anaerobes: grow only under anaerobic conditions (even a low partial pressure of O_2 is toxic for them);
- facultative anaerobes: have metabolic pathways for both respiratory and fermentative metabolism;
- aerotolerant microorganisms: have fermentative metabolism, grow in the presence of O_2 but do not use it;
- microaerophils: aerobic bacteria that require less oxygen than is present in the air and grow best under modified atmospheric conditions.

Oxygen is a component of water and oxygen compounds of the cell. Otherwise, it is mainly used as an electron acceptor for breathing processes, including energetic metabolism. Aerobic bacteria (*Micrococcus*, *Mycobacterium* genera, etc.) require the presence of oxygen. Their growth is generally dependent on the content of oxygen in the environment. For strictly anaerobic bacteria (*Clostridium*, *Desulfovibrio*, methane bacteria), oxygen is toxic or lethal. This is explained by the fact that oxygen inhibits the achievement of the required E_h value or leads to the formation of toxic hydrogen peroxide.

The presence of O_2 is an important factor for growth of aerobic bacteria. These microorganisms can have **cytochromes** and **cytochrome oxidase**, some **oxidoreductive enzymes**, which react with molecular oxygen to form a superoxide (O_2^-), a hydroxyl radical ($HO^{\cdot}$) or hydrogen peroxide (H_2O_2):

$$O_2 + \overline{e} \rightarrow O_2^-,$$

nonenzymatically: $\quad O_2^- + H_2O_2 \rightarrow O_2 + {}^{\cdot}OH + {}^-OH,$

and enzymatically: $\quad O_2^- + 2H^+ \xrightarrow[\text{\textit{superoxide dismutase}}]{} H_2O_2 + O_2.$

Hydrogen is contained in water, in reduced compounds, and in functional groups of most organic substances. Hydrogen, together with hydroxyl ions, influences the pH of the environment and participates in dehydrogenation and hydrogenation reactions directly associated with the catabolism and anabolism processes. Hydrogen also participates in the energy metabolism of the bacterial cell through the transport chain during respiration.

Sulfur is mainly used to synthesize amino acids (cysteine and methionine) and sulfur-containing proteins. These compounds are in a reduced state (–SH, –S–, –S–S–) only. The conversion of sulfhydryl groups into disulfide and vice versa is important in oxidative reduction processes associated with the change of redox potential. A suitable source of sulfur is sulfates and thiosulfates, from which sulfur is released during processes of assimilatory desulfurization. Organic compounds containing an oxidized sulfate group are mostly toxic (e.g., sulfonamides). Like nitrogen, oxygen, and hydrogen, sulfur can also be bonded directly to carbon atoms, while other elements are bonded to carbon through these four elements. Sulfur, sulfates, thiosulfates, and hydrogen sulfide are associated with the change of redox potential (rH) at oxide reduction processes.

Phosphorus (or salts of phosphoric acid) is a component of nucleic acids, including coenzymes and phospholipids. Some organic phosphorus compounds (ATP, phosphoenolpyruvate, acyl phosphate, etc.) can accumulate energy through macroenergetic bonds to provide energy for biosynthesis processes. The source of phosphorus is mainly phosphoric acid salts.

Microelements. Bacteria also require other minerals such as potassium, magnesium, calcium, iron, and chlorine as well as trace elements such as manganese, molybdenum, cobalt, copper, and boron. Most of these elements are used as components of the so-called metalloenzymes (Fe, Zn, Mn, Cu, Co, Mo, B, Ni, Se, etc.). Other trace elements have a reserved function for protein biosynthesis (K, Na), chlorophyll synthesis (Mg), osmoregulation (Na, Cl), and sporulation (Ca). Sodium is involved in the regulation of osmotic pressure and the transport system and affects the activity of some enzymes. Magnesium can take part in the synthesis of proteins and the synthesis or hydrolysis of ATP. Cobalt is a component of vitamin B_{12}.

The presence of iron is unconditional in the processes of respiration. Under anaerobic conditions, the iron in the cell is found as Fe^{2+} (soluble), while under aerobic conditions, it is in the form of Fe^{3+} (insoluble). The transport of iron ions into the bacterial cell occurs through special structures called siderophores. The iron hydroxamate complex is then transferred to the cell, the Fe^{3+} is reduced to Fe^{2+}, and the hydroxam-

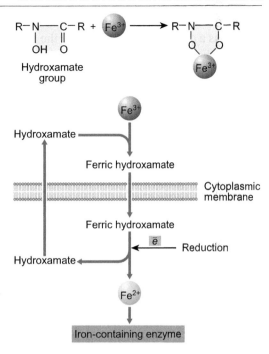

Fig. 3.3. Reduction of Fe^{3+} in the bacterial cell.

ate is excreted from the cell (**Fig. 3.3**). A number of bacteria can produce a complex of siderophores, which is referred to as **enterobactin** (*E. coli* and other enterobacteria), **bacillibactin** (*B. cereus, B. thuringiensis, B. anthracis*). The Fe^{3+}–enterobactin complex is transmitted across the membrane via active transport. Enterobactin (also known as enterochelin) is a high-affinity siderophore that acquires iron for microbial systems. Due to its high affinity, enterobactin is capable of chelating even in environments where the concentration of ferric ions is held very low, such as within living organisms. Enterobactin can extract iron even from the air. Pathogenic bacteria can compete for iron from other microorganisms using this mechanism, even though the concentration of iron is kept extremely low due to the toxicity of free iron. Bacillibactin is a catechol-based siderophore secreted by members of the genus *Bacillus*, including *B. anthracis* and *B. subtilis*. It is also involved in the chelation of ferric iron (Fe^{3+}) from the surrounding environment and is subsequently transferred into the bacterial cytoplasm via the use of ABC transporters.

The biosynthetic pathway of bacillibactin was first identified by Jürgen J. May *et al.* (2001) in the Gram-positive *B. subtilis*. The siderophore is synthesized through multimodular nonribosomal peptide synthetases (NRPSs), similar to enterobactin. However, unlike enterobactin, the genes responsible for encoding the bacillibactin synthetases are all located in one operon. This gene cluster is termed dhb — cognate to the catecholic structure of 2,3-dihydroxybenzoate (DHB) — and can be divided into specific genes responsible for encoding the enzymes. These are three genes *dhbE*, *dhbB*, and *dhbF*, which are translated into DhbE, DhbB, and DhbF synthetases. Notably, *DhbF* is characterized as a dimodular NRPS, unlike the monomodular EntF synthetase for enterobactin.

Sea bacteria produce a modified siderophore, which makes it possible to obtain iron even from a very poor environment. The involvement of some elements in the functioning of the bacterial cell is presented in **Table 3.1.**

Table 3.1. Involvement of trace elements in the functioning of the bacterial cell

Element	Function
Boron	Participates in the implementation of quorum sensing, components of polypeptides (antibiotics, mycotoxins)
Cobalt	Vitamin B_{12}, transcarboxylase (propionic bacteria)
Copper	Respiration, cytochrome c oxidase, photosynthesis, plastocyanin, some superoxide dismutases
Iron	Cytochromes, catalases, peroxidases, FeS proteins, nitrogenases
Manganese	Activator of a series of enzymes, some superoxide dismutases, phototrophy (photosystem II)
Molybdenum	Some flavin-containing enzymes, nitrate reductase, sulfite oxidase, urease, some formate dehydrogenases, coenzyme F_{430} (methanogens), carbon monoxide dehydrogenase
Selenium	Formate dehydrogenase, some hydrogenases, selenocysteine
Wolfram	Some formate dehydrogenase, oxotransferases of hyperthermophiles
Vanadium	Vanadium nitrogenase, bromoperoxidase
Zinc	Carbonic anhydrases, alcohol dehydrogenase, RNA and DNA polymerases, proteins of the DNA
Magnesium	Proteosynthesis, bacteriochlorophyll
Calcium	Sporulation
Nickel	Urease, hydrogenase, methyl-coenzyme M reductase, carbon monoxide dehydrogenase

Thus, the transport of iron in the bacterial cell can occur by using siderophores, enterobactin in enterobacteria, and bacillibactin in bacilli. Other elements involved in mineral nutrition can be transported by other transporters, which are various and depend on bacterial species.

3.5. Growth factors

Growth factor is a molecule (cell component) that cannot be synthesized by the cell from the nutrients presented in the external environment. Depending on the relationship to growth factors, the cells are:
- prototrophic: all necessary components are synthesized from nutrients;
- auxotrophic: some molecules cannot be synthesized and are obtained from the environment.

Auxotrophy is natural (the organism does not have the appropriate genetic equipment) or originates from a prototrophic organism by a loss of mutation. Essential vitamins, amino acids, and bases are required.

Growth factors are substances that usually act as precursors to the biosynthesis of macromolecular compounds or coenzymes that are necessary for cell development and growth. The cell cannot synthesize these substances itself, and therefore, they must be delivered in the finished state. Unlike the basic carbon sources, only a small amount of growth factors is required. Bacteria that lack the ability to synthesize one or more growth factors may be referred to as auxotrophic. In contrast, prototrophic bacteria can synthesize the necessary growth factors from simple components. The function of the growth factors and their role for bacterial cells are shown in **Table 3.2.**

Table 3.2. Function and role of growth factors for bacterial cells

Growth factor	Function	Bacteria
B_1 (thiamine, aneurin)	Precursor of thiamine pyrophosphate	Lactic acid bacteria, propionic bacteria, staphylococci
B_2 (riboflavin, lactoflavin)	Precursor of FMN, FAD	Lactic acid bacteria, propionic bacteria, clostridia, streptococci
B_3 (pantothenic acid)	Precursor of CoA	Lactic acid bacteria, hemolytic streptococci, *Proteus morganii*, *Corynebacterium diphtheriae*
B_4 (choline)	Donor of methyl groups	*Streptococcus pneumoniae*
B_5 (nicotinic acid)	Precursor of NAD^+, $NADP^+$	Lactic acid bacteria, *Gluconobacter oxydans*, *Proteus vulgaris*, *Staphylococcus aureus*, *Clostridium tetani*, *Shigella dysenteriae*
B_6 (pyridoxine)	Coenzyme of transferase	Lactic acid bacteria, *Streptococcus faecalis*
B_7 (biotin)	Participation in carboxylating processes	Lactic acid bacteria, propionic bacteria, staphylococci, β-hemolytic streptococci, *Bacillus coagulans*,
B_{12} (cyanocobalamin)	Component of coridine enzyme	Lactic acid bacteria, auxotrophic strains of *E. coli*
Folic acid	Precursor of coenzyme F	Lactic acid bacteria
Lipoic acid	Component of lipothylamide pyrophosphate	Lactic acid bacteria, corynebacteria
Vitamin C (ascorbic acid)	Regulation of redox potential	Some lactic acid bacteria, *Serratia marcescens*
Vitamin E (tocopherol) Vitamin K (phylloquinone)	Electron transport	Mycobacteria
Organic bases Adenine		Some lactic acid bacteria, *C. tetani, Shigella boydii*,
Xanthine and guanine	Synthesis of nucleotides	*Leuconostoc mesenteroides*, *Neisseria gonorrhoeae*
Uracil		Some lactic acid bacteria, *S. aureus, Shigella flexneri*, *C. tetani*, β-hemolytic streptococci

As growth factors, auxotrophic bacteria typically require the following:
- vitamins (a wide range of organic substances that form prosthetic groups of enzymes or an active center of a number of enzymes);
- purine and pyrimidine bases (for the synthesis of nucleic acids);
- amino acids (especially for protein synthesis).

Vitamins are primarily involved in the formation of coenzymes in the biosynthesis of nucleotides and other physiologically active components affecting the metabolic processes of bacteria. Lack of vitamins leads to the disruption of these processes, resulting in limited growth and proliferation of cells. For bacterial growth, B-group vitamins are mostly required, including the following compounds:

Thiamine or *aneurin* (vitamin B_1). It is a growth factor of lactobacilli, propionic bacteria, and staphylococci. For some bacteria (e.g., staphylococci), just one of two components of the molecule containing the pyrimidine and thiazole is enough to synthesize thiamine (**Fig. 3.4**). Pyrophosphate is a coenzyme of carboxylase, an enzyme that catalyzes the decarboxylation of pyruvic acid and other keto acids.

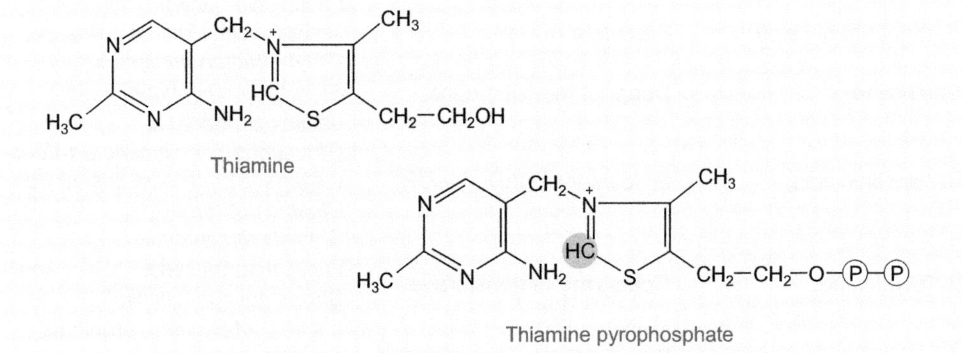

Fig. 3.4. The structure of thiamine and thiamine pyrophosphate.

Riboflavin or *lactoflavin* (vitamin B_2) is incorporated into flavin coenzymes: flavin mononucleotide (FMN) and flavin adenine dinucleotide (FAD) that mediate hydrogen transfer in respiratory processes (**Fig. 3.5**). FMN and FAD function as cofactors for a variety of flavoprotein enzyme reactions. In particular, they are required by lactobacilli (*Lactobacillus casei*), some pathogenic clostridia (*Clostridium tetani*), streptococci, and propionic bacteria.

FAD is necessary for the production of pyridoxic acid from pyridoxal (vitamin B_6) by pyridoxine 5′-phosphate oxidase. The primary coenzyme form of vitamin B_6 (pyridoxal phosphate) is FMN dependent. The oxidation of pyruvate, α-ketoglutarate, and branched-chain amino acids requires FAD in the shared E3 portion of their respective dehydrogenase complexes. Fatty acyl CoA dehydrogenase requires FAD in fatty acid oxidation. FAD is also needed to convert retinol (vitamin A) to retinoic acid via cytosolic retinal dehydrogenase. The synthesis of an active form of folate (5-methyltetrahydrofolate) from 5,10-methylenetetrahydrofolate by methylenetetrahydrofolate reductase is $FADH_2$ dependent. FAD is necessary for the conversion of tryptophan to niacin (vitamin B_3). The reduction of the oxidized form of glutathione (GSSG) to its reduced form (GSH) by glutathione reductase is FAD dependent.

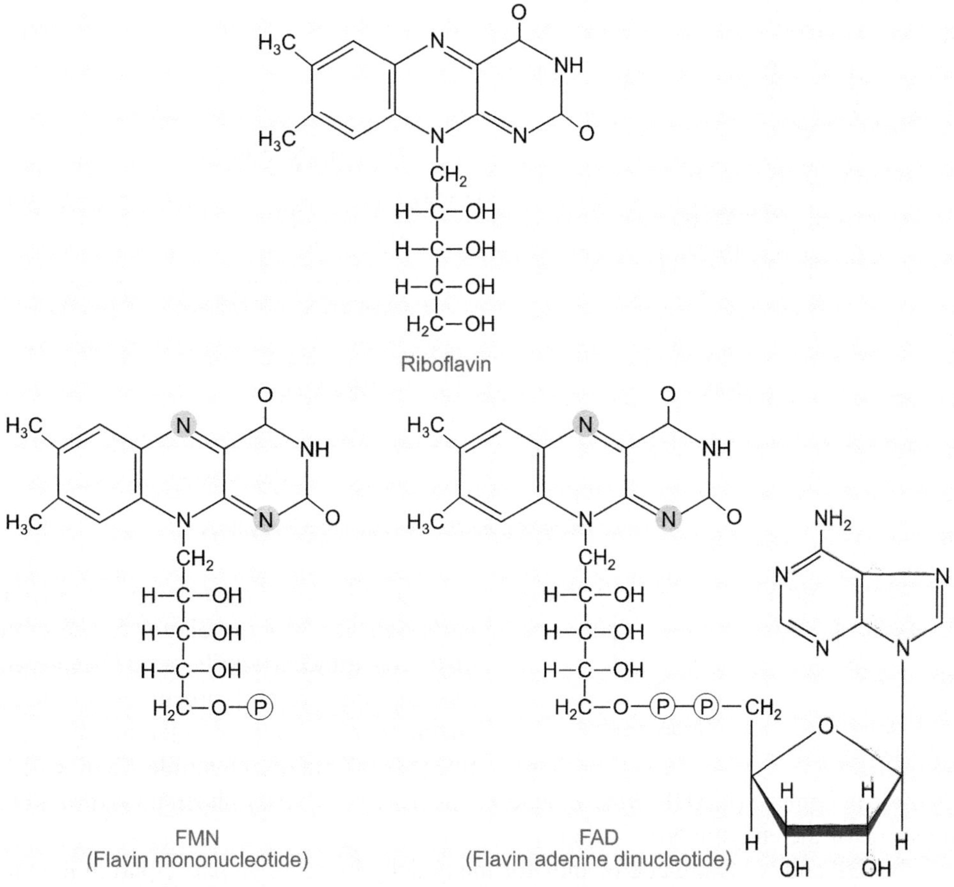

Riboflavin

FMN
(Flavin mononucleotide)

FAD
(Flavin adenine dinucleotide)

Fig. 3.5. The structure of main flavins.

Pantothenic acid. Its molecule consists of pantoic acid (β, δ-dihydroxy-γ, γ'-dimethylbutyric acid) and β-alanine, and is a growth factor for most lactobacilli (*Lactobacillus plantarum, L. casei*), hemolytic streptococci, *Proteus morganii*, and *Corynebacterium diphtheriae*. It is a component of coenzyme A (**Fig. 3.6**), which is involved in the transfer of acyl groups, for example, in the synthesis of higher fatty acids. A derivative of pantothenic acid, called pantetheine or pantethine, depending on whether it is in the reduced or oxidized form, is also used as another growth factor in many lactic bacteria (*Lactobacillus delbrueckii* ssp. *bulgaricus*). For some bacteria (corynebacteria), pantothenic acid may be replaced by one of its constituents, namely β-alanine.

$$HO-CH_2-\underset{\underset{CH_3}{|}}{\overset{\overset{CH_3}{|}}{C}}-\underset{\underset{OH}{|}}{CH}-\underset{\overset{||}{O}}{C}-NH-CH_2-CH_2-COOH$$

Pantothenic acid

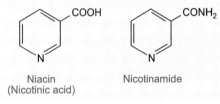

Coenzyme A

Pantoic acid   β-Alanine   Cysteamin

Pantothenic acid

Pantetheine

Fig. 3.6. The structure of pantothenic acid and coenzyme A.

Niacin
(Nicotinic acid)

Nicotinamide

Nicotinic acid (also known as niacin) is vitamin B$_3$. Niacin belongs to the group of the pyridinecarboxylic acid. Niacin and nicotinamide are both precursors of the co-enzymes nicotinamide adenine dinucleo-tide (NAD) and nicotinamide adenine dinu-cleotide phosphate (NADP) *in vivo*. NAD converts to NADP by phosphorylation in the presence of the enzyme NAD$^+$ kinase.

NADP and NAD are coenzymes for many dehydrogenases, participating in many hydrogen transfer processes (**Fig. 3.7**). NAD is important in the catabolism of fats, carbohydrates, proteins, and alcohol as well as cell signaling and DNA repair, and NADP mostly in anabolism reactions such as fatty acid and cholesterol synthesis.

Niacin was first described by chemist Hugo Weidel in 1873 in his studies of nico-tine. The original preparation remains useful: the oxidation of nicotine using nitric acid. For the first time, niacin was extracted by Casimir Funk, but he thought that it was thiamine and due to the discovered amine group, he coined the term "vitamine."

Nicotinic acid is used as a growth factor in *Lactobacillus plantarum, Leuconostoc mesenteroides, Staphylococcus aureus, Proteus vulgaris, Clostridium tetani*, and *Shigella dysenteriae. Proteus vulgaris* can use both nicotinic acid and nicotinamide, but *Pasteurella* requires only nicotinamide that cannot be replaced by nicotinic acid. The whole molecule of the coenzyme is the so-called V-factor of hemophilia, which cannot be replaced (in hemophilia) by individual components.

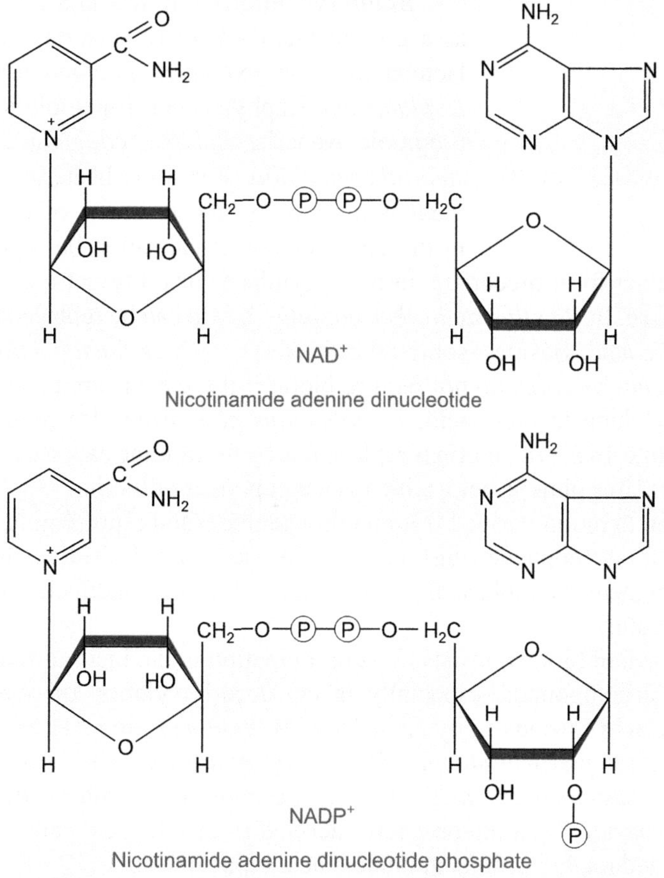

NAD⁺

Nicotinamide adenine dinucleotide

NADP⁺

Nicotinamide adenine dinucleotide phosphate

Fig. 3.7. Structure of nicotinamides.

Pyridoxin or adermin (vitamin B_6). It is a growth factor especially in lactoba-cilli (*Lactobacillus casei*, *L. delbrueckii*) and *Streptococcus faecalis*. Other bacteria commonly synthesize pyridoxin. It is a derivative of pyridine and occurs in three forms (pyridoxol, pyridoxal, pyridoxamine) with a similar effect (**Fig. 3.8**). The last two compounds interact with each other as an active ingredient of transaminases and decarboxylases of amino acids.

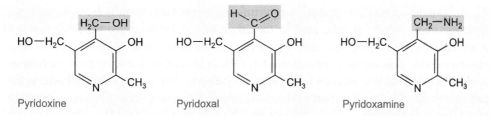

Pyridoxine   Pyridoxal   Pyridoxamine

Fig. 3.8. Structures of three forms of vitamin B_6.

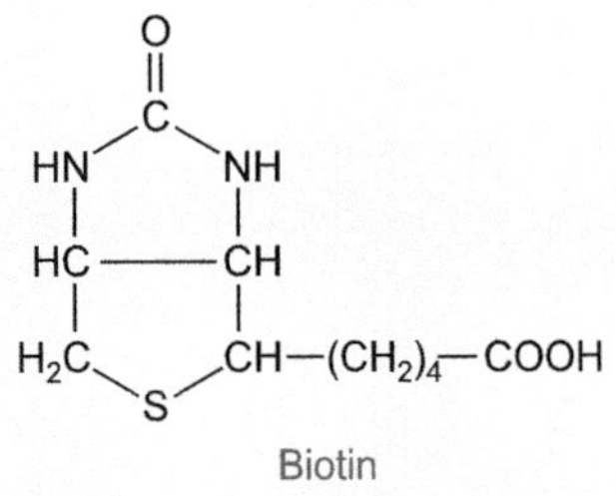

Biotin

Biotin (vitamin B$_7$). It acts in a low concentration as a growth factor of many bacteria, especially some lactobacilli, for example, *Lactobacillus helveticus*, *L. planarum*, staphylococci, β-hemolytic streptococci, propionic bacteria, *Bacillus coagulans*, and *Bacillus stearothermophilus*. For some bacteria, biotin requirement is affected by the presence of other substances in the environment. It can often be replaced by compounds that either form precursors in its biosynthesis or represent metabolites, where biotin is involved. In *Bacillus stearothermophilus*, biotin can be replaced by either oleic acid or aspartic acid. Besides, some other bacteria, such as *Lactobacillus delbrueckii* and *Streptococcus faecalis*, do not require biotin if these acids are present in the environment. In addition to oleic acid, *Lactobacillus plantarum* also requires acetate as a biotin substitute. In *E. coli*, biotin is replaceable by pyrimidine bases. *Corynebacterium diphtheriae* requires pimelic acid, which is a component of biotin.

In addition to biotin derived from natural sources (and representing a urea derivative with a natural thiophene ring), many synthetic biotin derivatives have been prepared. Some of them can replace biotin in bacteria, others are ineffective or antagonistic (e.g., desthiobiotin).

The function of biotin consists in its participation in the incorporation of CO$_2$ into certain organic compounds, especially in the decarboxylation of pyruvic acid and α-ketoglutaric acid carried out by *Lactobacillus plantarum* and *E. coli*.

Folic acid (or *pteroylglutamic acid*) is a growth factor for *Lactobacillus casei*, *L. delbrueckii*, *Streptococcus faecalis*, and *Leuconostoc mesenteroides* ssp. *cremoris* (**Fig. 3.9**). Coenzyme F (tetrahydric acid) derived from folic acid carries out the transfer of formyl and methyl groups in nucleic acid biosynthesis.

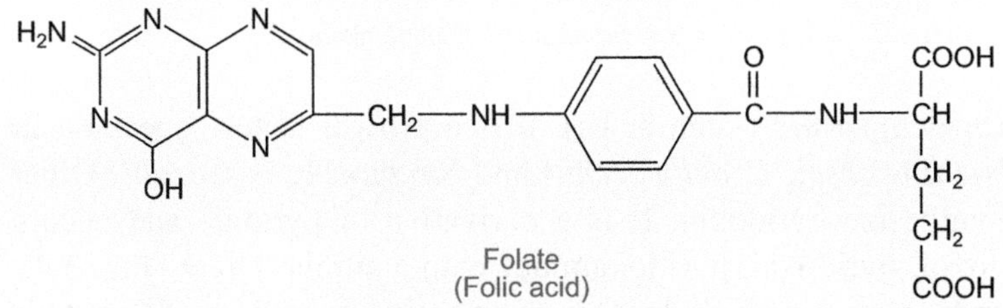

Folate
(Folic acid)

Fig. 3.9. Structure of folic acid.

It contains a pteridine ring with a pyrimidine nucleotide, *p*-aminobenzoic acid, and glutamic acid. A lack of folic acid in the environment leads to growth inhibition and a decrease of ribonucleic acid levels in cells. For some bacteria, such as *Lactobacillus plantarum and Leuconostoc mesenteroides*, folic acid is replaceable by *p*-aminobenzoic acid, which is a basic substance for folic acid synthesis. These are probably bacteria that can synthesize the pteridine part of the molecule. For *Streptococcus faecalis*, some non-physiological pteridines, such as rhizopterin (N10-formylpheric acid), replace folic acid because the pteridine ring cannot be synthesized by these bacteria (**Fig. 3.10**).

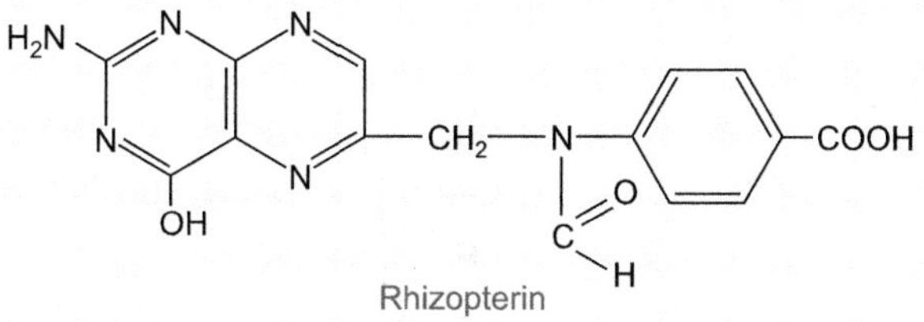

Fig. 3.10. Structure of rhizopterin.

Other bacteria, such as *Leuconostoc mesenteroides* ssp. *cremoris*, need folic acid derivatives, which are structurally related to active coenzymes, for example, leucovorin, for growth (**Fig. 3.11**).

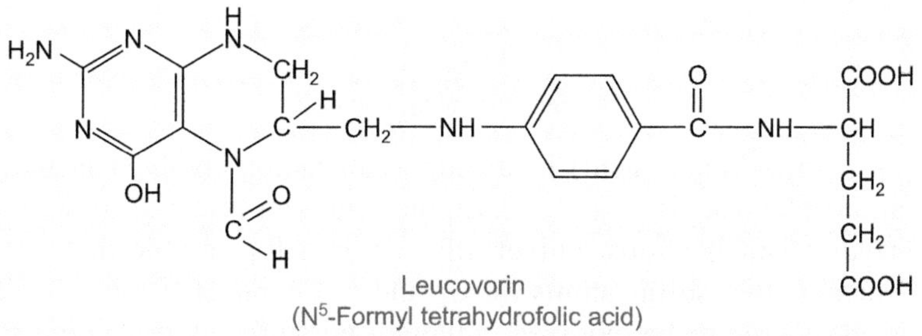

Fig. 3.11. Structure of leucovorin.

Cyanocobalamin (vitamin B_{12}) belongs to the cobalamin family of compounds, which are composed of a corrinoid ring and an upper and lower ligands (**Fig. 3.12**). The upper ligand can be an adenosine, methyl, hydroxy, or cyano group. Vitamin B_{12} is synthesized by prokaryotes and is a growth factor for *Lactobacillus lactis*, *L. leichmannii*, *L. acidophilus*, *L. delbrueckii*, and *L. helveticus*. From other species, the need for vitamin B_{12} was observed in some *E. coli* mutants and some soil bacteria. Cobalamin can be synthesized de novo in prokaryotes through two alternative routes according to the timing of cobalt insertion and the molecular oxygen requirement. These pathways are the aerobic pathway, which has been best studied in *Pseudomonas denitrificans*, and the anaerobic pathway, which has been best studied in *Salmonella enterica*, *Bacillus megaterium*, and *Propionibacterium shermanii*. Chemically, vitamin B_{12} is a complex compound, somewhat resembling a hem system, in which the cobalt is bound to the nitrogen of the pyrrole nuclei and to the cyanide group. For some B_{12} vitamin precursors, 5′-desoxyadenosine is present instead of the cyano group.

For some lactobacilli, it can be replaced by thymine and deoxyribosides that contain cytosine, adenine, and hypoxanthine or some deoxyribosides. Vitamin B_{12} can be replaced by reducing substances, such as ascorbic acid. For auxotrophic strains of *E. coli* dependent on vitamin B_{12}, this substance can be replaced by methionine. It is also worth noting that vitamin B_{12} is a growth factor for phototrophic microorganisms, which require it in the medium for biosynthesis of bacteriochlorophylls.

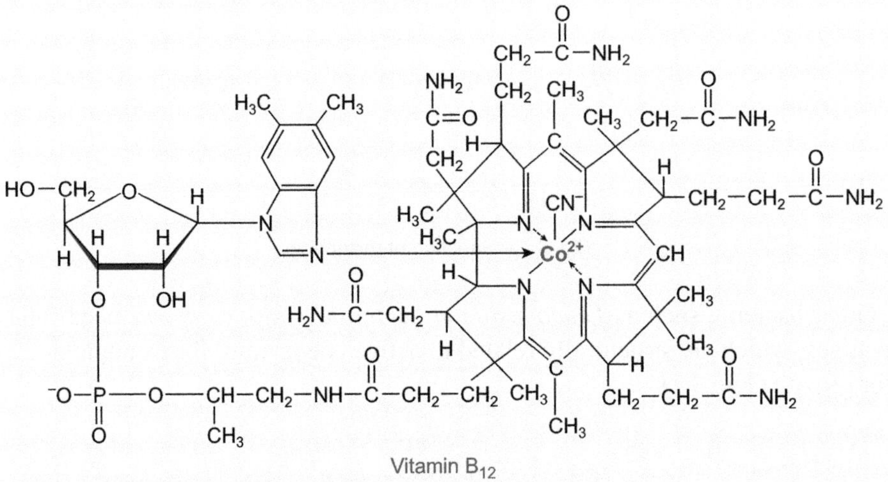

Vitamin B$_{12}$

Fig. 3.12. Structure of cyanocobalamin.

Lipoic acid (6,8-thioctic acid) is an important factor involved in the oxidative decarboxylation of keto acids (**Fig. 3.13**). Lipoic acid is an essential cofactor of dehydrogenase enzymes and is almost universally required for aerobic metabolism. This compound contains two sulfur atoms (at C$_6$ and C$_8$) connected by a disulfide bond and, thus, is considered to be oxidized, although no sulfur atom can exist in higher oxidation states. It is required, for example, in strains *Lactococcus lactis* ssp. *cremoris*, *Lactobacillus casei*, and corynebacteria. In bacteria, the lipoic acid forms a biologically active conjugate with thiamine (also with thymidine phosphate), which has been designated as thiamide, originally lipothiamine pyrophosphate (**Fig. 3.14**).

Lipoic acid (6,8-Thioctic acid)

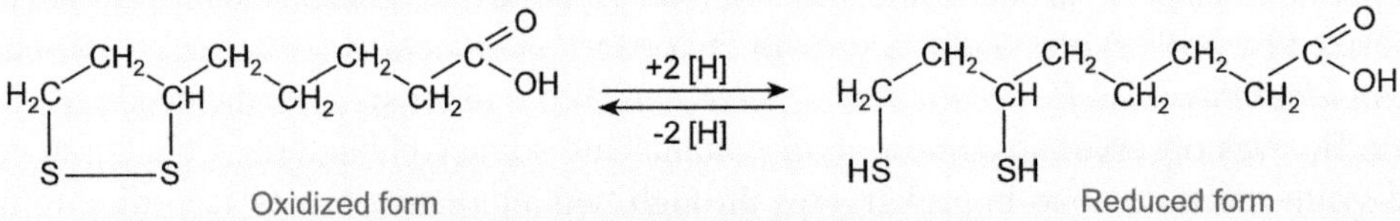

Fig. 3.13. Oxidized and reduced forms of lipoic acid.

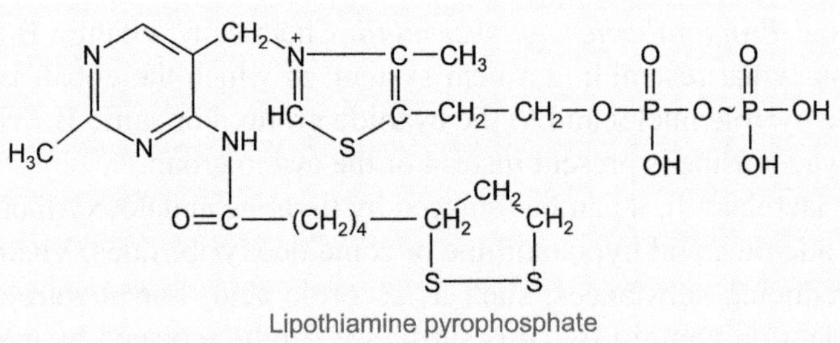

Lipothiamine pyrophosphate

Fig. 3.14. Structure of lipothiamine pyrophosphate.

Choline is a water-soluble vitamin-like essential nutrient. It is required in the exogenous form by *Streptococcus pneumoniae*. However, most bacteria can synthesize this substance and are, therefore, independent of its presence in the environment. It works in the transfer of methyl groups during the synthesis of some amino acids and further in the formation of lecithin and phospholipids.

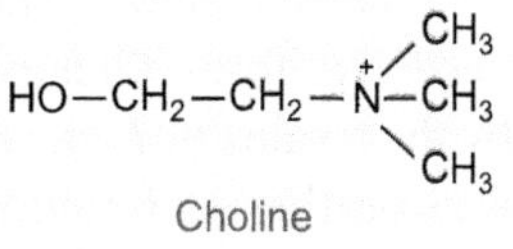

$$HO-CH_2-CH_2-\overset{+}{N}\overset{CH_3}{\underset{CH_3}{-}}CH_3$$

Choline

Other vitamins. The need for other vitamins is not so significant in bacteria. For example, vitamin C (ascorbic acid) is used substantially less because most organisms can synthesize it (**Fig. 3.15**). The exception is the strains of *Serratia marcescens* and some lactobacilli, for which it is required as a growth factor. The main function of ascorbic acid is the maintenance of a suitable redox potential in the cell.

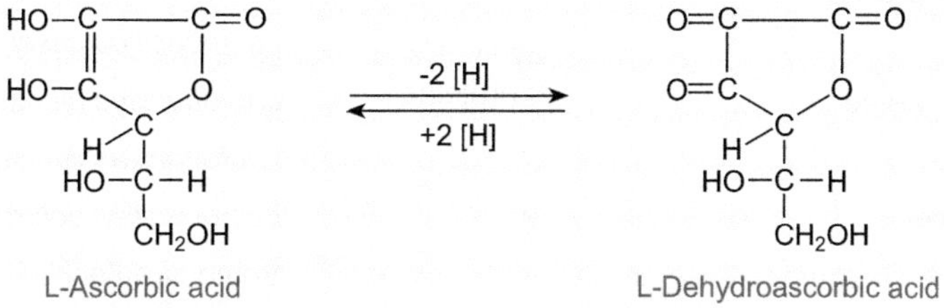

Fig. 3.15. Structures of L-ascorbic and L-dehydroascorbic acids.

Even *vitamin E* (**tocopherol**) and *K* (**phylloquinone**), which act as coenzymes of electron transfer in oxidative phosphorylation, can be used as growth factors in some bacteria. The issue of the importance of these vitamins for bacteria has not been resolved yet. They are required, for example, for *Mycobacterium paratuberculosis*.

Organic bases. Purine and pyrimidine bases are used by bacteria for nucleic acid synthesis. Intracellular concentrations of purines and pyrimidines do not undergo such changes as those of amino acids.

The proportion of the different types of nucleic acids varies during the bacterial cell growth, and thus they are fast recycled.

Adenine is essential for the growth of *Clostridium tetani* and may be replaced by hypoxanthine. It is also a growth factor in some lactic acid bacteria and *Shigella boydii*. It is believed to be used by bacteria even in the NAD^+ synthesis process. If present in sufficient quantities, it may reduce the bacteriostatic effect of certain sulfonamides.

Xanthine and guanine. The strains of *Leuconostoc mesenteroides* require these bases for growth, whereas *Neisseria gonorrhoeae* require hypoxanthine. Some purine nucleosides, such as adenosine, guanosine, and inosine, or nucleotides may also be used as growth factors in bacteria.

Uracil. This growth factor is required by the strain *Staphylococcus aureus* growing under anaerobic conditions. The same function is also found in *Shigella flexneri*, *Lactobacillus plantarum*, *Leuconostoc mesenteroides*, *Clostridium tetani*, and some hemolytic streptococci. Some strains of *Lactobacillus delbrueckii* ssp. *bulgaricus* use orotic acid (6-carboxyluracil) instead of uracil.

Amino acids. Many bacteria lack the ability to synthesize some amino acids needed to form other essential metabolites or to synthesize proteins. Alanine is required by the strains of *Leuconostoc mesenteroides* ssp. *cremoris* and *Pediococcus cerevisiae*. The synthesis of pantothenic acid in corynebacteria requires β-alanine, which cannot be replaced by α-alanine (**Table 3.3**).

Table 3.3. Amino acids as a growth factor for some lactic acid bacteria

Amino acids	Function	Bacteria
α-Alanine		*Leuconostoc mesenteroides* ssp. *cremoris*
Aspartic acid, cysteine, lysine, phenylalanine, proline		*Leuconostoc* genus
Arginine	Synthesis of proteins	*Lactobacillus casei*
Glutamic acid, valine		*Lactobacillus plantarum*
Tryptophan		*Lactobacillus xylosus*
Tyrosine, serine		*Lactobacillus delbrueckii*
Histidine		Some streptococci

Glutathione

Growth factors in *Leuconostoc* genus are aspartic acid, cysteine, cystine, lysine, phenylalanine, and proline. Lactic acid bacteria have particular requirements for amino acids. *Lactobacillus casei* requires arginine to grow, *Lactobacillus plantarum* needs glutamic acid and valine, tryptophan is required by *Lactobacillus xylosus*, and tyrosine and serine by *Lactobacillus delbrueckii*. Some streptococci need histidine to grow. For *Neisseria gonorrhoeae*, the glutathione tripeptide is a significant growth factor and is used to maintain the appropriate redox potential.

3.6. Sources of energy

Like all organisms, bacteria need energy, and they can acquire this energy through a number of different ways. In order to grow, bacterial cells must have an energy source, a source of carbon, and sources of other required nutrients. The nutritional requirements for bacteria can be grouped according to the carbon source and the energy source. Some types of bacteria must consume preformed organic molecules to obtain energy, while other bacteria can generate their own energy from inorganic sources.

Most bacteria acquire energy by decomposing inorganic or organic matter. The use of light quanta of energy is ***phototrophy***. Organic and inorganic substances are used in ***chemotrophy*** (energy is obtained by oxidoreduction processes, one is oxidized, the other is reduced).

An electron acceptor and hydrogen are:
- produced by donor catabolism (fermentation);
- molecular oxygen (aerobic respiration);
- oxygen in oxidized inorganic compound anaerobic respiration (sulfates, nitrates, CO_2, etc.).

Depending on genetic features, ***chemotrophic microorganisms*** can produce energy in one or more ways:
1. Only aerobic respiration (*Pseudomonas putida*).
2. Aerobic and anaerobic respiration, but not fermentation (*Paracoccus denitrificans*).
3. Aerobic, anaerobic respiration, and fermentation (*Escherichia coli*).
4. Aerobic respiration and fermentation (some enterobacteria).
5. Anaerobic respiration and fermentation (*Desulfovibrio*).
6. Only fermentation (*Clostridium*).
7. Only anaerobic respiration (*Methanobacterium*).

Forms of energy and its transformation in a bacterial cell. Due to the cleavage of chemical bonds, the energy released during degradation processes (exergonic reactions) is transformed into the energy of macroenergetic bonds mainly of ATP type and released into the environment. Bound energy is used primarily in endergonic reactions (biosynthesis processes).

Transformation of energy (ATP) into other forms:
- ***Kinetic energy*** allows the movement of bacterial cells and the flow of cytoplasm.
- ***Osmotic energy*** is applied to the diffusion of substances through the cytoplasmic membrane and osmoregulation.
- ***Electric energy*** occurs in the form of surface charge, mainly affecting sorption.
- ***Light (fluorescent) energy*** is radiation in the so-called bioluminescence of luminescent bacteria (*Photobacterium*, *Vibrio*).
- ***Thermal energy*** is unused released chemical energy that escapes into the environment in the form of heat.

Chemoorganotrophic bacteria utilize hydrogen and organic matter, which can be simultaneously a source of carbon, as donors. Organic or inorganic substances or molecular oxygen could be acceptors of hydrogen and electrons.

Chemolithotrophic bacteria gain energy from the oxidation of inorganic hydrogen and electrons of molecular oxygen. Oxygen is the acceptor of hydrogen and electrons.

Phototrophic bacteria. These bacteria utilize energy from sunshine, which turns into the energy of macroergic bonds of ATP. According to the nature of hydrogen and electron donors, they include:
- *Photoorganotrophic bacteria* utilize hydrogen and electrons from simple organic compounds or molecular hydrogen. CO_2 or organic matter could be an acceptor of hydrogen and electrons.
- *Photolithotrophic bacteria* require hydrogen and electron donors from reduced inorganic sulfur compounds, usually hydrogen sulfide. CO_2, which can be used as a source of carbon, is the acceptor of hydrogen and electrons.

3.7. Transport of compounds across the plasma membrane

The permeability of surface structures of bacterial cells can be typically various. While across the cell wall, ions and little molecules can permeate easily, the plasma membrane is semipermeable and provides a gateway mainly for water solutions.

Because the cell is an open system, there must be controlled, selective, and bidirectional transport between the cytoplasm and the external environment. The barrier is a semipermeable cytoplasmic membrane. There are three types of transport:

- nonspecific diffusion;
- specific transmission by a protein carrier;
- pinocytosis (which does not exist for bacteria).

Nonspecific diffusion is characteristic to only a small number of molecules and ions (water, fat-soluble molecules, Cl^-, HCO_3^-, NO_3^-, etc.). The rate is directly proportional to the concentration (or electrochemical) gradient of the substance and temperature. Schemes of diffusion of one and two substances are presented in **Fig. 3.16.**

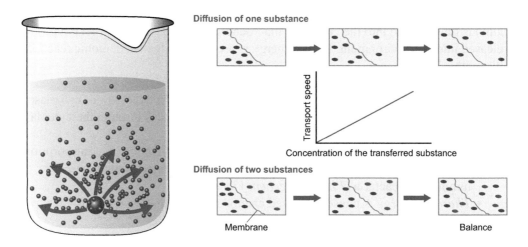

Fig. 3.16. Diffusion of substances.

Simple diffusion takes place in both directions (into and out of cells).

Transport by specific carrier proteins:

- *facilitated diffusion* with the participation of specific membrane proteins (such as channels and carriers), which involved in molecules transferring across the membrane (no energy supply);
- active transport via binding proteins; energy is supplied by ATP (***primary active transport***);
- active transport using the chemiosmotic gradient (proton gradient) of the cytoplasmic membrane (***secondary active transport***);
- group translocation is mediated by transport protein, and transmission is associated with the modification of the transmitted substance (***phosphorylation***).

Easy diffusion:
- The transport is carried out by a specific protein after the concentration of the substance decreases (**Fig. 3.17**).
- The rate depends on the concentration gradient, the number of transport proteins, and the temperature. One transport protein molecule can transmit up to 60,000 substrate molecules per second.
- Transport works both ways (into and out of cells).

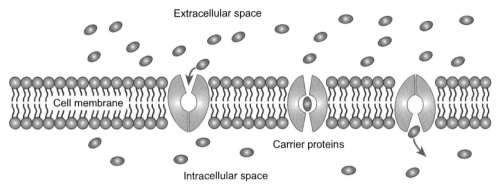

Fig. 3.17. Transport by a specific protein.

The semipermeability of the membrane is not perfect. Due to asymmetric structure set by the weird organization of molecules and lipids, membrane pores are created. Through these pores, some kinds of substances can permeate. The transport of these substances across the membrane can be passive (diffusion) or active depending on special energetic and transport systems.

3.8. Passive transport

This type of transport consists in diffusion (**Fig. 3.18**). Diffusion provides concentration balance, and electrochemical potential between the outside and the inside of a cell and energy are not required. Transported substances stay in basic form. Usually two types of diffusion are distinguished:

(1) simple diffusion;

(2) facilitated diffusion.

Simple diffusion is used mainly for transport of water or some ions (Cl^-, HCO_3^-, NO_3^-) and nonpolar molecules across the membrane.

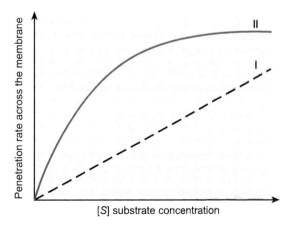

Fig. 3.18. Dependence of penetration rate across the membrane on substrate concentration (without transporter I, with transporter II).

The speed is directly proportional to the concentration or electrochemical gradient according to the molecule or temperature. The size of transported particles and their lipophilic character are also important.

Facilitated diffusion takes place using a protein or lipoprotein transporter, which is located in the membrane (**Fig. 3.19**). The presence of a specific binding place and changeable configuration enable binding of the transporter to the substrate and transport across the membrane. After releasing the substrate, the transporter is regenerated. The transport speed is dependent on the concentration gradient; however, after the saturation of the transport system by the substrate, the curve is not increasing anymore. The curve has hyperbolic shape according to this relation:

$$v_{in} = v_{max}^{in} \frac{S_m}{K_{S_b} + S_m}; \quad v_{ex} = v_{max}^{ex} \frac{S_b}{K_{S_m} + S_b}$$

where v is the transport speed of substrate: v_{in} into the cell and v_{ex} from the cell; v_{max} is the maximum transport speed of substrate: v_{max}^{in} into the cell and v_{max}^{ex} out of the cell; S is the concentration of substrate: S_m in the cell and S_b in the outer environment; K_S is saturated substrate concentration: K_{S_b} in the cell and K_{S_m} in the outer environment.

Facilitated diffusion is characterized called permease, whose binding place is complementary only for one substrate or for a group of structurally similar molecules. Permeases are membrane transport proteins, a class of multipass transmembrane proteins that allow the diffusion of a specific molecule into or out of the cell in the direction of a con

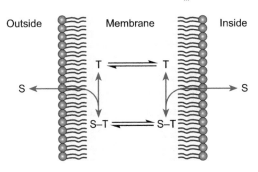

Fig. 3.19. Facilitated diffusion.

centration gradient. The permease binding is the first step of translocation. LacY protein from *Escherichia coli* is an example of a permease. In bacteria, this transport system has place for phosphate. Phosphate transport is necessary for the formation of special intermediated product, which transports some divalent cations, such as Mg^{2+}, Co^{2+}, Zn^{2+}, Mn^{2+}. Some bacteria can transport also sugar (*Pseudomonas aeruginosa*) or glycerol (*Enterobacteria*) by facilitated diffusion.

3.9. Active transport

Active transport is transfer of nutrients into the cell and metabolites to the external environment. It is always carried out by proteins. Some of them are present in the membrane (***constitutive***); some are synthesized only in the presence of a substrate on the outside of the membrane (***inducible***). There is always a certain amount of energy required for the active transport of substances across the cytoplasmic membrane. The transport (substrate transfer) rate can be limited by the number of stereospecific transport protein molecules and the amount of energy required to carry out the process.

Pathways for obtaining energy for the transport system:
- ATP cleavage: primary active transport;
- transport linked to the electrochemical ion gradient: secondary active transport;
- phosphoenolpyruvate-dependent phosphotransferase system: group translocation;
- the proton gradient can also be generated by light.

Permeases are membrane proteins that transport molecules across the membrane against the concentration gradient by consumption of energy. For example, the permease system creates protein transport and source of energy when transporting lactose in *E. coli*. The transporter called *M*-protein located on the membrane can occur in two conformations. In the direction to the outer environment of the cell, *M*-protein has affinity against lactose, and in the presence of β-galactoside, *M*-complex transports sugar into the cell. In the presence of an energy source, *M*-protein changes his conformation and loses affinity against sugar. Than sugar can be released from the complex and starts accumulating in cell. Supply energy is not used for transport but enables the accumulation of lactose inside the cell against the concentration gradient.

If energy source is eliminated, for example by adding sodium azide, active transport cannot occur, and the permease system enables only facilitated diffusion. The system's affinity to the substrate (lactose) or other β-galactoside and transport speed on both sides of membrane are similar. β-Galactoside permease is coded by the specific gen *LacY*. A mutation in this gene results in strains unable to transport lactose from the outer environment to the cell and then lactose transformation. These strains are called cryptic, because they do not use this substrate even if they have enzymatic equipment. The inability of some wilds strains of *E. coli*, which use citrate as a source of energy, can be also considered a cryptic state. All these strains could realize the production and transformation of citrate during metabolism.

Active transport takes place with almost all amino acids, which can accumulate in some bacterial cells in the concentration, which is higher than in the outer environment, and they create an amino acid pool. Transport of amino acids is realized by the facilitated diffusion of specific transporters — permeases. Transport requires a supply of energy from the cell. The whole process is realized in several phases according to the scheme in **Fig. 3.20.**

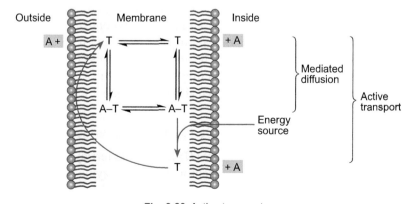

Fig. 3.20. Active transport.

The first two phases remind facilitated diffusion. During these phases, a complex amino acid–a carrier (TA) is made, and its transport across the osmotic barrier by a change in carrier conformation is realized. In the next phase, supply of energy increases the carrier's enzyme level (T^+), which leads to loss of its affinity to the substrate, and the amino acid is released inside the cell. By inhibition of this phase, for example, by azide or reduction of temperature, the carrier has high affinity to the substrate on both sides of the membrane, so transport can occur in both directions.

Next to the transport of individual amino acids, primarily proline, histamine, and methionine, transport systems in bacteria can transfer also groups of these compounds with different structures. The transport of associated groups of these amino acids includes glycine, valine, alanine, serine, leucine, isoleucine, phenylalanine, tyrosine, tryptophan, lysine, arginine, and aspartic and glutamic acids. In the mechanism of active transport in bacteria, also changes of hydrogenation activity and the ionic gradient are involved, except for a change of the transformation carrier protein (**Fig. 3.21**).

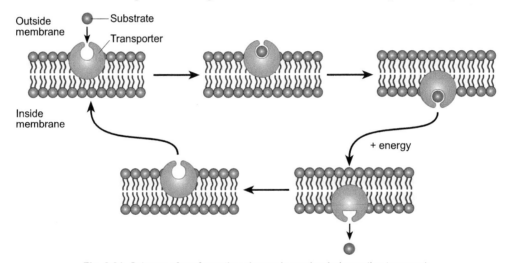

Fig. 3.21. Scheme of conformation change in carrier during active transport.

The connection between active transport and the process of dehydrogenation was proved in the membrane of preparations from *E. coli* and *Staphylococcus aureus* cells. In both cases, there was a noticeable increase in the consumption of some amino acids when *D*-lactate was added to suspension. The importance of this effect consists in cyclic reduction and reoxidation of disulfide and sulfhydryl groups of *D*-lactate dehydrogenase (**Fig. 3.22**). *D*-Lactate dehydrogenase (called also cytochrome, EC 1.1.2.4) is an enzyme that catalyzes the chemical reaction:

(*D*)-Lactate + 2 ferricytochrome $c \rightleftharpoons$ pyruvate + 2 ferrocytochrome c.

Thus, two substrates of this enzyme are (*D*)-lactate and ferricytochrome c, whereas its two products are pyruvate and ferrocytochrome c. This enzyme belongs to the family of oxidoreductases, specifically those acting on the CH–OH group of the donor with a cytochrome as an acceptor.

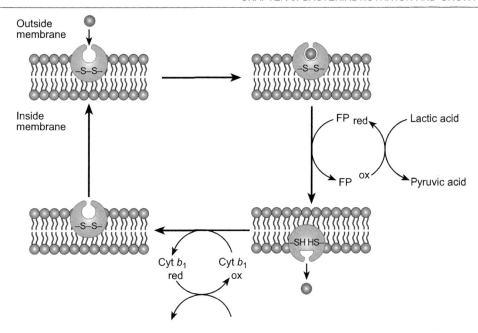

Fig. 3.22. Scheme of molecule transport across the membrane connected with lactate oxidation.

Affinity to the transported molecule is shown only by the oxidized form of the carrier. The following reaction connected with a conformation change in the carrier enables the transfer of substrate across the membrane and its release into the cell. Reoxidation of the reduced form occurs probably in the presence of cytochrome b_1. The curiosity of this system is that metabolic energy in the form of macroergic bonds is not required.

Facilitation of active transport by change of the ionic gradient. An electrochemical gradient is a gradient of electrochemical potential, usually for an ion that can move across a membrane. The gradient consists of two parts, the chemical gradient, or difference in solute concentration across a membrane, and the electrical gradient, or difference in charge across the membrane. When there are unequal concentrations of an ion across a permeable membrane, the ion will move across the membrane from the area of higher concentration to the area of lower concentration through simple diffusion. Ions also carry an electric charge that forms an electric potential across a membrane. If there is an unequal distribution of charges across the membrane, then the difference in electric potential generates a force that drives ion diffusion until the charges are balanced on both sides of the membrane.

This effect is explained by the presence of the oxidation chain through the membrane. Its activities are influenced by respiration processes or anaerobic dehydrogenation, which leads to the release of protons and creation of different pH and electric potential in the membrane. A difference in pH on the outside and inside membrane layers can be around 1 and in membrane potential about -150 mV. The increase of pH and electric potential is accompanied by the penetration of neutral molecules, primarily sugars and amino acids, or anions, primarily phosphates, across the membrane. It is supposed that anticarries are located in the membrane and transport cations out of the cell into the outer environment to maintain differences of potentials (**Fig. 3.23**).

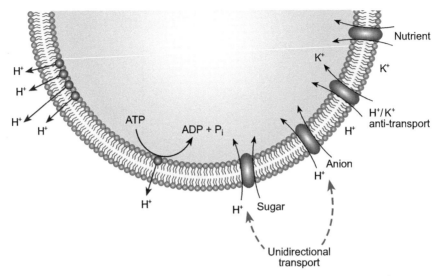

Fig. 3.23. Different types of proton and cation transport through the plasma membrane.

Thus, both passive and active transports are an integral part of the bacterial cell's life. In passive transport, the energy is not used, and small molecules are trapped in the cell by diffusion. However, in the active transport, there are other mechanisms and carriers of molecules, and the cell uses energy in this case.

3.10. Transport of iron and its regulation

Iron is a necessary element for most aerobic and facultative aerobic microorganisms. Several species, such as *Lactobacillus plantarum* and *Borrelia burgdorferi*, are an exception. Lactobacilli live in microaerophilic conditions and do not synthesize gem-containing proteins. *Borrelia burgdorferi* normally grows in iron deficient conditions; it does not contain **iron-derived proteins**, which were found in other bacteria and **iron-regulated genes**. It contains only 10 atoms of iron per cell. Although iron is one of five metals most abundant in nature, all organisms have problems with the provision of this metal. In aqueous solutions, iron is in two stable states of Fe^{2+} and Fe^{3+}. The equilibrium between these states strongly depends on the molecular environment of iron ions. In an aqueous solution and in most biological fluids in the presence of oxygen at pH about 7, iron exists in the form of iron hydroxide, the solubility constant of which is 4×10^{-38} M. Accordingly, the concentration of iron in the solution is 10^{-17} mol/L (at pH 7) and 10^{-8} mol/L (at pH 4). This is a much lower concentration than is needed to maintain the growth of microorganisms. Such an extreme iron deficiency is also found in seawater, and in humans and animals, where iron is associated with transferrin of blood serum or lactoferrin of secretory fluids. In anaerobic conditions, the provision of iron is not an issue for organisms, since compounds of bivalent iron are well soluble.

Iron containing heme and Fe–S proteins and directly bound to proteins performs important functions in the cell, namely transport of oxygen, mitochondrial energy metabolism, transport of electrons, and synthesis of deoxynucleotides. Iron contains active centers of reducing enzymes and directly participates in the reduction reactions associated with the transition of Fe^{2+} to Fe^{3+}. Membrane-bound respiration cannot occur without iron. Aerobic respiration with oxygen as the final acceptor of electrons as well as anaerobic respiration with nitrate, sulfate, sulfur, fumarate, and CO_2 or Fe^{3+} instead of oxygen occurs in different bacteria.

Redox enzymes include ribonucleotide reductase, the main enzyme for DNA synthesis, and aconitase of the tricarboxylic acid cycle. Iron-enzyme **catalase** and **superoxide dismutase** destroy active radicals, preventing damage to the cell. In addition, iron carries out a number of regulatory functions in the metabolism; in particular, it participates in the regulation of the synthesis of oxidoreductase, heme, bacterial proteases, diphtheria and other toxins, iron metabolism enzymes, and the biosynthesis of riboflavin (vitamin B_2).

On the other hand, free iron may catalyze the formation of high-reactive free radicals that can bind to sugars, amino acids, phospholipids, DNA, and organic acids and cause damage to membranes and DNA. Iron is one of the elements, which are often deficient, which leads to dramatic changes in the metabolism, namely, the reduction of biomass, the decrease in the content of enzymes containing iron-porphyrin groups, and the decrease in the activity of enzymes containing nonhemic proteins (NADH dehydrogenase, succinate dehydrogenase).

It is found that virulence is associated with the functioning of a specific transport system for iron. For example, in *Neisseria gonorrhoeae*, the frequency of changes in antigenic characteristics increases by five to nine times in the case of iron starvation. In addition, the frequency of DNA recombinations and the rate of repair processes are also changing. The *Neisseria meningitidis* genes, which provide iron transport, *tonB*, *exbB*, and *exbD*, also play an important role in the onset of the disease. *Neisseria* not only competes for transferrin iron with host cells, but also affects the function of transferrin receptors.

All of the above suggests that cells and the whole body should support the homeostasis of iron in such a way as to ensure their physiological needs, but to prevent the accumulation of excess iron. Several mechanisms are used to eliminate the toxicity of metal cells, which includes neutralizing enzymes and iron intracellular chelators, and the direct control of the transmembrane transport of the metal.

Transmembrane transport maintains a balance between the amount of metal required for biological processes and its amount that can be toxic. Cells contain a variety of transport systems, and this diversity is an important environmental factor as well as a human health factor, since changes in the intracellular iron pool are observed in many diseases, including neurodegenerative diseases, atherosclerosis, tumors, and microbial infections, and aging.

To get iron, bacteria and mushrooms have several strategies. Many microorganisms secrete ***siderophores*** — highly chelated Fe^{3+} chelators and specific transport systems that ensure the flow of the Fe^{3+}–siderophore complex into the cell. The siderophore is released in the medium and binds to Fe^{3+}; this complex binds to the siderophore-receptor protein; the iron is released from the complex or ferrireductase or by the cleavage of the etheric bond. The iron can be released either on the surface, or the entire complex is transported inside the ABC transporter.

Another strategy, which is characteristic not only for microorganisms but also for plants, is the recovery of Fe^{3+} on the cell surface by a transmembrane retrieval system. Iron-reductase (metal-reductase) restores iron to a divalent state. In this form, iron can be absorbed by cells. Some microorganisms release the reduced compounds into the environment, resulting in iron entering the cells in the form of soluble Fe^{2+} ions. Some pathogenic bacteria (*Neisseria gonorrhoeae*, *Haemophilus influenza*, etc.) can use eukaryotic proteins that bind iron-transferrin, lactoferrin, ferritin, or hemin, by attaching them to their surface structures, as a source of iron.

In the supply of cells with iron, involved exotoxins, which cause the death of host cells, free iron, which can be used by the infecting microorganism. In the regulation of genes that encode the synthesis of some exotoxins and siderophores, iron (shigo toxin, hemolysin, aerobactin) is involved.

The assimilation of iron includes several stages: the binding of a complex of Fe^{3+}–siderophore with special membrane proteins, metal transfer into cells, release of iron from the complex, and its inclusion in reserve proteins. Iron absorption systems have been well studied in *E. coli* and some other types of bacteria as well as in several types of fungi (*Ustilago sphaerogena*) and yeasts (*S. cerevisiae*).

Porins of the outer membrane cannot provide the penetration of siderophores into the periplasmic space. This requires proteins called receptor proteins. Siderophores bound to the receptors penetrate through the external membrane that requires energy and, in complexes with the binding proteins, form the periplasmic space.

Some bacteriophages (T1), antibiotics (albomycin), and colicin are bound on the surface of cells with polyfunctional receptors and can penetrate into cells using an iron transport system. The non-specificity of receptors allows the use of different ligands to study the transport process. For example, mutants with a damaged transport of iron resistant to the T1 phage (T-one) revealed Ton protein, one of the components of the iron transport system.

Some antibiotics are structural analogues of siderophores and can therefore penetrate into cells using iron transport systems. The transfer of the Fe^{3+}–siderophore complex through the membrane may be carried out by several ATP-dependent transport systems.

Siderophores (siderochromes, sideramines) are low-molecular compounds (0.5–10 kDa) with high affinity for Fe^{3+}. Complex formation constants for iron with siderophores is 10^{23}–10^{52}. Fe^{3+} ions bind to six oxygen atoms in the siderophore molecule. After the formation of the complex with iron, conformational changes occur in the siderophore molecule, and in this form, it is recognized by the membrane receptors of the cells.

Microorganisms synthesize siderophores that are diverse in their chemical nature. Information on the synthesis of siderophores by microorganisms appeared in the 1950s. The first described siderophores were mycobactin, arthrobactin, ferric chloride, and coprogen. Currently, in cultures of almost all studied species of aerobic and optional anaerobic microorganisms, substances that are involved in the binding of Fe^{3+} and transport it into cells are detected (Table 3.4).

Table 3.4. Binding constants for Fe^{3+} complexes at pH 7

Iron complex	Constants for Fe^{3+} complexes
Fe^{3+}–enterobactin	10^{52}
Fe^{3+}–ferrioxamine B	10^{32}
Fe^{3+}–ferrichrome A	10^{29}
Fe^{3+}–transferrin A	10^{24}
Fe^{3+}–EDTA	10^{25}

Siderophores have a different chemical structure, but most of them can be divided into two main groups: **phenolates** and **hydroxamates**. Ligands of the phenol type are found only in bacteria. These are derivatives of 2,3-dioxybenzoic acid. Bacteria of the intestinal group (*E. coli*, *Salmonella enterica*) form **enterobactin** (also known as **enterochelin**). This is a cyclic triester of *L*-serine, in which all three amino groups are acylated with residues of 2,3-dioxybenzoic acid (Fig. 3.24, I). Iron forms complex with enterochelin, in which one molecule of ligand has one Fe^{3+} atom.

Bacteria *Azotobacter vinelandii* synthesize azotocheline of *N,N'*(2,3-dioxy)-*N*-benzoyl-*L*-lysine. In addition, hydroborate siderophores containing citrate residues, aerobactin (Fig. 3.24, II), schizokinin, and arthrobactin were detected in bacteria. Aerobactin is the second siderophore enterobacterium that can be encoded by a plasmid or a chromosome.

Phenolic siderophores are formed from chorismic acid, which is a common precursor of aromatic acids and some vitamins. The biosynthesis of enterobactin consists of the following stages: chorismate → isochorismate → 2,3-dihydro-2,3-dioxybenzoate → 2,3-dioxybenzoate → enterobactin. In the synthesis of 2,3-dioxybenzoate from chorismate, isochorismate synthase (EC 5.4.4.2), 2,3-dihydro-2,3-dioxybenzoate synthetase, 2,3-dihydro-2,3-dihydroxybenzoate dehydrogenase (EC 1.3.1.28), which are encoded by the *entC*, *entB*, and *entA* genes, respectively, are involved. The complex of products of four other genes *entD*, *entE*, *entF*, and *entG* ensures the formation of enterochelin.

Some microorganisms form several siderophores, which differ little in their chemical structure. Siderophores in bacteria *Cryptococcus melibiosus* contain alanine. Actinomycetes and some bacteria (*Pseudomonas*, *Arthrobacter*) form siderophores of the group of ferrioxamines (Fig. 3.24, III). Linear or cyclic molecules of compounds of this group contain residues of acetate, succinate, and amino-*N*-oxyaminoalkyl groups.

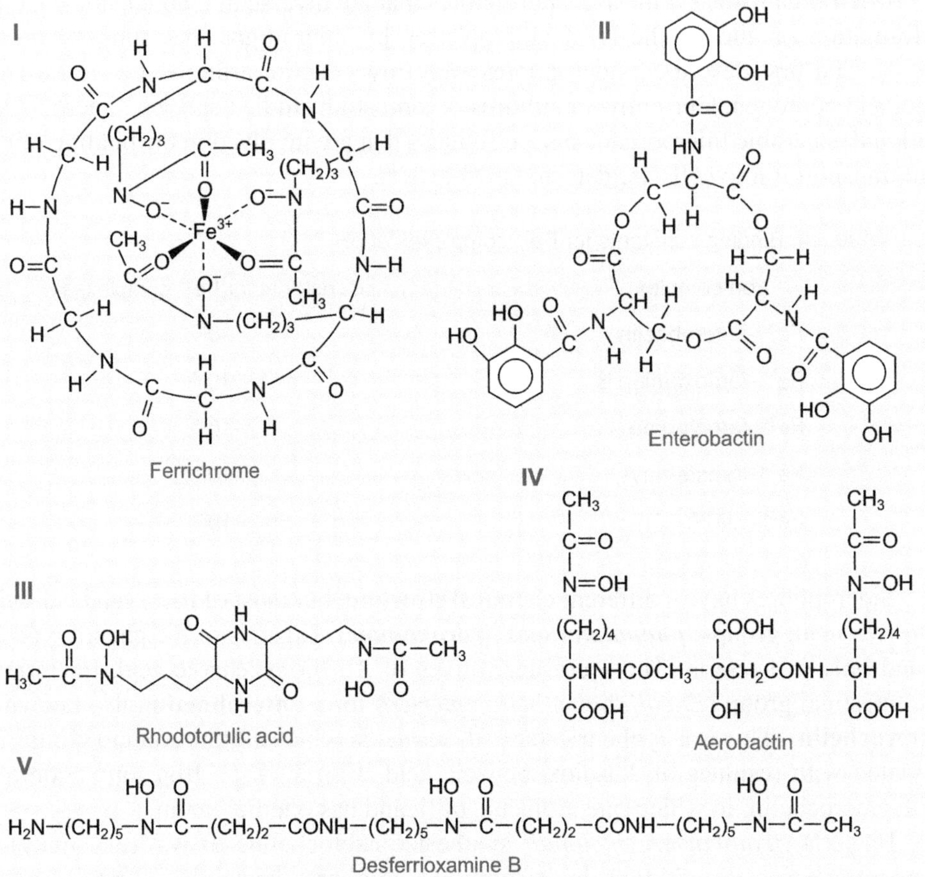

Fig. 3.24. Structure of some siderophores of microorganisms.

In bacteria, siderophores of hydroxamate type, containing the remainder of citrate, **aerobactin** (**Fig. 3.24, IV**), schizokinin, and arthrobactin, were detected. Some bacteria form mixed-type siderophores that contain both phenolic and hydroxamate groups. Pathogenic for plants, fluorescent pseudomonads *Pseudomonas syringae* pv. *syringae* form siderophore pyoverdine, unique in its composition. It contains β-hydroxyaspartic acid, threonine, serine, and lysine at a ratio of 2:2:2:1; the α-hydroxy group of each residue of β-hydroxyaspartic acid is a ferric-binding ligand. The assimilation of iron in pseudomonads, mediated by siderophores, has been intensively studied in recent years, due to its possible role in the supply of iron by plants and in the virulence of pathogenic strains.

Some microorganisms are able to use siderophores formed by other organisms. *E. coli,* which synthesizes enterochelin and aerobactin, has a citrate-dependent system of transport of iron ions and can use coprogen and rhodotorulic acid.

Mycobacterium smegmatis forms extracellular siderophores called **exochelins** and intracellular siderophores **mycobactin** and can use iron from the complex with citrate and rhodotorulic acid, but not with ferrioxamine B or ferrichrome. The photo-synthetic bacterium absorbs iron from complexes with citrate and parabactin, although it does not form its own siderophores.

Siderophores have an extremely high affinity for Fe^{3+}, but many also form relatively stable complexes with copper, molybdenum, aluminum, and some transuranium elements.

There are four mechanisms for the absorption of an iron–siderophore complex:

- Shuttle mechanism. The complex of Fe–siderophore enters the cell; the iron is released under the action of reductase, or the ligand is replaced; and the recipient siderophore performs the function of iron storage. The liberated siderophore can bind a new iron molecule. Such a mechanism is characteristic for siderophores of the family of ferric chromes and for coprogens.
- The mechanism of direct transfer. Iron is released from the complex with a siderophore without the introduction of the ligand into the cell. At the same time, iron is not restored, but there is a membrane-bound exchange of the siderophore on an intracellular chelating agent. This type of iron transfusion is described for the family of rhodotorulic acid.
- An esterase-reductase mechanism that functions for Fe-triacetyl fuzarinin C. After absorbing the complex, the ligand's radioactive bonds rupture; the fuzarinin portion is excreted; and iron ions are restored.
- The restorative mechanism described for some ferric chromes that are not part of the cell, and the recovered iron is transported in the form of Fe^{2+}. This type of transport is also characteristic for microorganisms that do not synthesize directly from siderophores but can use siderophores synthesized by other microorganisms.

Bacteria *E. coli* include five different absorption systems with the participation of siderophores. These systems are encoded by more than 20 genes located in several operas. They include:

- ferrichrome absorption system (*fhu*ACDB operon);
- system for the absorption of enterochelin (enterobactin) (*fep*A, *fep*B, *fes*, and *fep*CDG genes). The way of absorption of enterochelin degradation products is encoded by the *cir* and *fiu* genes;
- absorption system of coprogen (*fhu*E gene);
- a system for the absorption of complexes of Fe^{3+}-dicitrate, the components of which are codified by the operon *fec*IRRRRRABCD.

The *exb*BD and *ton*B genes, whose products are important for the translocation of Fe^{3+} complexes across the external membrane of Gram-negative bacteria, should also be included in this list. The TonB protein has an extremely prolonged sequence of [73-(Glu-Pro)$_5$-Ile-Pro-Glu-Pro-Pro-Lys-Glu-Ala-Pro-Val-Val-Ile-Glu-(Lys-Pro)$_7$-109]. It is assumed that it functions to link energy from the potential of the inner membrane to open channels in Ton-dependent outer-membrane proteins, including ferrisiderophore transport systems.

ExbB functions in the inner membrane and is activated or stabilized under the action of TonB. All known ferrisiderophore specific outer membrane proteins in *E. coli* are TonB-ExbB-dependent: enterochelin receptor (FepA), aerobactin (IutA), citrate (FecA), ferrichrome (FhuA), and rhodotorulic acid (FhuE).

Many of the above-mentioned proteins require a cell only in specific growth conditions when microorganisms are lacking in iron. Therefore, most of them are expressed as much as possible only when the iron is insufficient and strongly repressed when the cell feels an excess of iron.

The main siderophore of *E. coli* is enterochelin. **Desferrienterochelin** in the medium is bound to Fe^{3+}, after which the products of the *fep* genes are able to move ferrienterochelin back into the cell (**Fig. 3.25**).

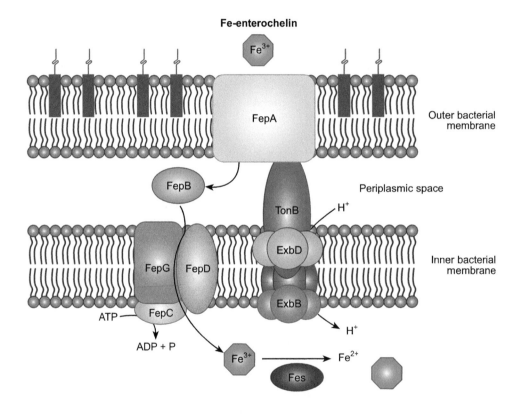

Fig. 3.25. Enterochelin (Ent) system in *E. coli*.

The outer membrane protein FepA (Mw = 79 kDa) functions as a monomer and is able to transfer the complex across the membrane. It contains 29 transmembrane regions and an amino acid sequence called **Ton box**. The energy potential of the cytoplasmic membrane is transmitted to FepA, and the transport of the siderophore across the membrane is started. Two other membrane proteins, ExbB and ExbD, form a heterohexamer with a central channel that changes the conformation of TonB. The periplasmic ferrienterochelin-binding protein (FebB) directs the complex to the intra-embodiment transport system, which consists of FepG, FepD, and FepC.

Enterochelin, using the FepA protein of the outer membrane, passes through a process coupled to TonB. The periplasmic FepB transmits Fe-enterochelin to FepG and FepD intramuscular proteins. The passage through the internal membrane requires the energy of ATP FepC. In the cytoplasm, ferric enterochelin esterase cleaves enterochelin, releasing the recovered iron. FepC is a membrane-bound ATPase that provides the energy needed to transport the complex through FepG and FepD to the cytoplasm. Fes esterase cleaves enterochelin, reducing affinity for iron.

In the regulation of the biosynthesis of phenol and hydroxamate siderophores, iron is involved. When cultivating the strain *Klebsiella aerogenes* in medium with a low iron content, the activity of enzymes that catalyze the formation of enterochelin increases by 15–40 times. The regulation of the formation of phenolate siderophores by iron is found in *E. coli*, *Azotobacter vinelandii*, *B. subtilis*, *K. aerogenes*, *Vibrio cholerae*, and other types of bacteria. Strains of *E. coli* and *B. subtilis* under conditions of insufficient iron supply can accumulate up to 75–150 mg of phenolate siderophores in 1 liter of culture medium.

The regulation of the biosynthesis of enterochelin in *E. coli* was studied in detail. The synthesis and absorption of this siderophore is encoded by a cluster of 14 genes *ent–fep*, which includes seven transcription units that transcribe in two directions. In the synthesis of 2,3-dioxybenzoate from the chorismate, isochorismate synthase, 2,3-dihydro-2,3-dihydroxybenzoate synthetase, and 2,3-dihydro-2,3-dioxybenzoate dehydrogenase, which are encoded by the *entC*, *entB*, and *entA* genes, respectively, are involved. To connect dihydroxybenzoate with serine and for the formation of a trimmer, a complex of products of four other genes, *EntD*, *EntE*, *EntB*, and *EntG*, is required. The synthesis and absorption of enterochelin are controlled by 13 genes, and another *entG* gene is identified near the 3′-end of the *entB* gene due to a mutation that retains the *entB* activity but eliminates the *entG* activity required for the enterochelin synthesis. Bifunctional peptide (EntBG) is also required for the synthesis of dihydroxybenzoic acid, and for the synthesis of enterochelin from dihydroxybenzoylserine.

To study the regulation of the expression of the genes responsible for the synthesis of enterochelin, the method of fusion of the *ent* genes with the *lac* operon that carries the β-galactosidase gene but without the regulatory region has been used. It was shown that with iron deficiency, there is a 5–15-fold increase in the expression of all investigated *ent* genes.

The influence of iron on the accumulation of various siderophores of hydroxamate type was found in many species of pro- and eukaryotic organisms, but it is not established which stages of biosynthesis are regulated by this metal. Iron suppresses the formation of fuzarinins, coprogen, and schizokinin.

In *A. aerogenes*, the biosynthesis of aerobactin is encoded by chromosomal genes, while in *E. coli*, the five-gaseous aerobatic operon is located on the ColV plasmid, which, moreover, determines the ability to virulence. Under conditions of insufficient iron supply, numerous genes, whose function is associated with the absorption of iron, are subjected to deregulation in *E. coli*. These genes encode absorption systems that have high affinity for iron (*fhu* — absorption of Fe-hydroxamates, *fec* — transport of Fe-citrate, *fep* — transport of Fe-enterochelin) and also enzymes associated with the synthesis of siderophores (*iuc* (syn. *aer*) — aerobactin, *ent* — enterochelin). With the participation of iron, other genes are regulated (*cir* — receptor of colicine; *slt* — shiga-like toxin). All these genes are repressed with growth in a medium with high iron content.

The aerobactin iron transport system in *E. coli* and *Salmonella enterica* is repressed by the product of the *fur* gene. It is shown that in order to form the Fe^{3+} complex and bind to the promoter of the aerobactin operon, the dimer form of the Fur protein is required, with the binding being metal dependent. In addition to Fe^{2+}, activators are Mn^{2+}, Cu^{2+}, Co^{2+}, Cd^{2+}, and partially Zn^{2+} in high concentrations.

The *fur* gene product carries out the control of the expression of iron transport genes by the negative type. In *E. coli*, the Fur-regulon includes at least five iron-transport systems as well as a number of other genes that are controlled by iron. The Fur protein is a repressor that is capable of blocking the transcription of genes regulated by iron binding to specific sequences in the promoter region. The product of the *fur* gene is a protein with Mr = 16.8 kDa, which contains histidine and cysteine. In the aerobactin operon, a nucleotide sequence (*fur*-box or iron-box) that is protected from the action of DNAase in the presence of the Fur protein and Mn^{2+} is identified. This DNA sequence from 19 bp: 5'-GAT (A/T) ATGAT (A/T) AT (C/T) ATTTTC-3'.

DNA sequences, like fur-boxes, are found in many genes whose expression is regulated by the Fur protein. The *fur* genes are cloned in *Yersinia pestis*, *Vibrio cholerae*, *Pseudomonas aeruginosa*, and *Neisseria*, which suggests the similarity of regulatory mechanisms in different bacteria. In *B. subtilis*, three fur-shaped proteins have been described.

The transport of iron in *E. coli*, *Vibrio anguillarum*, and *Pseudomonas*, in addition to negative Fur-dependent regulation, is subject to positive control. The Fur gene product regulates the transcription of at least 50 genes in *E. coli* and *S. enterica*, many of which are not related to iron transport. Some genes are regulated by the Fur-protein for a positive type. The most studied system is the ferruginous superoxide dismutase enzyme encoded by the *SodB* gene. The activity of this enzyme increases with the presence of iron and the product of the *Fur* gene. The regulation is posttranscription and stabilize the *SodB* mRNA.

Since many genomes of animal and plant pathogens have been sequenced, this made it possible to characterize their iron-transport systems. The presence of complex mechanisms for regulating the expression of iron transport genes ensures their inclusion in response to multiple signals only when they are absolutely necessary.

3.11. Transport of proteins

After synthesis, proteins should be included in the periplasmic membrane, periplasmic space, or external membrane. There are other proteins that are released from the cell into the environment. These proteins include hydrolases, exotoxins, proteins that provide resistance to antibiotics, and enzymes for cell wall synthesis. The integral proteins of the plasma membrane are enzymes, cytochromes, and special and transport proteins. In the external membrane of Gram-negative bacteria, porins and other carrier proteins are located. All these proteins are synthesized in the cytoplasm.

The transfer of proteins across the plasma membrane and their excretion from the cell to the environment is called secretion. Excretion is most common in Gram-positive bacteria. The Sec system of proteins across membranes in bacteria usually uses the Sec pathway. Many proteins secreted through a plasma membrane of bacterial cells have a signal sequence at the *N*-terminus, through which they travel to a secretion apparatus placed on the plasma membrane. This is typical for most proteins of Gram-negative bacteria, localized in the periplasm and in the outer membrane as well as proteins of Gram-positive bacteria that are released into the medium.

Genetic studies, mainly *E. coli*, have allowed identifying a number of genes and their products that mediate this process. **Table 3.5** represents genes involved in the main translocation protein system. Protein SecB, which is a specific protein–chiron, binds to many precursors, mainly due to binding to certain places of the mature part of the molecule.

Table 3.5. Components of the system of translocation of the cytoplasmic membrane secretory proteins in *E. coli* and their presence in other bacteria

Groups and characteristics	Name	Size	Localization	The presence in other bacteria
Secretion chaperones	SeB	18	Cytoplasm	Widespread in *Enterobacteriaceae*
	DNAK	69		
General chaperones	GroEL	62	Cytoplasm	Universal
	GroES	11		
Secretory ATPase	SecA	102	Cytoplasm, ribosome, peripheral cytoplasmic membrane	*Bacillus subtilis Enterobacteriaceae*
Translocase	SecD	67	Cytoplasmic membrane	*Brucella abortis, Archaea*
	SecE	14		*Bacillus subtilis*
	SecF	39		*Archaea*
	SecY	48		*Bacillus subtilis*
Signal peptidases	LepB	36	Cytoplasmic membrane	*Salmonella enterica, Pseudomonas fluorescens, Bacillus subtilis*
	LspPA	18	Lipoprotein, signaling peptidase, cytoplasmic membrane	*Pseudomonas fluorescens, Klebsiella aerogenes, Pseudomonas aeruginosa*
	Ppp	25	Cytoplasmic membrane	*Vibrio cholerae, Bacillus subtilis*
Others	sRNA	4.5	Cytoplasm	*Pseudomonas fluorescens*
	ffh	48	Cytoplasm	
	FtsY	54	Cytoplasmic membrane	*Bacillus subtilis*

Chaperones are proteins that stabilize the conformational state of other proteins in the process of their synthesis or after its completion. Such stabilization ensures the correct collapse of the protein molecule. In the case of sec-dependent secretion, SecB stores a prefix in a state suitable for movement. However, the secretion of certain proteins does not depend on SecB; its function is performed by the general action *chaperone* proteins (GroEL). The PreBox–SecB complex binds to the SecA protein, part of which is linked to the SecY/E/G integral membrane complex on the inner side of the plasma membrane. Next, the SecB–supramental–SecA complex interacts with a complex of SecM/E/G membrane proteins, which leads to the stimulation of the SecA protein of ATP activity. It is possible that the hydrolysis of ATP contributes to the release of the prefilm from the complex with SecA. The secretion of proteins through the plasma membrane is provided by a number of proteins — *SecD, SecE, SecG,* and *SecY* genes as well as a protein with the ATP activity of SecA.

Integral membrane proteins bind to a very hydrophobic N-terminal signal sequence that immerses the protein in the membrane. It consists of 15–30 amino acids and includes 11 hydrophobic amino acids, between which there are short hydrophilic residues. The signaling sequence ends with glycine and alanine residues, which determine the location of the signal peptidase. Splitting leads to the release of a mature protein. This mechanism of secretion completely depends on the presence of a signal sequence. Removing the signal sequence or mutation that results in its alteration prevents secretion.

Unlike integral membrane proteins, the proteins of the periplasm and the outer membrane split their signaling sequence when exported. The role of the signal sequence is to mediate the binding of the polypeptide to the membrane and change the conformation of the protein so that it can be dissolved in the membrane. Signal sequences themselves, however, cannot export proteins.

Some proteins are secreted regardless of Sec proteins or using additional proteins. For example, in Gram-negative bacteria, one protein (pullulanase, amylase) passes through the plasma membrane using the Sec pathway, and through the external one using another pathway, which involves at least 14 genes. Systems of this type are typical for proteins involved in the formation of flagella, fimbria, and bacteriocins.

Thus, the transfer of proteins in the cell and their secretion were considered, but for the growth and functioning of cells, a variety of mechanisms for the transport of complex lipids, in particular, external membrane components, carbohydrates, including extracellular polysaccharides, and nucleic acids, are also required.

3.12. Group translocation

Some kinds of carbohydrates are transported through the bacterial membrane in the phosphorylated form. The process of phosphorylation happens during the transport of different bacteria, including *E. coli, Salmonella enterica, Staphylococcus aureus, Klebsiella aerogenes, Bacillus subtilis,* and *Lactobacillus plantarum.* During the translocation, hexoses and other glycides are phosphorylated on the terminal carbon, except for fructose, where phosphate binds to C1. A phosphate donor in bacteria is usually **phosphoenolpyruvate (PEP)**. The transfer is realized by the phosphotransferase system. It consists of an active thermoresistant protein called phosphotransferase, which has a molecule weight about 9400, and two enzymes — cytoplasmic enzyme (En1) and membrane specific enzyme (En2). Transport takes place in two steps.

At first, En1 transfers phosphate from PEP to the carrier protein HPr, and HPr binds to the N1-histidine residue according to the equation:

$$PEP + HPr \rightarrow P–HPr + pyruvate.$$

Protein (HPr) and En1 are not in the cytoplasmic membrane; they are probably part of the cytoplasm. The mentioned mechanism of carrier protein phosphorylation is similar for all bacteria. On the other hand, the second step varies for individual bacteria in transported glycides.

During the second step, En2, which binds in the membrane, is catalyzed.

$$P–HPr + carbohydrate \rightarrow carbohydrate–P + HPr.$$

The different character of the phosphotransferase system on the second step confirms that in *E. coli*, En2-catalyzed transport consists of two proteins, En2 A and En2 B. Nevertheless, every protein 2 A is specific for a certain carbohydrate and is labelled after that (En2 A^{glu}). For the activation of En2 A, glycerol phosphate is required which is in a small quantity included in the *E. coli* membrane. On the other hand, the fraction of En2 B, whose content can be almost 10% of the membrane proteins, is part of the transport system for glucose, fructose, and mannose.

The second step in the group translocation of lactose in *S. aureus* has a different progress. The dominant difference is in En2^{lac}, which is bonded to the membrane. The soluble protein factor 3^{lac} is also part of the reaction according to the equation:

$$3 P–HPr + F3^{lac} \rightarrow tri–P–F 3^{lac} + 3 HPr$$

$$tri–P–F3^{lac} + 3 \text{ lactose} \rightarrow 3 \text{ lactose–P} + F3^{lac}.$$

Phosphate bonding to $F3^{lac}$ occurs through the histidine molecule in this protein.

The transport of purine and pyrimidine bases in enterobacteria and their transformation in appropriate mononucleotides have also the character of group translocation and take place according to the equation:

Purine + Phosphoribosyl pyrophosphate $\xrightarrow{\text{Phosphoribosyltransferase}}$ Purine Pyrimidine Mononucleotide + Pyrophosphate
Pyrimidine

(Outside) (Inside)

Osmotic barrier. Effects of facilitated diffusion and active transport prove that the semipermeable membrane has a selective character and, with the transport systems, creates an osmotic barrier, which protects bacteria against the penetration of undesirable molecules into the cell.

3.13. Bacterial growth and multiplication

In the right conditions, bacterial cells grow to a certain size and then divide. The enlargement of cell dimensions involves a number of biological processes that result in the formation of cellular components, especially ribosomes, most proteins, and nucleic acids. The summary of these processes is often referred to as the growth cycle.

The division itself is a complex process and its mechanism is not entirely clear. Most bacteria are divided transversely by first dividing the nucleoid (chromosome) and then forming a cell divider separating the parental cell into two parts. Some bacteria can multiply by budding. Chromosome replication can be divided into roughly three phases — initiation, replication itself, and termination.

In the first phase, the chromosome connects its origin of replication to the cytoplasmic membrane (mesosome), where the enzymes, polymers, and ligases are found. In the next phase, double helix develops, and chromosome replication occurs. At the end of the replication process, the complementary fibers are ligated into the circle by the enzyme ligase, and the chromosomes are sequentially separated (**Fig. 3.26**).

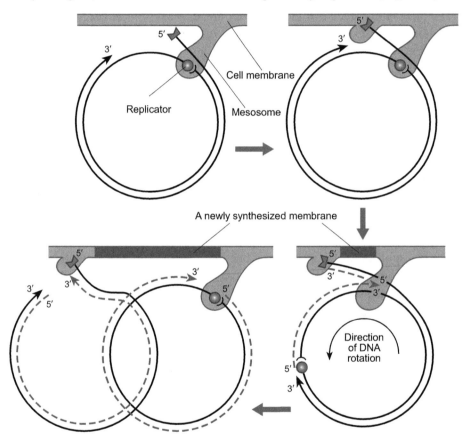

Fig. 3.26. Replication of bacterial chromosome by Jacob, Brenner, and Cuzin (1963).

An initial impulse to replication is most likely given by a certain amount of volume of the growing cell. The separation of the chromosome between the parental and daughter cells during the cell cycle is probably enabled by the growth of the cytoplasmic membrane touching the chromosome. The creation of the transverse septum separating the cells is conditioned by the completion of one replication cycle and the division of the chromosome.

The development of the septum begins by cleaving the cytoplasmic membrane-peptidoglycan layer complex in the parental cell. The resultant structure is gradually filled with the cell wall material. Gram-positive bacteria start producing cell wall material at the beginning of this process, while in Gram-negative bacteria, wall material synthesis is performed in the next phase. In these bacteria, the constriction of the cell happens during the initiation of cytokinesis directly in the dividing area (**Fig. 3.27**).

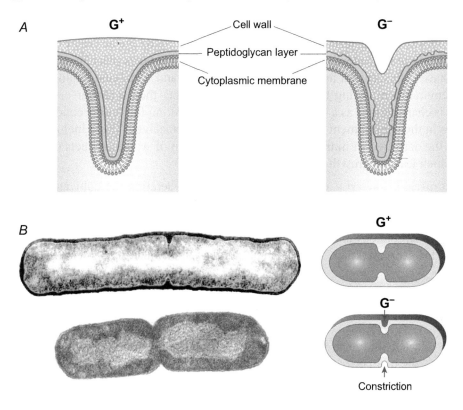

Fig. 3.27. Development of the transverse septum in Gram-positive and Gram-negative bacteria: differences in the synthesis of the cell wall (*A*), constriction of the cell (*B*).

The different nature of the cytoplasmic membrane and cell wall contact is apparently the cause of the variance of septum development in Gram-positive and Gram-negative bacteria. The activity of hydrolytic enzymes results in the separation of the daughter cell. In some bacteria, after the partition, the cells are not completely disconnected. They remain attached by a smooth bridge made of fibrous structures. Depending on the shape of the cell and the direction of the dividing plane, various formations such as diplococci, streptococci, sarcina, and staphylococci can be created.

In every environment, bacteria multiply as long as they find the conditions required for cell division, that is, nutrients, temperature, oxygen pressure, pH, and redox potential. Therefore, the rate, at which new cells are created, depends on the changes happening in the environment.

Bacteria and their environment form a comprehensive system. This system can be either closed or open depending on the cultivation method, which distinguishes:

(1) bacterial growth under conditions of static cultivation;

(2) bacterial growth under conditions of continuous cultivation.

3.13.1. Growth under conditions of static cultivation

The concept of static bacterial cultivation. In static cultivation, the properties and composition of bacterial environment vary depending on the activity of the bacterial cells. There is no steadily ascending trend here, but it takes place in the characteristic phases of ascension and decline. Since all cells and metabolic products remain in the environment during the whole cultivation period, the bacterium–environment system is characterized as a closed system. Growth-limiting factors are conditioned by changes in the environment caused by the bacteria themselves. These include, in particular, the depletion of nutrients and the accumulation of waste products of metabolism, which leads to growth and dying.

Submerged bacterial cultivation. By mixing or aerating the cultivated culture, improved nutrient utilization and, therefore, more intensive propagation can be achieved. This way is called submerged bacterial cultivation. It is mainly applied in the field of industrial microbiology aimed at acquiring certain important substances, including antibiotics. However, even this improved method is essentially a closed system.

Growth curve of bacteria. Changes in the static culture environment are reflected in the irregular, phasic character of bacterial growth. The sequence of growth phases depending on cultivation time can be graphically expressed by the so-called bacterial growth curve (**Fig. 3.28**).

It is clear that after the inoculum is transferred to the living environment, after stagnation, the number of cells in the static culture gradually increases, but after the climax, it begins to decrease again. An increase or loss of cell count is not uniform. In some time intervals, the growth is faster, in others slower. On the curve shown, these sections define the so-called growth phases. For most bacteria grown under static conditions, the following growth phases are found most often:

(1) The *lag* phase is usually the period following bacteria transition to a new environment. The cell count is practically constant; the number is often reduced by the loss of less viable cells. At this stage, surviving bacteria adapt to new conditions, generate necessary enzymes, and increase their volume mainly by synthesizing the cellular components necessary to initiate the reproduction process, i.e., nucleic acids and proteins. The duration of the lag phase depends on the size and age of the inoculum in addition to the nature of the environment.

(2) The accelerated growth phase represents the period in which the culture is already fully adapted to the environment. The cells begin to multiply with an increasing rate of division, and the intensity of their metabolism increases rapidly. The high physiological activity of the cells is accompanied by their increased sensitivity to unfavorable environmental factors.

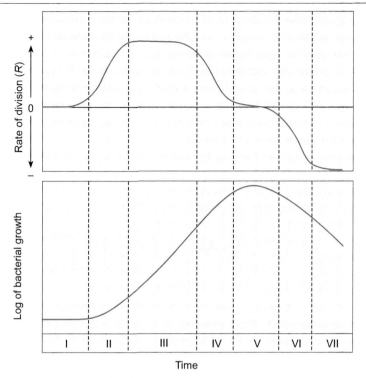

Fig. 3.28. Bacterial growth curve.

(3) The logarithmic or exponential phase is characterized by intense multiplication of bacterial cells increasing by geometric growth. The cutting speed is therefore constant. Intense division is accompanied by active cell metabolism, rapid substrate utilization, and metabolite production. The decrease in cell count due to death is minimal in relation to the increase in new subjects. Among the factors influencing the length of the log phase, there are the composition of the environment and the temperature of cultivation. Other factors are the characteristics of the cultivated species.

(4) The slow growth phase is characterized by a gradual slowdown in the processes of reproduction and total metabolism. The rate of division decreases as the number of dying cells increases. This is mainly due to the changes that occur in the environment by the gradual depletion of nutrients and the accumulation of toxic metabolic products. The adverse environmental factors accompanying these processes also include pH and redox potential.

(5) The stationary phase is a period of growth during which, due to the continuing adverse changes, the number of dying cells is increased by the growth of the culture. Therefore, the splitting speed is zero. At this stage, the number of cells is at its maximum. Certain bacteria grown in the same volume of the medium under certain conditions do not change this maximum concentration (the so-called M-concentration); therefore, it was suggested that the limiting factor

of cell proliferation is the lack of "biological space." This assumption, based on Malthus' population theory, was later criticized as one-sided. At the same time, it was shown that even in the same volume of environment, the number of cells can be substantially increased by changing conditions, for example, by aerating and, in particular, by introducing a continuous-flow culture. Bacterial growth is limited by several co-acting factors. These are primarily the nutrient concentration per unit area of the cell, the pH, the redox potential, and eventually some other environmental features.

(6) The phase of decline (or accelerated death) also corresponds to a time period characterized by a rising loss of cells. The rate of division drops below zero and gets negative values. This state is still influenced by deteriorating environmental conditions. The drop of the nutrient concentration below the critical level results in a decrease in metabolic activity, the gradual degradation of resources, and, finally, mass destruction of the cells. The course of this phase is, in addition to the properties of the culture itself, influenced by the nature of the accumulation of metabolic products.

Enzyme synthesis during bacterial growth. During bacterial growth, enzymatic apparatus is being modified. The beginning of growth is linked to the maximum creation and activity of enzymes. These enzymes, mainly catalyzing biosynthesis processes, include, for example, proteinase and amylase endoenzymes and some primary glycosylation enzymes. In the time of the intensive growth of the bacterial culture, these enzymes are used primarily as cell biosynthesis catalysts.

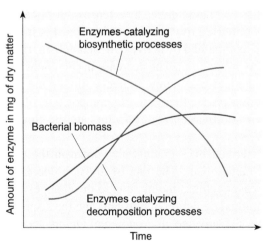

Fig. 3.29. Creation of bacterial enzymes in dependence on growth.

Other enzymes have their creation and maximum catalytic effect shifted to the phase of decline. Most of these enzymes are deaminases, exogenous amylases, and proteinases, etc. Besides the catalysis of some synthetic processes, their function is the catalysis of decomposition processes (**Fig. 3.29**).

Thus, under conditions of static cultivation, bacterial growth has a phase character and depends on the metabolic activity of cells and environmental conditions (presence of free substrate, concentration of terminal metabolites, population saturation). Every phase may differ in different bacteria and depends on the bacterial species. Depending on the growth phase, two types of enzymes that catalyze biosynthetic and decomposition processes are involved in bacterial growth.

3.13.2. Growth constants

The growth of bacteria in the logarithmic phase is usually characterized by the so-called growth constants. They include:
(1) bacterial cell division speed;
(2) generation time;
(3) lag time;
(4) specific growth rate.

All these constants can be deduced mathematically by simple considerations. Their values are to be understood mostly as relative values based on a simplified concept of bacterial growth and multiplication. Nevertheless, they become a significant help, especially in the field of comparative physiology and taxonomy.

Cell division speed. Assuming a bacterial cell divides into two by binary fission, the previous number of cells doubles with each new generation. We can express the gradual increase in cell count with this simple scheme:

Number of generations	0	1	2	3	4	n
Number of cells	1	2	4	8	16	2^n

If the initial number of cells is denoted as N_0, then the total number of cells generated at the end of the first generation will be $N_1 = N_0 \cdot 2$, at the end of the second generation $N_2 = N_0 \cdot 2^2$, at the end of the third generation $N_3 = N_0 \cdot 2^3$, and at the end of the nth generation, N reaches a value according to the formula:

$$N = N_0 \cdot 2^n.$$

The values of the initial (N_0) and the final (N) number of cells needed to quantify the relationship are usually determined using the direct (counting chamber) or indirect (based on the number of colonies) methods. The n value represents the number of generations in a given population; it expresses how many times an average cell was divided and is calculated using decimal logarithm.

$$\log N = \log N_0 + n \cdot \log 2, \text{ from where}$$

$$n = \frac{1}{\log 2} \cdot \left(\log N - \log N_0 \right).$$

By expressing the number of generations n depending on the growth time of the culture, we obtain the so-called average division rate (R):

$$R = \frac{n}{t} = \frac{1}{\log 2} \cdot \frac{\left(\log N - \log N_0 \right)}{t - t_0},$$

where the expression $t - t_0$ is the difference between the beginning (t_0) and the end of the time interval, where the number of cells was found. It is given in hours.

Generation time is the time needed to create a new generation ($n = 1$), that is, the time between cell divisions. Its derivation is very simple. If the division rate represents the average number of cell divisions per time unit, it is clear that the generation time representing the number of time units required for one cell division will correspond to the inverted value of the average division rate:

$$G = \frac{1}{R} = \log 2 \cdot \frac{t - t_0}{\left(\log N - \log N_0\right)}.$$

Both quantities, that is, the division rate and generation time, are objective in nature only if they are derived from the exponential growth of the bacterial culture. Their value depends on the properties of the bacterial species and on environmental conditions, as shown in the attached **Table 3.6**.

Table 3.6. Generation time in some bacterial species

Bacteria	Temperature (°C)	G
Escherichia coli	37	17 minutes
Bacillus thermophilus	55	18 minutes
Klebsiella aerogenes	37	21 minutes
Serratia marcescens	30	25 minutes
Staphylococcus aureus	37	27–30 minutes
Streptococcus lactis	37	48 minutes
Lactobacillus acidophilus	37	1.1–1.5 hours
Mycobacterium tuberculosis	37	13.2–15.5 hours
Treponema pallidum	37	33 hours

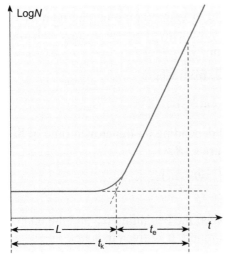

Fig. 3.30. Graphical derivation of the *lag* time (L) in bacteria.

The *lag* time. The actual proliferation of bacteria is often preceded by a period in which cells adapt to new conditions. The transition from this phase to the exponential phase is not always clear, and sometimes it is difficult to decide whether and how long the phase takes place. The *lag* time (L) is the growth constant used to determine the duration of the *lag* phase. It represents the time interval between the beginning of the inoculation and the start of the exponential phase. The end of that interval is defined by the intersection of the line, which is parallel to the time axis at the distance of the initial number of cells, with the line obtained by prolonging the exponential growth (**Fig. 3.30**).

The *lag* time can be derived from the difference between the total growth time observed (t_K) and the time the culture needs to pass to the exponential growth phase (t_e):

$$L = t_K - t_e,$$

where t_K is the end of the relevant measurement period included in the exponential growth phase; t_e can be determined from the relation for division speed or generation time:

$$t_e = \frac{1}{\log 2} \cdot \frac{\log N - \log N_0}{R} = \frac{1}{\log 2} \cdot G\left(\log N - \log N_0\right).$$

Specific growth rate. This constant represents the time increment per unit of growing biomass. It can be expressed by the number of cells, the optical density, the amount of dry matter, etc. It can be derived from the relationship characterizing the period of constant biomass growth over time:

$$\frac{dX}{dt} = \mu \cdot X,$$

where X = amount of biomass grown; μ = specific growth rate.

By integrating the formula above, we obtain a general equation of growth during which bacteria multiply at a constant rate:

$$X = X_0 \cdot e^{\mu \cdot t},$$

where X_0 is the initial amount of biomass (number of cells, dry weight, optical density, etc.).

From this equation, a specific growth rate can be determined using natural logarithms:

$$\mu = \frac{\ln X - \ln X_0}{t}.$$

After converting to decimal logarithms using conversion ratio $\ln X = m \cdot \log X$, where $m = 2.302$, and introducing the expression $t - t_0$ for a certain time interval, we obtain a relationship

$$\mu = 2.302 \cdot \frac{\log X - \log X_0}{t - t_0}.$$

If we express the amount of biomass (X_0, X) by the number of cells (n_0, n), we can determine the relationship between the specific growth rate of division (R) or between the specific growth rate and the generation time (G). These relationships are given by the following formulas:

$$\mu = \frac{2.302}{3.32} \cdot R = 0.693 \cdot R, \quad \mu = \frac{0.639}{G}.$$

Maximum growth rate, saturation constant. While studying the use of nutrients by bacterial cells, it was found that the value of the specific growth rate in the exponential growth phase is directly dependent on the substrate concentration (essential nutrients). This dependence can be expressed by the formula:

$$\mu = \mu_m \frac{S}{K_S + S}.$$

where μ_m = the maximum growth rate; S = the substrate concentration; K_S = saturation constant.

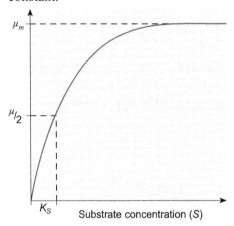

Fig. 3.31. The dependence of the specific growth rate on the substrate concentration.

Hyperbolic curve (**Fig. 3.31**) is the graphical expression of this relationship. The asymptote is the value of μ_m. The saturation constant K_S is numerically identical to the substrate concentration, for which the specific growth rate corresponds to half $\mu_m - (\mu_m/2)$. Actual saturation constant values are low. They range in the order of tens of mg per liter of medium for carbon sources and in the range of units of mg per liter of medium for amino acids. The func tional relationship between specific growth rates and substrate concentrations can be used, in particular, to study the nutritional value of different nutrient sources. These substances, usually in a suboptimal concentration, are used by bacteria at a different rate, which is reflected in different but specific conditions of specific growth rates.

3.13.3. Deviations from normal growth curve

Sometimes, bacterial growth in static culture does not follow the growth curve shown above, but takes on a different course. The phenomena that differ significantly from the normal sequence of growth phases include mainly growth in the multiple logarithmic phase and growth characterized by double growth curve (diauxie).

Multiple logarithmic phase. It is a type of multiplication process, which reminds several self-dependent exponential phases with different values of the division rate (R). In general, this phenomenon can be explained mainly by changes in the composition of the environment occurring in the course of cultivation. This is the case, for example, when one of the essential nutrients is exhausted and the transition to another substrate happens or when the temporary accumulation of metabolites acts as secondary sources of nutrition. In this way, the growth of *E. coli* occurs in a suboptimal concentration of carbon dioxide. The presence of CO_2 allows the metabolic biosynthesis of stockpile metabolites, thanks to the catalytic effect of carboxylation enzymes. Their gradual depletion allows the bacteria to multiply in several subsequent phases (**Fig. 3.32**).

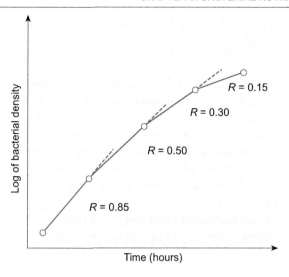

Fig. 3.32. Multiple exponential phase.

The double growth curve — Diauxic growth. Diauxic growth or diauxie is any cell growth that takes place in two phases and can be illustrated with a diauxic growth curve. Diauxie is characterized by growth, during which the cells return to the exponential growth phase after the logarithmic phase (**Fig. 3.33**). This reproduction pattern also corresponds to the dividing rate values associated with the individual growth phases. The described method of growth occurs in an environment that contains a mixture of two different organic compounds. During growth, bacteria use one of the components of the mixture only. Only after its exhaustion, the second component can be applied as a substrate. The successive or immediate successive use of the two substances is facilitated by the fact that the substrate, which is assimilated preferentially, inhibits the formation of the enzymes necessary to convert

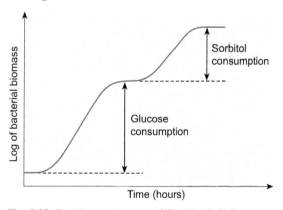

Fig. 3.33. Double growth curve of *E. coli* cells in the presence of glucose and sorbitol.

the remaining substrate. Therefore, the biosynthesis of these enzymes can take place only after using the first carbon source. This is only possible during the new *lag* phase. An example of diauxie is the growth of *E. coli* cells in a medium containing a mixture of glucose and sorbitol as a carbon source.

Thus, diauxic growth is a growth ability in the medium containing a mixture of two different substrates. Bacteria first use one substrate, after which they metabolize another.

3.14. Multiplication of microorganisms under conditions of continuous (*dynamic*) cultivation

The cause of changes in metabolic activity and the growth rate of bacterial cells growing under static cultivation are mainly changes in the environment generated by bacteria itself. The result of their activity is conditioned by the collaboration of a number of factors that cannot be separated or eliminated in a closed system.

To suppress environmental changes as a limiting growth factor, it is possible to use such a cultivation method, in which a state of dynamic equilibrium is created between the bacteria and the environment. These conditions are met by a continuous or flow-through method (**Fig. 3.34**).

The principle of this method is quite simple. A fresh environment is continuously poured into the bacterial culture container. Since the volume of the cultivator does not change, the same amount of environment that has been partly utilized by the bacteria has to flow out of it, as the fresh environment is being transferred in. During this process, part of the bacterial cells is washed away from the cultivator. If flow rate is the same as the rate of new cell growth and thus the rate of cell loss, the number of bacteria in the cultivator remains constant.

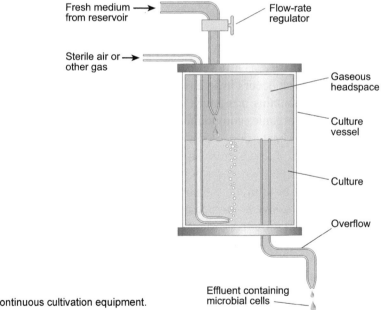

Fresh medium from reservoir

Flow-rate regulator

Sterile air or other gas

Gaseous headspace

Culture vessel

Culture

Overflow

Effluent containing microbial cells

Fig. 3.34. Diagram of continuous cultivation equipment.

A steady influx of fresh media and a flow of metabolic products ensure the growth of bacteria under virtually the same conditions, that is, in a physiologically stable state. Thus, the properties of the culture do not change unless we consider the changes underwent by the cells between dividing. The bacterium–environment system can be considered open in this case, because in this state of dynamic balance, the bacteria can be maintained indefinitely (**Fig. 3.35**).

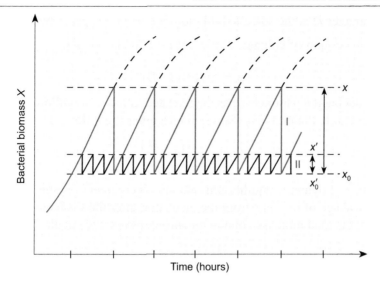

Fig. 3.35. Dependence of biomass growth on the interval between repeated influx of fresh medium using flow-through method (X_0, X: in 1-hour intervals, X'_0, X': in 15-minute intervals).

Chemostat. The growth rate of bacteria grown using the flow-through method can be regulated in essentially two ways. The first so-called external regulation mode depends on the change in the concentration of the essential nutrient in the culture medium. This is done by means of a device called chemostat. This device allows regulating the flow rate of an environment, in which the essential nutrient is in suboptimal concentration, while other nutrients are present in excess. Changing the rate of flow changes the concentration of the basic nutrient in the cultivator and thus also the rate of cell proliferation. The bacteria grow faster when the flow rate of fresh medium is increased.

Turbidostat. The second, internal mode of control is done by a device called turbidostat. There is a constant rate of multiplication as a regulating factor, which itself controls the inflow rate of a fresh, full-fledged environment. Upon reaching a predetermined upper cell density in a cultivator controlled by, for example, a photocell, the inflow rate increases, and the cell number decreases. On the other hand, the decrease in density at the selected lower limit causes the flow to stop or slow down, hence the population expands to the original density. Then the process is repeated.

Mathematical determination of bacterial growth rate in a flowing environment. Bacterial growth in a flowing environment is characterized by the specific growth rate of μ, contained in the relationship already known:

$$X = X_0 \cdot e^{\mu \cdot t}.$$

When using the flow-through method, the relationship above can be expressed as follows:

$$X = X_0 \cdot e^{\mu \cdot t} - D \cdot X \cdot e^{\mu \cdot t},$$

where X = total amount of biomass; $e^{\mu \cdot t}$ = biomass increment; $D \cdot X \cdot e^{\mu \cdot t}$ = biomass loss.

The parameter D is the so-called dilution rate, which is given by the rate of flow of the medium per unit of volume $\left(D = \dfrac{F}{V} \right)$. If $t = 1$ (hour) we get:

$$X = X_0 \cdot e^{\mu \cdot t}(1-D).$$

By introducing the parameter n as the repeated inflow and outflow of the environment for a time unit, the previous expression becomes the following:

$$X = X_0 \cdot e^{\mu \cdot t}\left(1-\frac{D}{n}\right)^n.$$

In the state of dynamic equilibrium, biomass increment is immediately compensated by the leakage of the cells from the incubator, after the fresh environment flows in. Therefore, the total biomass volume remains constant and equals the original state $(X = X_0)$:

$$1 = e^{\mu}\left(1-\frac{D}{n}\right)^n.$$

If the number of inflow and outflow intervals (n) becomes infinite, the expression $\left(1-\dfrac{D}{n}\right)^n$ changes to e^{-D}, that is:

$$1 = e^{\mu} \cdot e^{-D} \text{ so that } \mu = D.$$

The same result may be achieved by derivation of this equation according to the time. For the biomass increase, may be getting the following relationship:

$$\frac{dX}{dt} = \mu \cdot X.$$

Because in a state of dynamic equilibrium, each increase in the amount of biomass must be immediately compensated for by its loss, that is, by a leakage, the time increment of biomass must be virtually zero. This relationship can be expressed by the equation:

$$\frac{dX}{dt} = \mu X - DX = 0,$$

where D is the already known dilution rate. By adjusting this relationship, we get:

$$\frac{dX}{dt} = X(\mu - D) = 0 \text{ so that } X(\mu - D) = 0 \text{ and } \mu = D$$

Dependence of specific growth rate on dilution rate. In both cases, we deduced that the specific growth rate (μ) must be equal to the dilution rate (D), when using continuous cultivation. We can prove this by the following considerations:
- Let us assume that the dilution rate D exceeds the specific growth rate μ. The original formula for the biomass increase will then be as follows:

$$\frac{dX}{dt} = \mu \cdot X - D \cdot X < 0.$$

- The value of the increment (dX/dT) thus becomes negative, and the amount of biomass, that is, the cells in the system, will gradually decrease. As the number of cells is reduced, substrate concentrations increase. If, under these conditions, the value of μ is smaller than the value of μ_m, this difference may, according to the increasing concentration of the substrate, be offset according to the known relationship:

$$\mu = \mu_m \frac{S}{K_S + S}.$$

If the dilution rate (D) is not greater than the maximum specific growth rate, a new equilibrium characterized by a higher substrate concentration and a lower cell count occurs. However, if the value of D exceeds the value of μ_m, then all cells from the cultivator will be washed out.

Another condition can occur if the specific growth rate exceeds the dilution rate. Under these conditions, bacteria can use the substrate perfectly and increase the number of cells. On the other hand, reducing the substrate concentration by its better use will cause its deficiency and thus a reduction in the specific growth rate. Therefore, a new steady state will be created in the system in which the values of both constants equalize, that is, $\mu = D$. This characteristic of the continuous process, consisting in equalizing the values of μ and D up to the range of μ_m, is referred to as the self-regulatory capacity of the considered system.

Under the steady-state conditions, therefore, the specific growth rate and the dilution rate are equal. However, it is necessary to know which D values allow to achieve a steady state. The steady state of a system or a process is in a steady state if the variables (called state variables) that define the behavior of the system or the process are unchanging in time. In order to determine them, the relationship between the dilution rate and the substrate concentration is an important factor. In the cultivator, there is a substrate of concentration S_a used by bacteria. It flows out at concentration S. The change of the substrate concentration is therefore given by the relationship:

Substrate
concentration = Substrate influx – Substrate reflux – Substrate consumption.
increase

Substrate consumption can be expressed as the ratio between cell growth (μX) and yield (Y), that is, the mass amount of biomass produced per unit weight of the substrate used. Then the equation is:

$$\frac{dS}{dt} = DS_a - DS - \frac{\mu X}{Y}.$$

This equation is similar to the equation expressing the time increment of the biomass depending on the dilution rate:

$$\frac{dX}{dt} = \mu X - DX.$$

Both equations contain the quantity μ, which is a function of the substrate concentration $\left(\mu = \mu_m \dfrac{S}{K_S + S} \right)$. If we place an expression showing this functional dependence, we obtain equations:

and

$$\frac{dX}{dt} = X \left[\mu_m \left(\frac{S}{K_S + S} \right) - D \right] \qquad (1)$$

$$\frac{dS}{dt} = D(S_a - S) - \frac{X}{Y} \mu_m \left(\frac{S}{K_S + S} \right). \qquad (2)$$

These relationships perfectly define the behavior of the culture in continuous cultivation conditions. If S_a and D are maintained at a constant level and if D does not exceed a certain critical value, there are only X and S values, for which $\dfrac{dX}{dt}$ and $\dfrac{dS}{dt}$ are equal to zero, when the system is in dynamic equilibrium. If we solve both equations under these conditions, we get for the first equation:

$$X \left[\mu_m \left(\frac{S}{K_S + S} \right) - D \right] = 0, \text{ from where } X = 0$$

and

$$\mu_m \left(\frac{S}{K_S + S} \right) = D, \text{ from where } S = K_S \frac{D}{\mu - D}$$

and for the second equation:

$$DS_a - DS - \frac{DX}{Y} = 0 \text{ or } D \left(S_a - S - \frac{X}{Y} \right) = 0, \text{ where } D = 0$$

and

$$S_a - S - \frac{X}{Y} = 0, \text{ from where } X = X(S_a - S).$$

If μ_m, K_S and Y for a given bacterium and environment are known, we can deduce the biomass and substrate concentration for any D and S values using these equations. If we neglect the X and S values depending on D, we get a graphical picture of the activity of the given system (**Fig. 3.36**).

From this graph, it can be deduced that at most possible D values, the substrate concentration in the cultivator in the state of dynamic equilibrium is very low. This means that the substrate is completely consumed by the bacteria. Only when D reaches the value D_m, that is, when cell leaching occurs, the unused substrate appears in the cultivator. At the critical value of D_k, when the cells are practically flushed, the substrate concentration becomes equal to S_a.

The amount of biomass reaches the maximum at zero dilution rate ($D = 0$) and minimum substrate concentration. This condition corresponds to the final stage of the static culture, in which the growth was terminated because of substrate depletion. With increasing D, the substrate concentration increases gradually, but the number of cells decreases, reaching zero at D_k. Thus, the critical dilution speed D_k has a considerable practical significance and, according to the relationship $\frac{dX}{dt} = \mu X - DX = 0$, is equal to the maximum value of $\mu(\mu_m)$ achieved if it reaches its highest S_a value.

$$D_k = \mu_m \frac{S_a}{K_S + S_a}$$

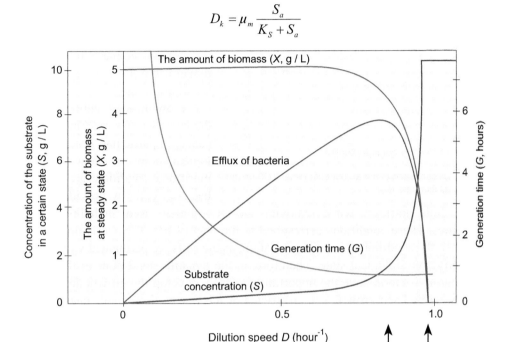

Fig. 3.36. Dependence of substrate concentration (S), biomass amount (X), and generation time (G) on dilution rates.

If $S_a < K_S$, which is usually true, D_k corresponds to the approximate value of μ_m. Whenever D exceeds D_k, $\frac{dX}{dt}$ will be negative, and for $X = 0$, a steady state cannot be achieved because bacteria will flow out of the cultivator too fast to multiply there.

From the graph above, the outflow of bacteria is also clear. It is expressed as DX, where X is a function of D. This curve reaches the maximum at the value D_m, which is optimal for reproduction. The last curve shows the dependence of the generation time (time of doubling the amount of biomass) on the dilution rate. The course of this curve shows that the generation time is the shortest at D_k, then it is gradually extended until it reaches its peak at $D = 0$.

The relationship between the specific growth rate and the dilution rate also influences the choice of the cultivators for continuous cultivation. At a significantly lower D value compared to μ value, the system can be kept in equilibrium by controlling the substrate concentration in the incubator, that is, using a chemostat. Conversely, at the value of the specific rate close to μ_m, a small change in the dilution rate results in a significant change in the substrate concentration and hence in the amount of biomass. It is

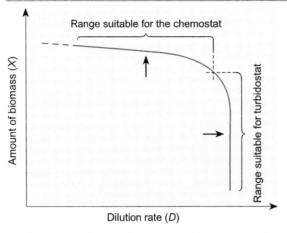

Fig. 3.37. Dependence of the amount of biomass (X) on the dilution rate (D).

therefore more appropriate to regulate this system by constant cell concentration, which corresponds to the turbidostat. The possibility of using one or the other device is shown in **Fig. 3.37** depicting the amount of biomass in a state of dynamic equilibrium depending on the dilution rate.

Flow-through cultivation methods have far-reaching significance, in particular, from a physiological point of view. Dynamic nutritional conditions are offered to bacterial cells by providing bacteria with fluid nutrition, while removing the environment used, including harmful metabolites. These conditions correspond to the way of life of bacteria undoubtedly better than the conditions of static cultivation, as the cells are permanently maintained in an active physiological state. Therefore, in the dynamic conditions of the flowing environment, bacteria have the opportunity to fully develop the growth and metabolism processes. Their growth is not restricted by adverse environmental changes, as in the case with static cultivation.

The benefits of continuous cultivation conditions are also reflected in the productivity of this process, given by the following relationship:

$$p = DX = n\left(1 - \frac{1}{e^{\frac{\mu}{n}}}\right)X$$

This relationship also indicates the productivity of the static cultivation process. Since the resulting cell concentration (X) is the same for both continuous and static cultivation, it is possible to determine the productivity of both culture methods from the expression:

$$p = n\left(1 - \frac{1}{e^{\frac{\mu}{n}}}\right).$$

Considering, for example, that static cultivation is completed in 10 hours, so $n = 1/10$ and $\mu = 0.3$, the productivity of this process is $p = \frac{1}{10}\left(1 - \frac{1}{e^3}\right) = 0.095$.

On the other hand, for continuous cultivation, where $\mu = D$, the productivity is given by the value of μ and is therefore 0.3. At the same cell concentration, the productivity in flow-through cultivation is about three times greater than in static cultivation.

3.15. Synchronous multiplication

Monitoring biochemical and physiological changes in bacterial cells during growth often encounters difficulties resulting from the fact that separation processes do not occur in all bacteria at the same time. This is because not all cells are in the same growth phase and are equally active or viable. Therefore, the culture passing through the individual growth phases is not homogeneous. Obtaining a developmental and hence physiologically cell homogeneous culture allows the method leading to the so-called synchronous cell division. Synchronization consists in preventing cells from dividing in a suitable manner, but leaves them with the possibility of further growth. During this time, individual bacterial cells can continue to grow normally, that is, they primarily carry out appropriate cell biosynthesis processes, including preparation for division, but cannot be divided. Once the original conditions are restored, all the cells are divided. This synchronous division continues, as a rule, during further growth of bacteria. The graphical representation of a synchronous growth gives rise to a characteristic stepped curve (**Fig. 3.38**).

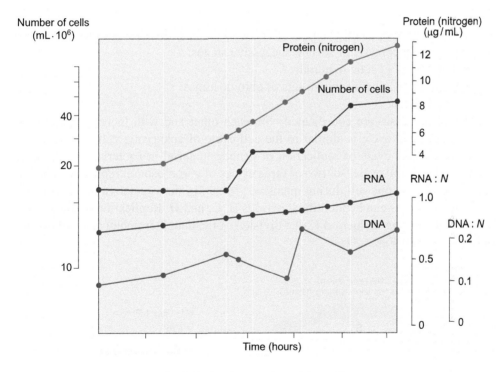

Fig. 3.38. Synchronous bacterial growth.

Induction of cell division synchronization. Synchronization of cell division can be achieved in several ways. The most commonly used method is the exposure of the culture to the effect of a cold shock. This is done by a sudden temporary reduction in temperature, so the cells are prevented from dividing, but not from growth. Then the original temperature conditions are restored. A similar effect can be achieved by temporarily transferring cells from a full-fledged environment to a deficient one, that is, lacking any nutrient or growth factor. It is a so-called starvation shock. Synchronization can also occur if cells of the same size obtained by filtration or centrifugation are seeded into the medium.

Importance of cell division synchronization. The importance of cell division synchronization consists in providing the opportunity to work experimentally with a homogeneous culture, for example, one in which the morphological and physiological properties of each cell correspond to the entire population. By synchronization, it has been proven that individual components of bacterial cells are not produced at the same rate during the generation period. For example, DNA synthesis in the *Azotobacter vinelandii* culture is stopped during synchronous division, whereas the synthesis of free amino acids increases at this stage. On the other hand, RNA and protein content increases steadily.

3.16. Bacterial cell cycle

The life cycle of a cell from cell division to the next division is called the vegetative cell cycle. It consists of several successive stages:
- doubling of genetic material;
- the difference between two sets of chromosomes;
- cell division.

These processes are coordinated with each other and with increasing cell mass. The bacterial cell cycle is similar to the cell cycle of eukaryotic cells, but has some peculiarities. One cycle of replication of genetic material in bacteria can take much less time than a cell cycle, so, two or three cycles of chromosomal replication can pass simultaneously in one cell during intensive growth. In the vegetative prokaryotic cell cycle, three phases can be identified: periods **B**, **C**, and **D**. Replication of the chromosome occurs during the period **C**, the division of the daughter nuclei and cell division during period **D** (**Fig. 3.39**).

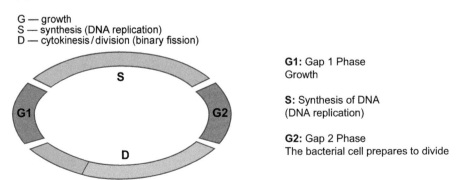

G — growth
S — synthesis (DNA replication)
D — cytokinesis / division (binary fission)

G1: Gap 1 Phase
Growth

S: Synthesis of DNA
(DNA replication)

G2: Gap 2 Phase
The bacterial cell prepares to divide

Fig. 3.39. Bacterial cell cycle.

In the cycle of eukaryotic cells, there is the phase of **Gap1**, abbreviated G1 (from gap), during which the mass of the cell increases. In bacteria, the corresponding phase is called period **B**. With rapid growth, a separate period is absent and replication cycles overlap. During the next phase, the replication of the genetic material occurs (period **C** or phase **S**).

In the eukaryote, after the completion of the **S phase**, the **Gap2** or **G2 phase** begins. The division of the daughter nuclei, so that each cell receives a complete gene, passes through the first part of period **D** in the prokaryote and during the phase of the mitosis (**M**) in the eukaryote. In the process of mitosis in the eukaryotes, a complex structure (**spindle separation**) that serves to move the daughter chromosomes is involved. The division of cells in the prokaryote occurs during the second half of the period **D**, whereas in the eukaryote it is during the phase of cytokinesis (**Table 3.7**).

Table 3.7. Stages of the cell cycle of prokaryotic and eukaryotic cells

Stage	Prokaryotes	Eukaryotes
Mass growth	Period B	Phase G1
DNA synthesis	Period C	Synthesis of DNA (phase S)
Mass growth	–	Phase G2
Distribution of daughter chromosomes	Period D	Mitosis (phase M)
Separation of cells	Period D	Cytokinesis

The beginning and duration of the stages of the cell cycle are studied by the following methods:

(1) elution from the membrane;

(2) analysis of cell size distribution;

(3) flow cytometry method.

In the ***method of elution from the membrane***, the culture in the exponential phase of growth is filtered through a nitrocellulose filter. Daughter cells are separated in the process of growth on the filter with the nutrient medium. The analysis of these cells, formed at different intervals of time, allows us to study the sequence of the cell cycle. This method has been used to determine the length of chromosome replication (period C) and the period of the completion of cell division (period D).

The cells of the growing culture can be fractionated in size (age), centrifuging them in a gradient of density (for example, in a sucrose gradient). By selecting individual fractions, a population of cells of a certain age can be obtained. The degree of DNA replication in the fractions can be determined if pulsed labeling of an exponentially growing culture is carried out. This method allows distinguishing periods of a cell cycle without DNA replication from the periods where replication occurs.

Flow cytometry is used to analyze the distribution of cells by size and the content of DNA. The fluorescently labeled DNA cell suspension is passed through an optical device that can measure light scattering and fluorescence of each cell. The size of the light scattering of cells is approximately proportional to their size, and the photomass of the labeled DNA is proportional to its content in the cell.

Synchronous cultures are also used to study the cell cycle. Such a culture can be obtained by using the melting elution method at 0 °C or by separating the smallest cells by centrifugation in the sucrose density gradient. Population growth is initiated by adding a fresh warmed medium. In synchronous culture, all cells are divided simultaneously. Culture synchronization can be also achieved by creating a hunger strike decreasing the amount of amino acids, but at the same time, cells are exposed to severe stress, which can cause a change in their physiological state.

The duration of the periods **C** and **D** of the cell cycle is almost independent of the growth rate, while the duration of the period **B** increases with increasing generation time. With rapid growth in the cell, there are two or three replication chromosome cycles that overlap. The replication of chromosomes consists of three processes: *initiation*, *elongation*, and *termination*. After replication of the chromosomes, processing of the nucleotide begins; two subsidiary chromosomes diverge and move to the centers of daughter cells. The division of the parent cell and the differentiation of the daughter cells occur at the end of the period D. In the process of division, there is a change in the direction of growth of the peptidoglycan layer, resulting in cell wall and plasma membrane invagination inside the cell. This requires the activation of one of the penicillin-binding proteins, *Pbp3*, which is involved in the synthesis of peptidoglycan transverse walls.

In the center of cells available at the end of the period D a specific structure, a periseptal ring, was found. It is a site that covers the region of invasion of the cell membrane. In the process of division, a moment comes, after which the division continues even if there is no protein synthesis. This moment coincides in time with the completion of the replication of the chromosome.

Many mutants with cell division disturbances have been obtained. Most of them are lethal, so genes have been identified by analyzing mutants that are viable at 30 °C and unable to grow at 37 °C.

Some defects in these genes are so damaging to the cell division that they change the shape of the bacteria. The analysis of the phenotypes of such mutants allowed the identification of two stages of division: septum formation and cell division. Mutants with a division of the *fts* group (from filamentation temperature-sensitive) have a blocked division. The *env* and *cha* mutations result in the formation of long chains of cells. The cytoplasm of these mutants is completely separated, but the cell membranes do not diverge. It was found that when the activity of the enzyme *N-acetylmuramoyl-L-alanine amidase* that cleaves bonds in the layers of peptidoglycan is reduced, daughter cells are covalently linked to the peptidoglycan layer surface.

Typically, cell division occurs in the middle of the cell, but in the mutant group (in the *minB* locus), the cell division occurs near one of the cells' poles, resulting in the formation of a mini-cell that does not contain DNA and an elongated cell containing the DNA of the parent cell.

All three processes of the cell cycle, including *chromosome replication*, *nucleotide processing*, and *cell division* are subject to fine regulation in accordance with the growth of cell biomass and are in some way coordinated. Several models of cell division regulation are proposed, but none of them is based on molecular biological data. In most hypotheses, an important role is given to such factors as the level of calcium ions and guanosine tetraphosphate.

A classic replicon model (**Fig. 3.40**) is proposed by F. Jacob, S. Brenner, and F. Cuzin in 1963, according to which the cell cycle is associated with a specific structure in the plasma membrane. At some point in the cell cycle, an unknown signal triggers replication. Two daughter chromosomes are attached adjacently; a membrane is formed between them. With the growth of the membrane, the chromosomes are removed from each other. Segmentation of the cell begins in the middle of the cell, between the daughter chromosomes. This model does not explain how the duality of the membrane structure is regulated and how the replication initiation takes place.

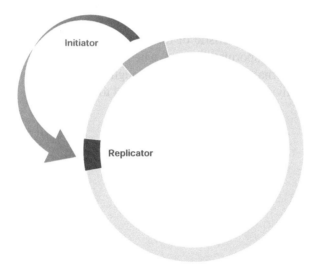

Fig. 3.40. Replicon model by Jacob *et al.* (1963): a *trans*-acting protein encoded by the initiator gene was proposed to recognize a *cis*-acting sequence (the replicator) that controls the initiation of DNA replication in the replicon.

There is also a model proposed by R. Prichard. It involves the participation of the following three components in the negative regulation of the cell cycle: the point of origin of the replication, the initiator protein, and the repressor acting on the point of origin of the replication. At the initiation of replication, the synthesis of repressor is stimulated. When interacting with the point of origin of the replication or with the initiator of the protein, the repressor inhibits the initiation until the cell volume doubles. Thus, the initiation of replication is associated with cell growth. However, so far, there is no data on the existence of a repressor to initiate replication.

The origin of chromosomal replication and cell division requires protein synthesis. However, after the start of the initiation, these processes continue until they are complete, even if the synthesis of the protein is inhibited.

The prokaryote cycle is not a real cycle, in which all subsequent steps require the end of the previous one. If the cell division is suppressed (cephalexin), the growth and replication of the chromosome continue and the daughter nuclei diverge, which results in the formation of long strands of cells with 4, 8, and even 16 clearly separated nucleoids.

In fast growing bacteria, replication cycles overlap and the completion of the chromosomal discrepancy is not a prerequisite for the initiation of chromosomal replication.

Data have been obtained indicating that the separation process has a separate regulation. The genes involved in regulation of the cell cycle development have not been detected in bacteria. All prokaryotic cell cycle genes identified by this time are not related to regulation but to the implementation of key events in the cell cycle.

Thus, the vegetative cell cycle of *E. coli*, which leads to the formation of two identical daughter cells, was examined. In bacteria, many other types of cell division are known, which are characteristic of differentiation processes. The cell division in *Bacillus subtilis* can be a sporulation stage, when one daughter cell becomes premature, and the other is a maternal cell. Two different daughter cells are formed in *Caulobacter*: one of them is free-living, and the other is attached to the surface.

CHAPTER 4

PROCESSES OF CELL DIFFERENTIATION

In this chapter, a short characteristic of the processes of bacterial cell diffe-rentiation is presented. These processes are complex and depend on certain environmental conditions, which cause long-term functional and structural changes in bacterial cells. A polar differentiation in species of *Caulobacter* genus is described. A special attention is paid to the differentiation of photosynthetic membranes in facul-tative phototrophic bacteria and the formation of heterocysts in cyanobacteria.

4.1. Characteristics of differentiation processes

The processes of differentiation include complex and long-term changes at the functional and structural levels under the influence of certain environmental condi-tions. The object of research of cell differentiation and development of multicellular structures using molecular biology methods is spore-forming bacteria. The subject of modern research is the questions:

- What events at the level of gene expression trigger processes of cell differentia-tion in response to stress?
- How do the cells communicate with each other for coordinated behavior in the formation of a complex multicellular structure?
- How is differentiated expression of genes provided for the formation of daugh-ter cells from one parent cell with different directions of development?

A simple form of differentiation is sporulation. It begins with an asymmetric cell division, resulting in the formation of two compartments: progressive, developing

Bacterial Physiology and Biochemistry
http://doi.org/10.1016/B978-0-443-18738-4.50004-8,

into the endosary, and the maternal cell, which, after the completion of the process of spore formation, undergoes lysis, releasing the endospore. Regulatory factors that activate the expression of genes specific for sporulation in response to certain signals play an important role in the initiation of spore formation. The nature of the signals is unclear, but the conditions for their occurrence are known.

Three factors that cause the initiation of spore formation are important:

(1) exhaustion in the environment of nutrients;

(2) high density of cell population;

(3) DNA synthesis (replication of DNA) necessary for the formation of mother cell chromosomes and spores.

The initiation of spore formation in *B. subtilis* occurs as a result of the integration of numerous signals that activate the histidine protein kinase, which, in turn, activates the phospho-signaling networks. In this way, several products of *spoO* genes are involved: *SpoOF*, *SpoOB*, and *spoOA*. Under the influence of signals arising from adverse growth conditions, the intracellular level of phosphorylated SpoA protein increases, which acts as a transcription regulator. The positive and negative control of gene expression is provided. The formation of cells of two different types during sporulation depends on the differential expression of genes in the corresponding compartments, including the spore and the parent cell. This programmed expression is determined by the sequential appearance of five sigma factors associated with differentiation.

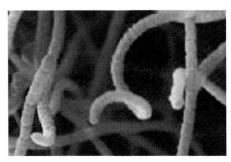

Fig. 4.1. Airy mycelium of *Streptomyces* sp. with exospores.

As was mentioned in **Chapter 2**, the species of *Streptomyces* genus form exospores during sporulation. Representatives of this genus have an unusual, for bacteria, ability of complex differentiation with the formation of a multicellular structure (**Fig. 4.1**).

The process of differentiation includes two phases: growth in the form of vegetative mycelium in optimal conditions and growth in the form of air micelles in fasting conditions. The growth of air micelles ends the formation of spore chains (sporangia) at the ends of the air hyphae. The amount of spores can reach fifty or more. When differentiating at the ends of the hyphae of air mycelium, regularly placed compartments containing one copy of the genome are formed. These compartments are differentiated (they change their shape and the thickness of the cell wall) and turn into mature spores with a hydrophobic surface. Exospores of *Streptomyces* sp. are similar in size to bacterial endospores. They are relatively resistant to drying but are sensitive to high temperatures.

The sensor mechanism that indicates the nutritional deficit is unknown. There is an assumption that it is associated with a change in the level of guanyl nucleotides. At the initiation of the formation of air mycelium, a decrease in the level of GTP and an increase in the content of guanosine tetraphosphate are observed.

Under conditions of nutrient deficiencies that cause cell differentiation, many *Streptomyces* form secondary metabolites (most known antibiotics are among them). In addition, autoregulatory factors, such as **pheromones** and proteins that regulate the morphological shape of hyphae of the airway mycelium, are formed. The diffusing factors (pheromones) or microbial hormones are capable of inducing sporulation. The pheromone **A-*factor*** and its derivatives are similar to auto-inductors involved in quorum sensing.

4.2. Polar differentiation in species of *Caulobacter* genus

Caulobacter crescentus belongs to the Gram-negative bacteria that have a polar flagellum (**Fig. 4.2**). The cells of this bacterium have stalks (prosthecae), which are structures 0.5–3.0 μm long, covered with the plasma membrane and the cell wall and containing the cytoplasm, but having neither ribosomes nor DNA. The stalk is formed at the pole of the cell, where the flagellum was located before. At the apex of the stalk, a small amount of adhesive material is released, which serves to attach cells to different surfaces. After the cell division on the free polar of the daughter cell, the flagellum and a few pills are formed. Here are also receptors of bacteriophage φCbK and chemosensors for chemotaxis. Sagittarius consists of three main components: the basal body of a mobile hook and the fiber. The basal body has several rings located inside the outer membrane (L-ring), the peptidoglycan layer (P-ring), the periplasmic space (E-ring), and the plasma membrane (MS-ring that fixes the flagellum in the plasma membrane and acts as a rotor). The rings and the hook are fastened with a central rod. The fiber consists of three different flagella proteins.

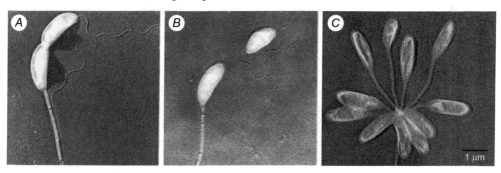

Fig. 4.2. Morphology and reproduction of *Caulobacter* sp.: the cell after division, visible prostheca and flagellum (*A*); the swarmer cell with flagellum (*B*); the cells in an electron microscope (*C*).

As was mentioned in **Section 2.7**, pili are threads with a diameter of 4.0 nm and a length of 1–4 μm. They consist of molecules of pilin, a hydrophobic protein with a molecular weight of 8.0 kDa, which is accumulated in the cytoplasm at the pole of the cell. This accumulation of pilin continues until the formation of pili begins before the cell division. In the process of vegetative growth of *C. crescentus*, there is an asymmetric binary cell division with the formation of daughter cells of various types: a moving swirler cell and a stationary stalk cell (**Fig. 4.3**).

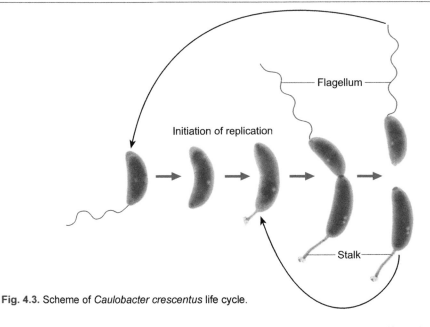

Fig. 4.3. Scheme of *Caulobacter crescentus* life cycle.

The stalk cell is divided again with the formation of a swarmer cell at the stalk distal pole. The swarmer cell with the flagellum at the old pole is not capable of DNA replication and cell division. For a short time, it is mobile, and then it loses the flagellum, mobility, and the pili and is differentiated into the stalk cell. With the onset of the formation of the stalk in the area where the flagellum was located, the cell initiates DNA synthesis and enters the S-phase of the cell cycle. Cell division begins during the replication of the chromosome and lasts almost half the time of the cell cycle. This division of stalk cells differs from the division of cells of eubacteria, in which it passes through the rapid formation of septum after the completion of chromosome replication.

The study of differentiation dependent on the cell cycle is supported by the differentiation of polar structure markers in *C. crescentus* cells as well as the fact that it is easy to obtain synchronous cultures in these organisms.

The biogenesis of polar structures depends on the initiation of a new DNA replication cycle and not on the completion of cell division. The bacterial swarmer cell contains a compact nucleotide 6,000S, which at the later stage of the cell cycle (when the swarmer cell is differentiated into the stalk) is converted into a normal nucleotide (30,000S).

Before division, the cell contains one compact nucleotide for a daughter swarmer cell and one normal nucleotide for the stalk cell. The transition from the compact nucleoide to the normal occurs very fast. The differential regulation of DNA synthesis in nursing cells, the swarmer, and the stalk cell in *C. crescentus* is one of the most striking examples of regulatory mechanisms related to development. In the stalk cell, DNA synthesis begins immediately after the division of the daughter cells, while the sister swarmer cell enters the S-phase of the cell cycle only when the stalk is started.

When replicating the chromosome in *C. crescentus*, a specific replication point of *oriC*, which does not function in *E. coli*, is used. It contains a site for *DnaA* protein binding and a strong *hemE* promoter. Replication and chromosome segregation require sequences of expression of genes, whose products are required for both of these processes. Such genes that are temporarily expressed include the *gyrB* gene (encoding the subunit of gyrase B) and *orf-1*. These genes are silent when located in the swarmer pole of cell before the division but are expressed when the swarmer cell is converted into a stalk. The origin of replication in cells before division also depends on the activity of the promoter *hemE*. Directly before the cell division in *C. crescentus* cells, DNA-methyltransferase is expressed. Conservative sites of the GANTC type in the swarmer cells are fully methylated and only partially methylated in the stalk cells. In the signal transduction, which coordinates the mobility and the formation of the stalk with the completion of the cell division cycle, the histidine kinase *PleC* and the protein controller *DivK* are involved. They are also involved in the signal transmission necessary for cell division.

Thus, the cycle of chromosome replication in *C. crescentus* initiates uneven segregation of mRNA into the daughter cells. The expression of genes on the swarmer and stalk cells is differentiated. Possibly, it is as a result of the selective "silence" of various groups of genes in the chromosomes localized at these two poles. The processes of DNA replication and cell differentiation depend on the time regulation of DNA methylation.

The process of formation of flagella goes through several stages and is regulated at the level of transcription. The formation of flagella in *C. crescentus*, its function, and the reaction of chemotaxis are determined by about 50 genes. These genes are organized into a hierarchical structure that has four levels. The genes of individual levels are expressed sequentially in the order determined by the sequence of the formation of flagella structures (basal body — hook — fiber). The expression of genes of each level depends on the products of the genes of the previous hierarchy level. Genes of the same level, even if they are involved in different transcription units, are read simultaneously. The transcription time is controlled by the cascade mechanism of transactive regulatory proteins.

Pili are formed after cell division at the pole of a swarmer cell from pilin protein (molecular weight: 8 kDa). When the stalk is formed, the pili are pulled inside the cell. Pilin subunits are synthesized in the stalk cells and in the cells before division and accumulated in the cytoplasm. The posttransient time control of pili formation is related to the division of cells. Cell division and stalk formation also depend on *SecA* protein, which is involved in the transfer of many proteins through the plasma membrane.

Other examples of polar differentiation of bacterial cells are the formation of actin at the poles of *Listeria monocytogenes* cells, the formation of the proteins attached to the cell surface in *Bradyrhizobium japonicum,* and the secretion of iron hydroxides on the concave surface of *Gallionella* bean-shaped cells.

4.3. Differentiation of photosynthetic membranes in facultative phototrophic bacteria

The differentiation of photosynthetic membranes in facultative phototropic bacteria depends on the light and the partial pressure of oxygen. Under conditions of dark hemotrophic growth, these bacteria use oxygen as the final acceptor of electrons for respiration. At low partial pressure of oxygen, they form a photosynthetic apparatus. The synthesis of pigments and proteins of the photosynthetic membrane is regulated in a coordinated way. The reduction of the partial pressure of oxygen in the dark initiates a coordinated synthesis of bacteriochlorophyll, carotenoids, and proteins that bind pigments. When increasing the partial pressure of oxygen, NADH dehydrogenase, cytochrome oxidase, and other components of the respiratory chain are synthesized. Simultaneously with the synthesis of the photochemical reaction center, when the partial pressure of oxygen is lowered, membrane inflammation and species-specific membrane structures are formed (e.g., vesicles in *Rhodobacter sphaeroides* and *Rhodobacter rubrum* or stacks of lamellae in *Rhodopseudomonas viridis*.

The formation of membrane structures is accompanied by the synthesis of phospholipids and other components of membranes and their inclusion in a growing membrane. Phospholipids integrate into the membrane during cell division, proteins of the photosynthetic apparatus throughout the entire cell cycle. The components of the photosynthetic apparatus are included into certain areas of the plasma membrane and, after its invagination, are localized in the intraplasmic membrane structures.

The intensity of light affects the formation and composition of the photosynthetic apparatus in facultative phototropic bacteria in anaerobic conditions. Reducing the intensity of light causes an increase in the number of photosynthetic units, the size of the antenna relative to the reaction center, as well as the area of the intracellular membranes. With the increasing intensity of light, the concentration of electron carriers (cytochromes and quinones) and the rate of phosphorylation per reaction center are increased. The action of light overlaps with the stronger influence of the partial pressure of oxygen. The photosynthetic genes are controlled by the regulatory genes, which coordinate all processes from signal transmission to membrane formation. Most operons of the photosynthetic genes are clustered. The genes encoding the enzymes for the synthesis of bacteriochlorophyll and carotenoids as well as regulatory proteins are localized in one cluster. The expression of these genes at a low level begins with weak promoters and encompasses superoperon DNA structures. The promoters of some genes are located in the upright operon. The transcription of genes can begin with various promoters. With low partial pressure, there is a transcribed transcription of several operons and genes that are less pronounced. The superoperon organization provides readings of several operands and a low level of expression of photosynthetic genes in aerobic and microaerobic conditions.

The partial pressure of oxygen and the intensity of light are stimuli that control the expression of photosynthetic genes through a common regulatory system (**Table 4.1**). The existence of the complex regulatory system for the differentiation of the photosynthetic apparatus is established mainly by genetic methods.

Table 4.1. Regulatory systems for the control of transcription of photosynthetic genes with changes in the intensity of light or partial pressure of oxygen

Stimulus	Sensor	Regulator	Operons
Low light intensity	?	Hrv (+) Activator	puhA pufQBALMX
High oxygen content	?	Hrv (–) Repressor	bchCXYZ bchH, bchD, crtl pucBACDE
Low oxygen content	RegB	Reg (+)	pufQBALMX puhA pucBACDE

Thus, the differentiation and formation of photosynthetic membrane structures is a complex process accompanied by the synthesis of phospholipids and other components of membranes and their inclusion in the growing membrane.

4.4. Formation of heterocysts in cyanobacteria under bound nitrogen deficiency

In some cyanobacteria, vegetative cells can be differentiated into heterocysts, akinetes (**spores**), or short threads without a cover, the shape of which differs from that of the vegetative cells. *Akinetes* are resting cells with a thick cell wall, which are formed when there is a shortage of sources of nutrition or energy. These structures provide survival in cold or drought conditions (**exospores**). *Heterocysts* are specialized cells, the function of which is the fixation of molecular nitrogen. They are formed in the species *Anabaena*, *Nostoc*, and *Cylindrospermum*, and only these cells are capable of fixing nitrogen. Heterocysts are placed along the threads of cyanobacteria in a certain order. Mature heterocysts are surrounded by three additional covers. Above the cell wall, there is a glycolipid layer followed by a polysaccharide and a fibrous polysaccharide layers. Only photosystem I functions in mature heterocysts. The components of photosystem II — phycobilisomes, phosphoribulokinase, and ribulose phosphate carboxylase — are lost during the formation of the cyst. *Thylakoids* are modified. At the sites of fixation of nitrogen in the heterocysts, there is active respiration, which provides the removal of residual oxygen. The synthesis of ATP passes through cyclic phosphorylation in photosystem I and oxidative phosphorylation in the respiratory chain of electron transport. The donor of electrons for nitrogenase is ferredoxin. Nitrogenase is homologous to other nitrogenases. The sugar of heterocysts is obtained from adjacent vegetative cells that capture CO_2. Heterocysts release nitrogen in the form of glutamine, which is transferred from cysts to vegetative cells along the strands immediately and is metabolized (**Fig. 4.4**).

The differentiation of the vegetative cell to heterocysts lasts for 20 hours. Outer layers of polysaccharides and a glycolipid layer are formed. Within a few hours, two proteases, one of which cleaves phycobiliproteins and second Ca-dependent protease, which cleaves the ribulose 5-phosphate kinase enzymes and ribulose diphosphate carboxylase (Calvin's cycle components), as well as other proteins are activated.

Morphological differences between heterocysts and vegetative cells can be observed in 5–10 hours. After this period, heterocysts are no longer reverted to vegetative cells. Differentiation of vegetative cells in heterocysts is a complicated multistage process that runs through the cascade of sensory-regulatory systems. Only some components of these systems are identified. It is unknown how the formation of heterocysts is initiated. It is shown that in 30 minutes after the nitrogen source is exhausted, a general signal of the amino acid fasting — the synthesis of polyphosphorylated nucleotides — occurs, but the classical response to amino acid fasting is not observed.

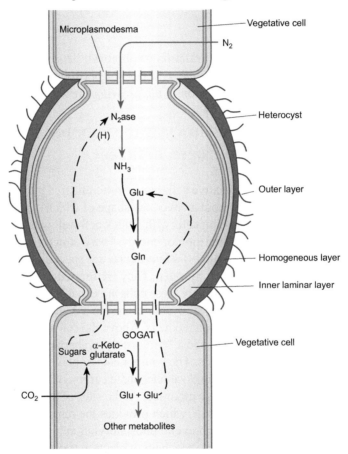

Fig. 4.4. Structure of a heterocyst and a way of nitrogen fixing and transferring it from the heterocyst.

The regulation of transcription plays an important role in differentiating heterocysts. Some genes encoding glutamine synthase, ATP synthase, and sigma factor are transcribed in heterocysts and vegetative cells from various promoters. Some promoters, in particular those responsible for the synthesis of nitrogenase, are expressed in heterocysts only. Different sigma factors are found at hunger for nitrogen. In heterocysts, about 1000 genes are expressed. Some of the products of these genes are not related to differentiation and belong to the families of known regulatory proteins.

CHAPTER 5

BACTERIAL METABOLISM

In this chapter, energy of biochemical reactions, carriers of hydrogen, the role of ATP, acceptors of electrons, and the types of phosphorylation of the bacterial cells are described. A special attention is paid to the types of metabolism, including catabolism and anabolism. The processes of catabolism of carbon compounds and fermentation are shown. Anaerobic respiration and the processes of oxidation of various compounds are presented. The anaerobic degradation and metabolism in different bacterial groups are also described. Another part of this chapter represents the processes of anabolism, including the biosynthesis of saccharides, lipids, amino acids, nucleotides, nucleic acids, and proteins. The regulation of metabolic processes such as enzyme synthesis and enzymatic activity and the specific of regulation mechanisms and energetic metabolism are represented. The metabolism of photolithotrophic and photoorganotrophic bacteria is also described.

Metabolism is a process including chemical transformation, which serves for gaining basic building material and energy for the synthesis of cell compounds and for other vital processes of bacterial cells. Depending on the character of these processes, metabolism can be divided into:

(1) **catabolism**—breaking down specific substrate into simpler products, while energy is released;

(2) **anabolism**, which represents biosynthesis of building blocks (glycides, amino acids, fatty acids, nitrogen bases, etc.) and macromolecular substances done at the expense of energy and fission products gained during catabolism.

Bacterial Physiology and Biochemistry
http://doi.org/10.1016/B978-0-443-18738-4.50005-X,

Both these metabolic processes take place simultaneously and they cannot exist without each other. The typical feature of biochemical reactions which take place during metabolism is group transfer (transglycosidation, transamination, transacylation, transphosphorylation, etc.). The group transfer is important especially when structural substances are being made, because their de novo creation is quite limited. Another important feature of metabolism is that reactions take place in firmly defined sequences called metabolic pathways. These pathways are often organized into a cycle. Depending on the conditions, many reactions can be carried out in alternative sequences with the same or similar function but using a different mechanism. All metabolic processes and their speed are controlled by specific regulatory systems, which, depending on the need and conditions, influence the synthesis of appropriate enzymes or their activity.

5.1. Energy of biochemical reactions

Free energy. All chemical reactions taking place inside the cell can be considered as thermodynamic reversible changes. However, during the reversible isothermal process, only a part of the total internal energy of the system is used for the realization of a reaction. This part is called free energy. The rest represents energy loss or entropy, which can be used for whatever work and is usually expressed as heat released into the surroundings.

The absolute value of free energy is unmeasurable. However, it is possible to determine its change, which occurs during the transition of the reacting system from one state to another. This change is expressed by the following equation:

$$\Delta G = \Delta H - T\Delta S,$$

where ΔG is the change of free energy, ΔH is the change of total (internal) energy of the system, T is absolute temperature, and ΔS is change of entropy.

During the chemical reaction, the change of free energy is closely related to the equilibrium constant of the reaction. If substances A and B enter the reaction, they undergo the conversion to reaction products C and D, therefore:

$$A + B \rightleftharpoons C + D.$$

The equilibrium constant K of this reaction is expressed by the relation:

$$K = \frac{[C][D]}{[A][B]}.$$

The relation between this value and the change of free energy results from the equation:

$$\Delta G = \Delta G^0 + RT \ln K,$$

where ΔG^0 is the standard change of free energy happening during standard conditions, which means the concentration of one mole of reacting substances and products of the reaction, temperature of 25 °C, and pH 7; R is the gas constant (8.32 kJ/mol/grad); K is the equilibrium constant of the reaction.

Exergonic and endergonic reactions. In a state of equilibrium, free energy is minimal, and no changes occur. Therefore $\Delta G = 0$ and previous equation looks like this

$$\Delta G^0 = -RT \ln K, \ \text{ or } \ \Delta G^0 = -2{,}303 \ RT \log K$$

This equation says that when we know the value of the equilibrium constant of a reversible chemical reaction, we can derive also the standard change of free energy, which then happens to be a characteristic thermodynamic quantity for this reaction. When the value of equilibrium constant is high, the reaction is about to be finished and the standard change of free energy is negative. Reactions like this, in which energy is being released, are called exergonic. On the other hand, when the value of the equilibrium constant is low, the reaction goes in the opposite direction and energy must be delivered into the system. In this case, the standard change of free energy is positive, and these reactions are endergonic.

Changes of free energy during oxidoreductive processes. The principle of chemical processes taking place inside the cell are oxidoreductive reactions. Thanks to that, we can determine the standard change of free energy from the difference between the redox potentials of reacting systems by the following relation:

$$\Delta G^0 = n \cdot F \cdot E'_0,$$

where n is the number of transmitted electrons; F is Faraday's constant (96.500 Coulombs or 96.56 kJ per volt gram equivalent); E'_0 is the difference between the standard redox potentials of the hydrogen acceptor and donor related to the physiological conditions of the system, which means all substances react in a concentration of one mole at a temperature of 25 °C and pH 7.0.

This equation says that during a conversion of one mole and transmission of one electron, the standard change of free energy with a one-volt difference between potentials will be equal to 96.56 kJ. If a couple of electrons is transmitted, the standard change doubles to 193.12 kJ.

Macroergic bonds. Energetic metabolism takes place connecting exergonic and endergonic reactions together. For synthesis to happen, the sum of changes of free energy of both these reactions must reach negative or at least zero value. In biological systems, an energy-rich compound, which can accumulate energy via the so-called macroergic bonds, usually takes part in these reactions. Macroergic bonds hold a high level of free energy, and therefore, the compounds containing this bond, can activate reactive substrate and enable its connection with a specific group or another fission product during biosynthesis.

The most important energy-rich compound is called adenosine triphosphate (ATP) (**Fig. 5.1**). Biochemically usable energy is bound in its molecule in two terminal phosphate bonds with about 30.56 kJ in each. The specialty of ATP is the ability to transfer phosphate groups, which have been split off from energy-rich compounds (such as phosphoenolpyruvic acid) to compounds with poorer bonds in terms of energy. The total concentration of AMP, ADP, and ATP in microorganisms is about 10–12 μmol in 1 g of dry matter, while the amount of ATP usually exceeds the total concentration of AMP and ADP.

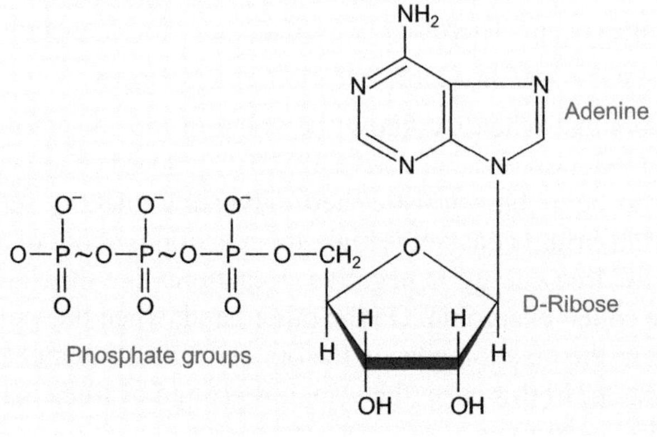

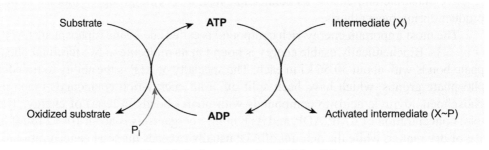

Fig. 5.1. Chemical structure of adenosine triphosphate.

As an example of a reaction connected with ATP, we can mention the phosphory-lation of glucose to glucose 6-phosphate. This reaction is endergonic, because it needs an energy of $G^0 = +12.56$ kJ to happen. If ATP takes part in this reaction, its macroergic bond ($\Delta G^0 = -29.31$ kJ) provides not only the required energy, but by splitting off the phosphate, it also enables the phosphorylation of glucose:

$$\text{Glucose} + P_i \rightarrow \text{Glucose 6-phosphate} \ (\Delta G^0 = +12.56 \text{ kJ})$$

$$\text{Glucose} + \text{ATP} \rightarrow \text{Glucose 6-phosphate} \ (\Delta G^0 = -16.75 \text{ kJ})$$

The regeneration of ADP to ATP demanding energy $G^0 = +29.31$ kJ is possible when connected with another reaction which contains a different energy-rich compound, such as phosphoenolpyruvic acid, whose hydrolysis releases energy $G^0 = -50.24$ kJ.

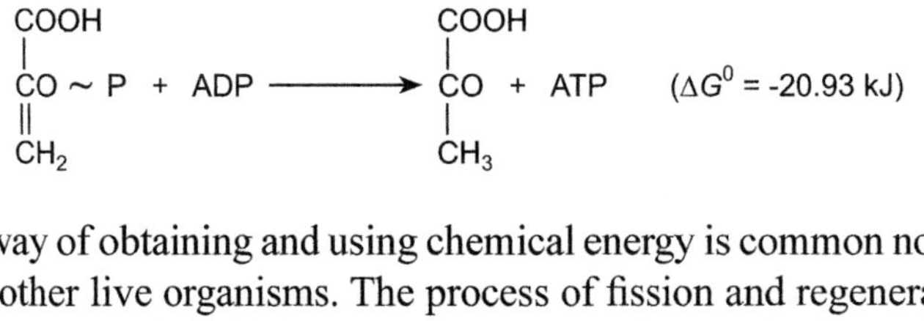

$$\begin{array}{c}
\text{COOH} \\
| \\
\text{CO} \sim \text{P} \\
\| \\
\text{CH}_2
\end{array} + \text{ADP} \longrightarrow
\begin{array}{c}
\text{COOH} \\
| \\
\text{CO} \\
| \\
\text{CH}_3
\end{array} + \text{ATP} \quad (\Delta G^0 = -20.93 \text{ kJ})$$

A similar way of obtaining and using chemical energy is common not only in bacte-ria, but also in other live organisms. The process of fission and regeneration of macro-ergic bonds in compounds, which accumulate energy, is commonly cyclic (**Fig. 5.2**):

Fig. 5.2. Cyclic regeneration of ADP to ATP.

Except ATP and phosphoenolpyruvic acid, there are also other compounds important for the energetic balance of reacting systems, which can create and split macroergic bonds. We can mention, for example, 1,3-biphosphoglyceric acid, acetyl phosphate, and especially acetyl-CoA (**Table 5.1**).

Table 5.1. Summary of the most important energy-rich compounds

Compounds	ΔG^0 (kJ/mol)
Phosphoenolpyruvate	-61.965
1,3-Biphosphoglycerate	-49.404
Acetyl phosphate	-42.287
Acetyl-CoA	-32.238
Adenosine triphosphate ($\rightarrow$ ADP + P$_i$)	-30.564
Glucose 1-phosphate	-20.934
Glucose 6-phosphate	-13.164
Glycerol 1-phosphate	-9.211

Creation of compounds with macroergic bonds in chemotrophic bacteria. Chemotrophic bacteria create compounds with macroergic bonds, most often the ones of ATP type, by phosphorylation of ADP at substrate level or by oxidative phosphorylation.

ATP creation at substrate level. In this case, ADP reacts with compounds with a macroergic bond, whose standard free energy (ΔG^0) of hydrolysis is higher than that of ATP. Phosphorylation at substrate level occurs during fermentation processes, when appropriate enzymes are present, and thanks to that, an organic phosphate is transmitted to ADP, while ATP is being created:

$$1\text{,}3\text{-Biphosphoglyceric acid} + \text{ADP} \xrightarrow{\enzyme} 3\text{-phosphoglyceric acid} + \text{ATP}$$

$$\text{Phosphoenolpyruvic acid} + \text{ADP} \xrightarrow{\enzyme} \text{pyruvic acid} + \text{ATP}$$

Because only a limited amount of energy can be obtained this way, fermentation processes are not considered as energetically effective. Therefore, organisms that carry out these processes need to maintain their energy levels high enough by processing sufficient amounts of substrate.

5.2. Carriers of hydrogen

Hydrogen is a very small and nonpolar molecule (paradoxically, a H$_2$ molecule is smaller than a single hydrogen atom), for which it is very difficult to work with enzymes. There are special hydrogen carriers to solve this problem in the cell. Examples of such carriers are nicotinamideaden inedinucleotide (NAD) and nicotinamideaden inedinucleotide phosphate (NADP). Hydrogen and electron donors can be released in the process of oxidation of organic compounds. Universal carriers of hydrogen are coenzymes (**Table 5.2**). Coenzymes are nonprotein bioorganic compounds that are necessary for the enzymatic action, that is, the conversion of the substrate into a product. Coenzymes can be combined with the protein part (apoenzyme) by non covalent physico-chemical or covalent bonds (in the latter case they are prosthetic groups of the enzyme protein: flavin coenzymes, pyridoxal phosphate, lipoic acid, etc.); sometimes coenzymes form complexes with apoenzyme only during the catalytic process of NAD and NADP.

Table 5.2. Carriers of hydrogen and electrons

Coenzymes	Reaction type catalyzed by coenzyme	Bioorganic compound derived from coenzyme
NAD (nicotinamideadenine dinucleotide)	$NAD^+ \xrightleftharpoons[-2H]{+2H} NADH + H^+$ Hydride ion transfer (:H⁻)	Vitamin PP (nicotinic acid, nicotinamide)
NADP (nicotinamideadenine dinucleotide phosphate)	Hydride ion transfer (:H⁻)	
FAD (flavin adenine dinucleotide)	 $FAD^+ \xrightleftharpoons[-2H]{+2H} FADH_2$	Vitamin B₂ (riboflavin)
FMN (flavin mononucleotide)	Transfer of hydrogen atoms (2 H⁺ + 2 ē)	
Coenzyme Q	Transfer of hydrogen atoms (2 H⁺ + 2 ē)	Ubiquinone
Glutathione	Hydrogen transfer in peroxidase and reductase reactions	Tripeptide: γ-Glutamyl-cysteinyl-glycine
Hemin coenzymes	Transfer of electrons	Metal porphyrins

By chemical nature, coenzymes are divided into:
- derivatives of vitamins, in particular, vitamin B (thiamine diphosphate), vitamin B₂ (flavinmononucleotide, FMN), vitamin B₆ (pyridoxal phosphate, pyridoxamine phosphate, pantothenic acid — coenzyme A), vitamin B₁₂—methylcobalamin, deoxyadenosylcobalamin, vitamin H (biotin) — carboxybiotin, folic acid — tetrahydrofolic acid;
- dinucleotides (nicotinamide derivatives: NAD, NAD(P); riboflavin derivative — FAD);
- nucleotides — derivatives of purines and pyrimidines (ATP, ADP, CTP, CDP, UTP, UDP);
- complexes of porphyrins with metal ions.

Role of NAD and NAD(P) in metabolism. NAD(P) is a universal transmitter of hydrogen and electrons (**Fig. 5.3**). Its reduced and oxidized form circulates freely in the cytoplasm. The NAD(P) + molecule is a redox system:

$$NAD(P)^+ + 2\,H^+ + 2\,\bar{e} \rightleftharpoons NAD(P)H^+ + H^+, \quad E_0' = -0.32\ \text{V}.$$

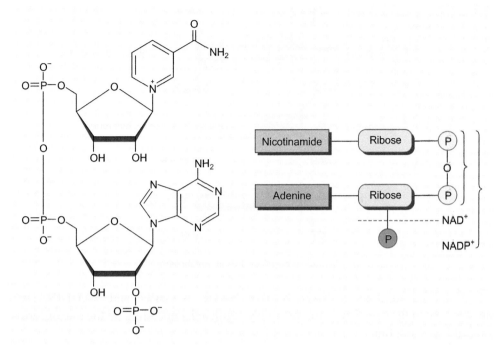

Fig. 5.3. Nicotinamide adenine dinucleotide phosphate structure (left) and the scheme of units of NAD⁺ and NADP⁺ (right).

NAD(P) is the second substrate of cytoplasmic dehydrogenases (lactate dehydrogenase, glyceraldehyde 3-phosphate dehydrogenase, etc.). NAD^+ is used as the second substrate by some dehydrogenases, while other use $NADPH^+$.

Bacteria (and some eukaryotic microorganisms and animal cells) contain the enzyme $NAD(P)^+$ transhydrogenase, catalyzing the reaction:

$$NADPH + NAD^+ \rightleftharpoons NADP^+ + NADH$$

from left to right, this is an **exergonic process**, but from right to left, it is **endergonic** (ATP or a proton gradient should be used). NADH is primarily intended to reduce oxygen and produce ATP at the membrane level (**exergonic reactions**). NADPH is primarily intended for reductions in biosynthetic processes (**endergonic reactions**).

Transhydrogenase is localized in bacteria on the cytoplasmic membrane, in eukaryotes in the mitochondria (the inner side of the inner membrane). The standard electron potentials of pairs **NAD⁺/NADH** and NADP⁺/NADPH are practically the same, -0.32 V.

In living cells, this enzyme primarily operates in the direction consuming NADPH and producing NADH, as the physiological ratio of NADPH/NADP⁺ is much higher than the ratio of NADH/NAD⁺.

Thus, the two substrates of this enzyme are NADPH and NAD⁺, while its two products are NADP⁺ and NADH. This enzyme participates in nicotinate and nicotinamide metabolism. The general role of NAD(P) in the metabolism is shown in **Fig. 5.4**.

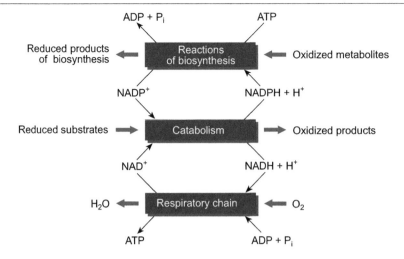

Fig. 5.4. Role of NAD(P) in metabolism.

The concentrations of **NAD⁺/NADH**, **NADP⁺/NADPH**, and **NADH/NADPH** are constant in the cell. They are sensitively regulated, depending on the rate of formation and consumption.

5.3. Role of ATP and its formation in bacterial cells

The energy released during the decomposition of organic matter is not immediately used by the cell but is stored in the form of high-energy compounds, usually in the form of adenosine triphosphate (ATP). ATP is a universal energy carrier and universal phosphate donor in anabolic reactions. By its chemical nature, ATP belongs to mononucleotides and consists of a nitrogenous base of adenine, a ribose carbohydrate, and three phosphoric acid residues (see **Fig. 5.1**).

$$\text{AMP} \xrightleftharpoons{\approx 42\,\text{kJ}} \text{ADP} \xrightleftharpoons{\approx 42\,\text{kJ}} \text{ATP}$$

The energy released by ATP hydrolysis is used by the cell for all kinds of metabolic pathways. Considerable amounts of energy are spent on biological synthesis. ATP is a universal source of energy for the bacterial cell. The supply of ATP in the cell is limited and replenished by the process of phosphorylation with different intensity during respiration, fermentation, and photosynthesis. ATP is replenished at an extraordinary rate.

System of **ADP ↔ ATP** is a universal coupler of catabolism and anabolism. Any exergonic reaction can drive any endergonic reaction. Therefore, metabolism as a whole is highly reliable. The linking role of ATP in the living cell metabolism (catabolism and anabolism) is presented in **Fig. 5.5**.

The central role of ATP in the synthesis of biological macromolecules is presented in **Fig. 5.6**.

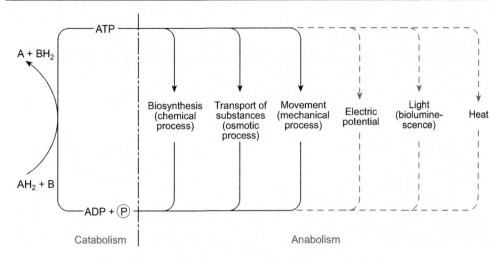

Fig. 5.5. Role of ATP in processes of catabolism and anabolism.

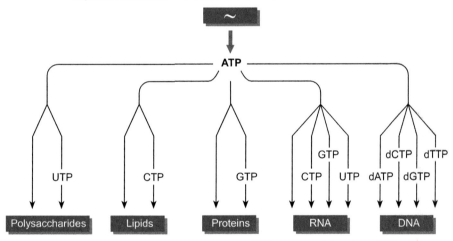

UTP: uridine triphosphate, CTP: cytidine triphosphate, GTP: guanine triphosphate, UTP: uridine triphosphate, dATP: deoxyadenosine triphosphate, dCTP: deoxycytidine triphosphate, dGTP: deoxyguanosine triphosphate, dTTP: deoxythymidine triphosphate

Fig. 5.6. Central role of ATP in bacterial cell.

The formation of ATP can be on the level of substrate (**Fig. 5.7**) and the membrane. ATP is accumulated under cytoplasmic membrane by the force of the H^+ diffusing through the ATPase in prokaryotes and on the inner membrane of mitochondria in eukaryotes.

Generally, O_2 is the most preferred terminal hydrogen and electron acceptor. Other substances can be used depending on the reacting couples. For example, $2\ H^+/H_2$ (-0.42 V) and fumarate/succinate (+0.02 V):

$$H_2 + fumarate^{2-} \rightarrow succinate^{2-}$$
$$succinate^{2-} + NO_3^- \rightarrow NO_2^- + H_2O$$
$$succinate^{2-} + \frac{1}{2}\ O_2 \rightarrow fumarate^{2-} + H_2O$$

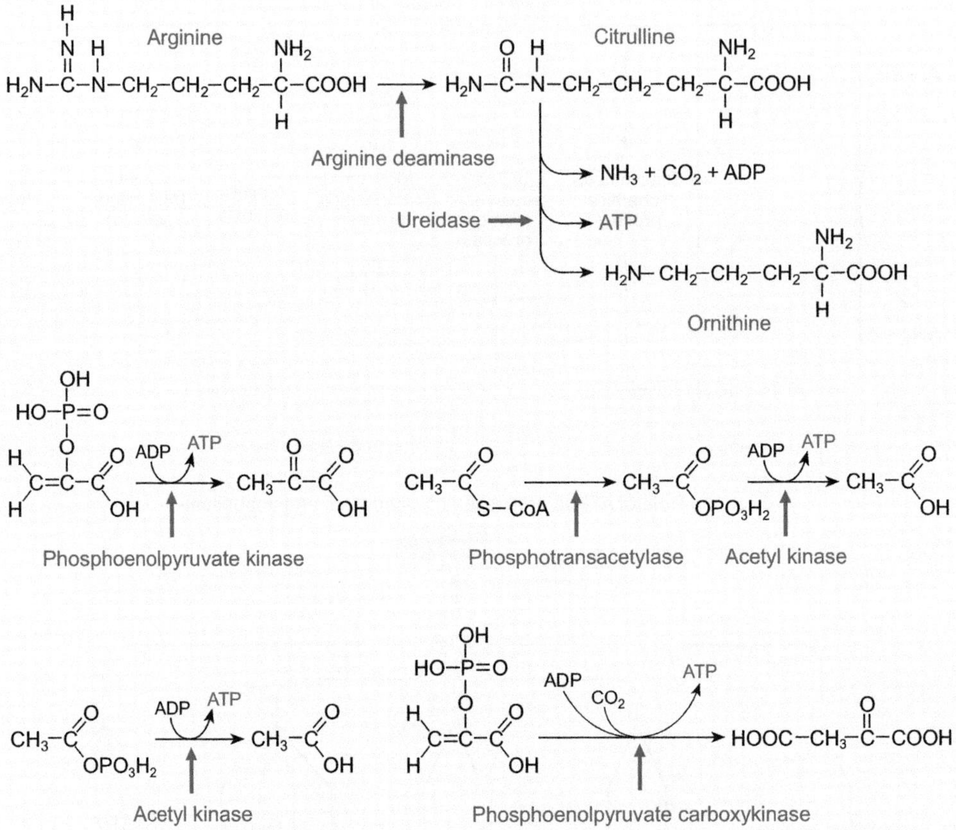

Fig. 5.7. ATP formation at substrate level.

The standard oxidation potentials of some substances and compounds are presented in **Table 5.3**. As was mentioned above, hydrogen is a universal electron donor in the bacterial metabolism. An electron donor is a chemical entity that donates electrons to another compound. It is a reducing agent that is, by virtue of its donating electrons, oxidized itself in the process.

The electron donating power of a donor molecule is measured by its ionization potential, which is the energy required to remove an electron from the highest occupied molecular orbital.

The overall energy balance (ΔE), that is, energy gained or lost, in an electron donor-acceptor transfer is determined by the difference between the acceptor's electron affinity and the ionization potential. Examples of reactions where H_2 is an electron donor are demonstrated in **Table 5.3**.

The terminal acceptor of hydrogen and electrons of chemoorganotrophic bacteria, which is able to recover energy by oxidation of H_2, can be O_2 (*Pseudomonas*), NO_3^- (*Paracoccus denitrificans*), SO_4^{2-} (*Desulfovibrio desulfuricans*), and CO_2 (*Methanobacterium*). These microorganisms have different types of dehydrogenases, for example:

- Hydrogen: NAD^+ oxidoreductase (hydrogen dehydrogenase) is soluble and membrane-bound (formation of reducing equivalents for biosynthesis and oxidation of hydrogen to form ATP).
- Hydrogen: ferricytochrome c_3-oxidoreductase (cytochrome c_3-hydrogenase) is membrane-bound (reacting at ubiquinone-cytb level via cytc (*Paracoccus denitrificans*).
- Hydrogen: ferredoxin oxidoreductase (ferredoxin hydrogenase) occurs in many facultative anaerobic bacteria (*E. coli, Proteus vulgaris*, and etc.).

Table 5.3. Standard oxidation potentials of some substances and compounds

Couple	Reaction	E_h (V)
NAD / NADH		-0.32
FAD / FADH$_2$		-0.22
FMN / FMNH$_2$		-0.19
Fumarate / succinate	$+ 2 H^+ + 2 \bar{e}$	+0.03
Menaquinone		-0.07
Ubiquinone		+0.11
2 Cyt b_{ox} / Cyt b_{red}		+0.07
2 Cyt c_{ox} / Cyt c_{red}		+0.25
2 Cyt a_{ox} / Cyt a_{red}		+0.38
½ O_2 / H_2O		+0.82
SO_4^{2-} / SO_3^{2-}	$+ 2 H^+ + 2 \bar{e}$	+0.20
$2NO_3^-$ / N_2O_4		+0.80

Examples of reactions where H_2 is an electron donor

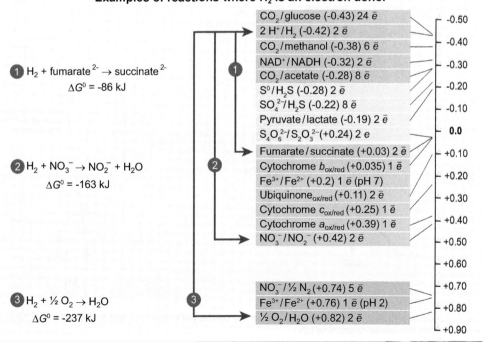

As is shown in **Fig. 5.8**, the accumulation of ATP in the bacterial cells can be by several pathways: in the process of fermentation and in the process of respiration. It depends on the physiological group of microorganisms and the conditions where they live (anaerobically or aerobically).

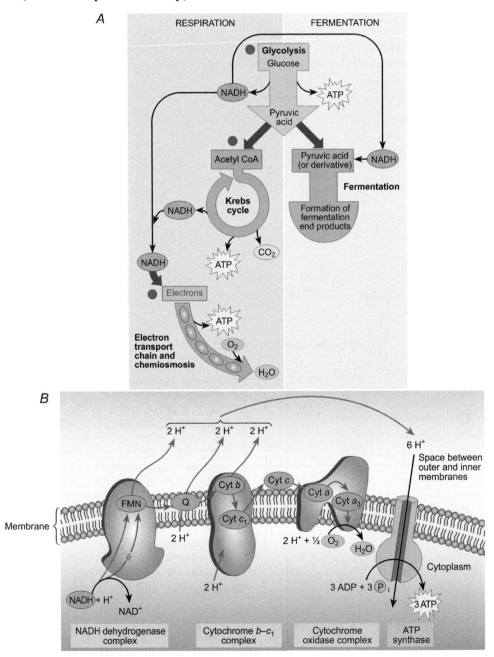

Fig. 5.8. Accumulation of ATP in bacterial cells: in the process of respiration and fermentation (*A*) and on the cytoplasmic membrane (*B*).

ATP is generated by creating a proton gradient and potential on the membrane by crossing the cytochrome chain. The respiratory chain in prokaryotic organisms has several types that are presented in **Fig. 5.9**.

General type including NAD, FAD, ferredoxin, quinone, and cytochromes

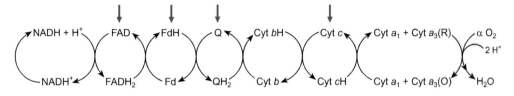

Branched linear chain with menaquinone and two types of cyt b

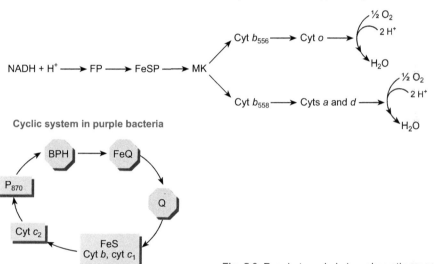

Fig. 5.9. Respiratory chain in prokaryotic organisms.

The efficiency **of living systems** during the oxidation of $NADH + H^+$ and complete oxidative phosphorylation with O_2 as a terminal hydrogen and electron acceptor is as follows:

$$\Delta G^0 = -nF \cdot \Delta E_h,$$

where n is number of electrons, F is Faraday's constant, and ΔE_h is redox potential between reactions:

$$NAD/NADH + H^+ \qquad -0.32$$
$$2\,H^+ + \tfrac{1}{2}\,O_2/H_2O \qquad 0.82$$

$\Delta G^0 = -2\,(96.500)\,(0.82 - (-0.32)) = \mathbf{-220}$ kJ/mol (theoretical),
ΔG^0 for ATP hydrolysis = -30.5 kJ/mol.
When three moles of ATP/mol NADH are synthesized, $3 \times (-30.5) = 91.5$ kJ.
Efficiency then is

$$E = \frac{91.5}{220} = 42\%.$$

The formation of ATP from ADP and P_i is energetically unfavorable and would normally proceed in the reverse direction. In order to drive this reaction forward, ATP synthase couples ATP synthesis during cellular respiration to an electrochemical gradient created by the difference in proton (H^+) concentration across the plasma membrane in bacteria. During bacterial oxygenic photosynthesis (*Cyanobacteria*), ATP is synthesized by ATP synthase using a proton gradient created in the thylakoid lumen through the thylakoid membrane and into the chloroplast stroma. ATP synthase consists of two main subunits, F_0 and F_1, which have a rotational motor mechanism allowing for ATP production. Because of its rotating subunit, ATP synthase is a molecular machine.

An ATPase first came to be recognized as a firmly bound component of bacterial cytoplasmic membranes from studies of *S. faecalis* protoplast ghosts. It was suspected at that time that the enzyme might be involved in ATP-dependent energy transduction processes occurring in the permeability barrier of the cell (**Fig. 5.10**). F_0 is a water in-soluble protein with eight subunits and a transmembrane ring. The ring has a tetramer shape with a helix–loop–helix protein that goes through conformational changes when it is protonated and deprotonated, pushing neighboring subunits to rotate, causing the spinning of F_0, which then also affects the conformation of F_1, resulting in switching of states of α and β subunits.

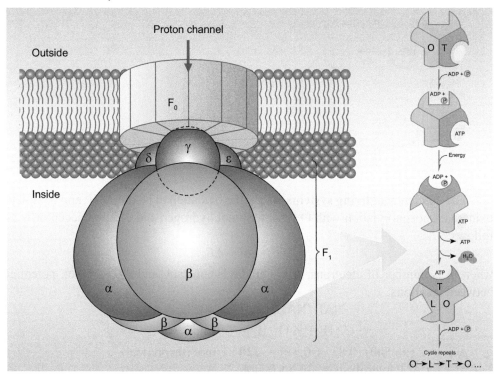

Fig. 5.10. Structure of ATPase in bacterial cells.

The F_1 portion of ATP synthase is hydrophilic and responsible for hydrolyzing ATP. The F_1 unit protrudes into the mitochondrial matrix space. Subunits α and β make a hexamer with six binding sites. Three of them are catalytically inactive and bind ADP. Other three subunits catalyze ATP synthesis. The other F_1 subunits γ, δ, and ε are a part of a rotational motor mechanism. Subunit γ allows β to go through conformational changes (i.e., closed, half open, and open states) that allow for ATP to be bound and released once synthesized.

5.4. Types of phosphorylation

Creation of ATP by oxidative phosphorylation takes place during the processes of respiration. These processes are related to hydrogen and electron transmission by chain of enzymes called "transmitters" on an inorganic compound during anaerobic respiration or on oxygen during aerobic respiration.

The transport chain, representing a pathway of hydrogen and electron transmission, consists of a series of reversible oxidoreductive energetic systems with a graduated redox potential (**Fig. 5.11**). Segments of this chain are very diverse in bacteria. According to the generally acknowledged and experimentally confirmed idea, hydrogen is transmitted by NAD dehydrogenases in the first degree of oxidation from substrate to flavoprotein dehydrogenases, which contain coenzymes FMN and FAD. From here, electron transmission continues through cytochromes to a final acceptor. The presence of cytochromes and their composition depend on the generic properties of bacteria and on the conditions of the surroundings.

The energetic efficiency of oxidative process depends on the amount of released energy and on the number of oxidative degrees. It is usually expressed by the quotient P/O, which represents the number of moles of esterified phosphate in the form of ATP to an atom of oxygen. Phosphate esterification occurs during oxidative phosphorylation, which takes place at the same time as a substrate oxidation. If a complex transport system takes part in the oxidative process, then the total gain of energy is three ATPs. When this system is not complete, the amount of created macroergic bonds is lower. This can be seen, for example, during the oxidation of succinate, which is not started by the first segment of the chain of transmitters, but by the second one, which is dehydrogenase containing flavoprotein. The amount of bound energy is in this case only two ATPs. It is similar to anaerobic respiration, where the last segment of the chain is missing — cytochrome oxidase containing cytochrome a, which accepts electrons from cytochrome c and transmits them to oxygen, while creating ATP.

ATP creation in phototrophic bacteria. Phototrophic bacteria (*Rhodospirillum*, *Chromatium*, and *Chlorobium* genus) create macroergic compounds of ATP type during the process of conversion of light energy into energy of chemical bonds. ATP creation is carried out by cyclic or non-cyclic phosphorylation.

Cyclic phosphorylation leads to ATP creation during the process of electron transmission in the closed system of transmitters with a gradually increasing redox potential. The process of cyclic phosphorylation results from the summary equation and scheme in **Fig. 5.12**. Transmitted electrons are released from bacteriochlorophyll by the activity of light quantum, which causes the transfer of electrons to the excited state. This is characterized by a high energetic potential. Electrons are then most likely transmitted from bacteriochlorophyll to ferredoxin by a yet unknown primary acceptor. In purple sulfuric bacteria (*Thiorhodaceae*), its molecule is made up from protein containing seven to eight atoms of iron and a sulfuric group. The molar mass is about 10,000 Da, absorption maximum at 390 nm. Very low standard redox potential -0.49 V, which is by 100 mV lower than in the $NAD^+/NADH + H$ is characteristic. Electrons then go from ferredoxin to quinone type transmitters, most often ubiquinone, that transmit them to cytochromes. Electron transmission to cytochromes b and c is, thanks to the strongly exergonic

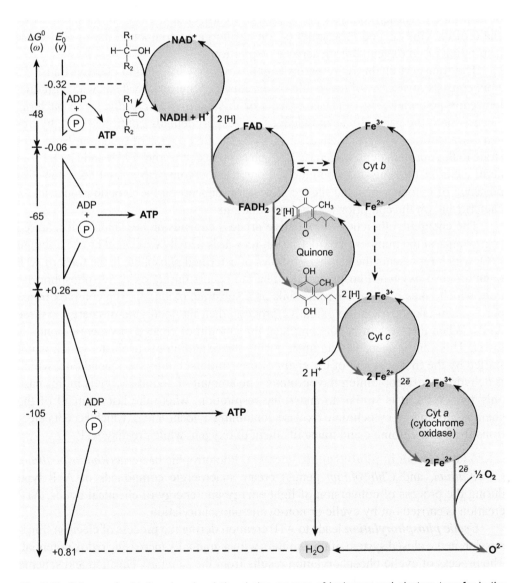

Fig. 5.11. Scheme of oxidative phosphorylation during process of hydrogen and electron transfer in the respiratory chain.

character of reaction, accompanied by energy release, which is used for the creation of ATP in the presence of ADP and an inorganic phosphate. The photophosphorylation of ADP occurs in two places simultaneously, which means that two molecules of ATP are created. Electrons then return from cytochrome c back to the bacteriochlorophyll, whose molecule goes from the temporary electropositive state to the original electroneutral one.

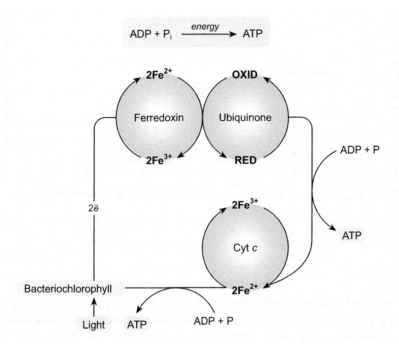

Fig. 5.12 Cyclic phosphorylation.

Noncyclic phosphorylation takes place in the presence of an external hydrogen and electron donor and enables the creation of not only ATP but also so-called reductive equivalents, which are reduced coenzymes necessary for CO_2 reduction or for proper working of other processes of biosynthesis. The beginning of noncyclic phosphorylation is same as with the cyclic one. However, electrons released from bacteriochlorophyll are not transmitted by ferredoxin to ubiquinone, but to flavin coenzymes, which simultaneously accept hydrogen from its external donor and go to a reduced state. The function of hydrogen and electron external donor in photolithotrophic bacteria is fulfilled by hydrogen sulfide, thiosulfate, or hydrogen gas; in photoorganotrophic bacteria, it is fulfilled by a simple organic substance such as succinic acid. Hydrogen then goes from flavin coenzymes to $NADP^+$, which is reduced to $NADPH + H^+$. Ferredoxin oxidation accompanied by FAD reduction is catalyzed by ferredoxin oxidoreductase. Chlorophyll returns to its original state thanks to electrons, which have been released during hydrogen fission from the external source. This transmission is carried out by cytochromes. The whole process of noncyclic phosphorylation results from the summary equation and scheme in **Fig. 5.13**.

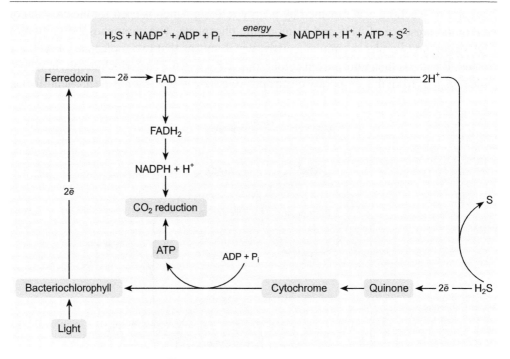

$$H_2S + NADP^+ + ADP + P_i \xrightarrow{\text{energy}} NADPH + H^+ + ATP + S^{2-}$$

Fig. 5.13. Noncyclic phosphorylation.

Current knowledge tells us that both processes, cyclic and noncyclic phosphorylation, take place in all phototrophic bacteria and complement each other. While the cyclic phosphorylation focuses on ATP creation and energetically secures biosynthesis processes thereby, noncyclic phosphorylation takes care of creating a sufficient number of reductive equivalents, necessary for CO_2 reduction.

5.5. Processes of catabolism

The main purpose of catabolic processes is the gain of energy necessary for bacterial growth. From this point of view, the catabolism must be understood as a number of processes that lead to decomposition or dissimilation of the substrate. These processes are accompanied by energy release and by the production of metabolic products. Both these components are used for proper biosynthesis and other vital functions by the bacteria cell, such as growth or multiplication.

The nature and progress of dissimilation processes vary a lot in different bacteria. When dissimilation takes place depends on the abilities of organism, on the nature of substrate, and on the surrounding conditions. Depending on the nature of the substrate, this process can be divided into the catabolism of:
- carbon compounds;
- nitrogen compounds;
- heterocyclic compounds.

5.5.1. Catabolism of carbon compounds and fermentation

The decomposition of nitrogen compounds in bacteria is important for gaining proper nitrogen nutrition and for energetic metabolism. The progress of these processes is conditioned not only by its originators, but also by the nature of hydrogen and electron donors and acceptors as well as environmental conditions. Depending on previous aspects, nitrogen compound catabolism can be divided into the following processes:

- fermentation;
- anaerobic respiration;
- aerobic respiration.

Fermentation. By the term "fermentation," it means processes of anaerobic dehydrogenation carried out by chemoorganotrophic bacteria, in which organic substances take part as hydrogen donors and acceptors. Fermented organic substrate is usually, after its previous activation, split to simple intermediates, which then enter oxidoreductive reaction in order to create final fermentation products. Substrate activation is caused by phosphorylation, where ATP and appropriate enzymes take part. Hydrogen transmission during oxidoreductive reactions is carried out by NAD dehydrogenases.

Considering energy, fermentation is a process that releases only a little amount of energy. The reason is that anaerobic dehydrogenation processes of substrate transformation stop creating simple organic compounds, which cannot be further decomposed by the cell. This is caused by the lack of enzymatic systems, which catalyze processes of terminal oxidation. The total amount of energy released by the substrate transformation during fermentation gets significantly lower by creating incompletely oxidized organic metabolites.

The most common substrate for the fermentation processes in bacteria are carbohydrates. However, some other substances, such as amino acids and heterocyclic compounds, can also be fermented.

The most common carbohydrates entering the fermentation process are hexoses, especially glucose. Glucose fermentation can run along several metabolic pathways, whose most important intermediate is pyruvic acid. Considering the nature of final products, which are created by the transformation of this acid (**Fig. 5.14**), and depending on the originator and conditions of the fermentation process, we can distinguish these types of fermentation:

(1) ethanol fermentation;
(2) lactic acid fermentation:
   - homofermentative;
   - heterofermentative.
(3) propionic acid fermentation;
(4) butyric acid fermentation and production of solvents;
(5) mixed fermentation;
(6) polysaccharide fermentation.

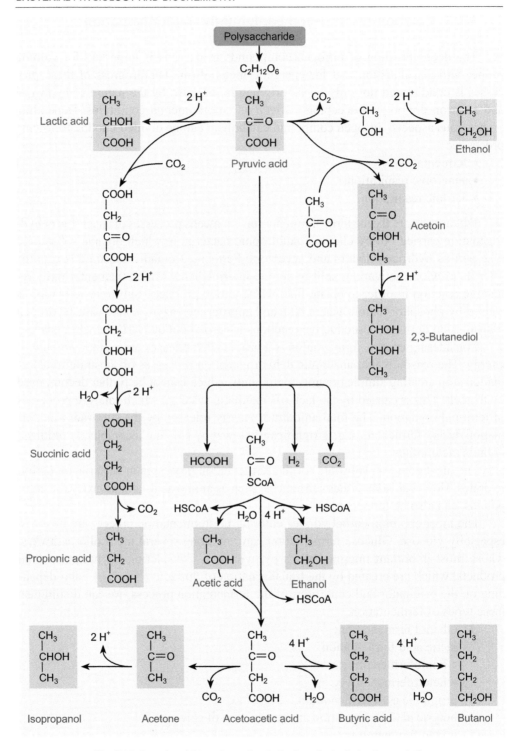

Fig. 5.14. Pyruvic acid transformation to final products during fermentation.

5.5.1.1. Ethanol fermentation

The starting substrate for ethanol fermentation are hexoses, which are transformed to pyruvic acid, from which ethanol and CO_2 are later created (**Fig. 5.15**). Originators are especially yeasts of *Torula* and *Saccharomyces* genera. As an example of bacteria carrying out this process, it can mention *Sarcina ventriculi*. Other bacteria with the ability to create ethanol run its production along different metabolic pathways.

Ethanol fermentation can be expressed by the summary equation:

$$C_6H_{12}O_6 \rightarrow 2\ CH_3CH_2OH + 2\ CO_2\ (\Delta G^0 = -234.5\ kJ)$$

The whole process captured in **Fig. 5.15** runs along the pathway called **h**exose **d**iphosphate (HDP) or Embden–Meyerhof–Parnas (EMP) pathway until pyruvic acid is created. Its individual phases include these reactions:

1. **Hexose phosphorylation.** If glucose is the substrate, in the beginning, it is phosphorylated to glucose 6-phosphate with the participation of ATP and hexokinase. Glucose 6-phosphate is then transformed by isomerization and by the effect of hexose phosphate isomerase to fructose 6-phosphate, which is further phosphorylated to fructose 1,6-diphosphate.

2. **Fission of fructose 1,6-diphosphate to two trioses, 3-phosphoglyceralde-hyde and dihydroxyacetone phosphate.** The reaction is catalyzed by aldolase, balance between trioses is displaced to the side of 3-phosphoglyceraldehyde and is kept by triosephosphate isomerase.

3. **Dehydrogenation of 3-phosphoglyceraldehyde to 3-phosphoglyceric acid.** At first, 1,3-diphosphoglyceric acid with a macroergic bond is created with the participation of glyceraldehyde 3-phosphodehydrogenase and in the presence of an inorganic phosphate. By the effect of phosphoglycerokinase, phosphate is transmitted from the macroergic bond to ADP, which leads to the creation of ATP and 3-phosphoglyceric acid.

4. **Transformation of 3-phosphoglyceric acid to pyruvic acid.** 2-Phosphogly-ceric acid is the first one to be created, and this happens by the rearrangement reaction of 3-phosphoglyceric acid with the participation of phosphoglyceromu-tase. By the effect of enolase, the dehydration of 2-phosphoglyceric acid occurs and creates phosphoenolpyruvic acid, which contains a macroergic bond. By splitting off the phosphate during ATP creation and thanks to the catalytic effects of phosphopyruvate kinase, this acid transforms to the pyruvic acid.

In the final phase, pyruvic acid is split by pyruvate decarboxylase to acetaldehyde and CO_2. The following reaction is catalyzed by NAD ethanol dehydrogenase, causing the transformation of acetaldehyde to ethanol.

The energetic balance of ethanol fermentation is two consumed and four new macroergic bonds of the ATP, created on the substrate level. It means a total gain of two ATPs. The efficiency of this process can be expressed as a percentage ratio between the free energy of both macroergic bonds and the total amount of released energy.

$$E = \frac{2 \times 30.6}{234.5} \times 100\ (kJ) = 26.1\,\%.$$

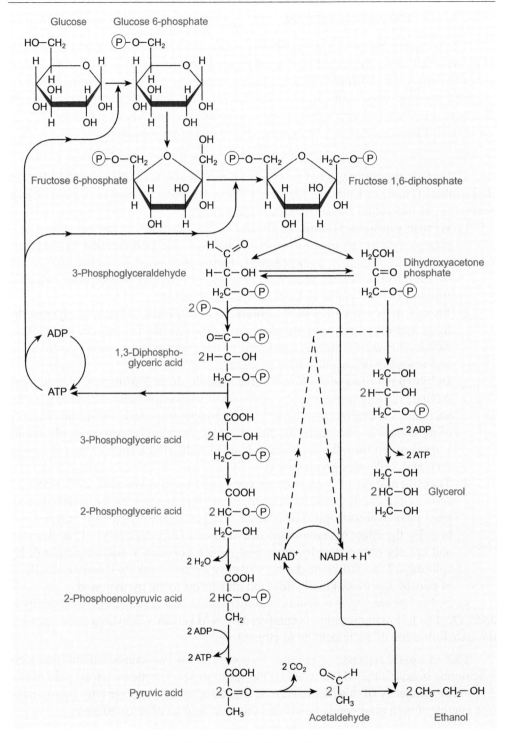

Fig. 5.15. Ethanol fermentation.

Ethanol fermentation is very important for practical purposes, especially for the production of ethanol and alcoholic beverages such as wine and beer. The hexose diphosphate pathway of transformation of glucose to pyruvate acid is not used strictly by anaerobic bacteria, but also by some facultative anaerobic bacteria.

Glycerol fermentation. The reduction of acetaldehyde to ethanol is suppressed by adding sulfite, which makes bonds with acetaldehyde. During this reaction, hydrogen is transmitted by reduced coenzyme ($NADH + H^+$) to hydroxyacetone phosphate, and glycerophosphate is created. At the same time, the balance between both trioses moves to the side of dihydroxyacetone. Glycerol is created from glycerophosphate after splitting off the phosphate. This change of ethanol fermentation to the glycerol one is used for glycerol production. The bond between the sulfite and acetaldehyde can be expressed by the equation:

$$C_6H_{12}O_6 + NaHSO_3 \longrightarrow H_3C - \overset{\overset{\displaystyle H}{|}}{\underset{\underset{\displaystyle OH}{|}}{C}} - SO_3Na \ + \ \overset{\overset{\displaystyle CH_2OH}{|}}{\underset{\underset{\displaystyle CH_2OH}{|}}{CH-OH}} + CO_2$$

Ethanol production by Entner–Doudoroff pathway. Bacteria species *Zymomonas mobilis* ferments glucose and creates ethanol not along the EMP pathway, but along the pathway called Entner–Doudoroff (ED). This pathway follows the so-called hexose monophosphate or pentose cycle. During this cycle, glucose 6-phosphate is not transformed by isomerization to fructose 6-phosphate as in the EMP pathway, but it is oxidized by dehydrogenase to gluconolactone 6-phosphate, which later undergoes hydrolysis because of the effect of gluconolactonase, and 6-phosphogluconic acid is then created. While in the EMP cycle, this acid undergoes decarboxylation with pentoses as a result, in the ED pathway, it is transformed in the presence of phosphogluconate dehydrogenase to 2-keto-3-deoxy-6-phosphogluconic acid, which is later split off by aldolase to glyceraldehyde 3-phosphate and pyruvic acid. Glyceraldehyde 3-phosphate is then glycolytically transformed into pyruvic acid, which provides ethanol and CO_2 for the terminal phase, same as with the ethanol fermentation.

Due to the fact, that only one molecule of triose phosphate, which can be used for ATP creation at the substrate level after entering the glycolysis, is created during the process of the ED pathway, the amount of gained energy is half compared to the EMP pathway, and the ratio is one mole of glucose to one mole of ATP. Glucose transformation in the ED pathway is schematically shown in **Fig. 5.16**. Ethanol can be also created as a side product of the metabolism of other bacteria (such as some *lactic acid bacteria*, *clostridia*, and *enterobacteria*). Acetaldehyde, which is a precursor to ethanol, is not created from pyruvic acid by the effect of pyruvate decarboxylase in these bacteria, but by the reduction of acetyl coenzyme A by aldehyde dehydrogenase:

$$CH_3CO - S\text{-}CoA + HADH + H^+ \rightarrow CH_3CHO + HS\text{-}CoA + HAD^+.$$

Emerging acetaldehyde is then reduced to ethanol by alcohol reductase:

$$CH_3-CHO \xrightarrow{\quad NADH + H^+ \quad NAD^+ \quad} CH_3-CH_2-OH$$

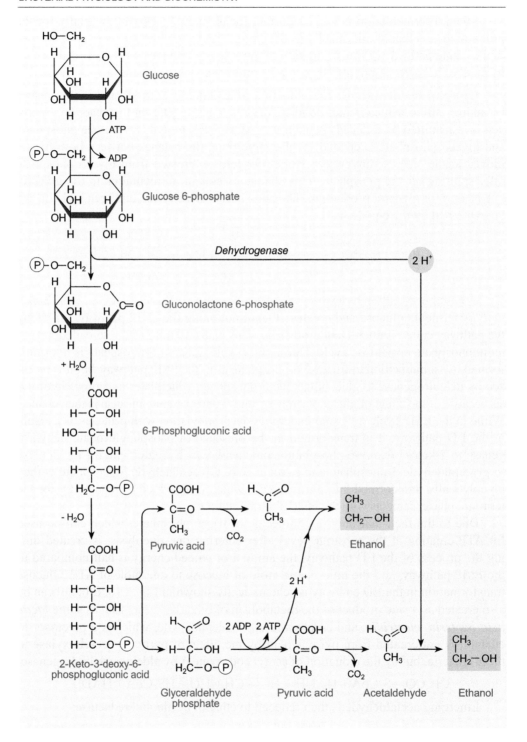

Fig. 5.16. Ethanol production by the Entner–Doudoroff pathway.

5.5.1.2. Lactic acid fermentation

This fermentation is carried out by a big number of bacteria. Most of them belong to the *Lactobacillaceae* family. According to the products, which are created during the fermentation, we can usually distinguish:

(1) **homofermentative lactic acid fermentation**, during which the main and only product created is lactic acid, and

(2) **heterofermentative lactic acid fermentation**, during which lactic acid and other substances are created.

Homofermentative lactic acid fermentation is carried out by species *Lactobacillus delbrueckii* ssp. *bulgaricus*, *L. casei*, *L. leichmannii*, *Streptococcus lactis*, etc. The substrate are mainly hexoses, whose fermentation runs along the glycolytic pathway. The terminal phase of this process, which is same as in the EMP pathway until the pyruvic acid is created, is represented by the transformation of this acid to the lactic acid. This reaction is catalyzed by NAD lactate dehydrogenase, which transmits the hydrogen withdrawn during triose phosphate dehydrogenation to pyruvic acid. The process of this reaction can be expressed by the following scheme:

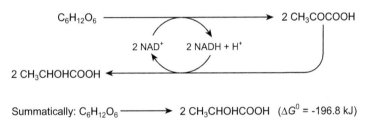

Summatically: $C_6H_{12}O_6 \longrightarrow$ 2 $CH_3CHOHCOOH$ (ΔG^0 = -196.8 kJ)

Depending on the stereospecificity of lactate dehydrogenase and in the presence of lactate racemase, D(-) and L(+) or DL-lactic acids can be formed. Because homofermentative lactic fermentation takes place, the energy yield of glucose forms **two ATPs**, and its efficiency is about 32%.

Heterofermentative lactic fermentation. It is characterized by the fact that not only lactic acid but also other final products are formed (**Fig. 5.17**). Most common are acetic acid, ethanol, hydrogen, and CO_2. Most originators of this fermentation, such as *Lactobacillus fermenti*, and *Leuconostoc mesenteroides*, are found to miss the basic enzymes of the EMP pathway (fructose-diphosphate aldolase and triose-phosphate isomerase). The fermentation of glucose takes place in the so-called phosphoketolase pathway. This pathway also originates from a hexose monophosphate or pentose cycle, during which glucose 6-phosphate undergoes conversion via 6-phosphogluconolactone to phosphogluconic acid. Its decarboxylation results in ribulose 5-phosphate, which passes through the action of epimerase to xylulose 5-phosphate. While the HMP pathway proceeds through pentose metabolism to form a number of other by-products, in the phosphoketolase pathway, xylulose 5-phosphate is cleaved in the presence of thiamine pyrophosphate by the action of phosphoketolase to glyceraldehyde 3-phosphate and

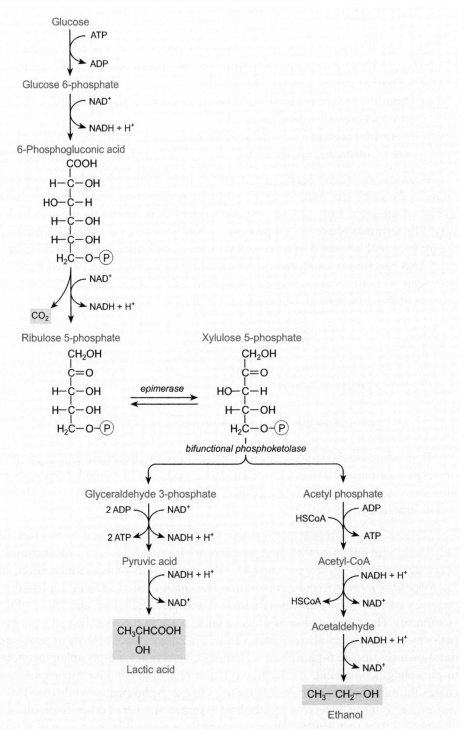

Fig. 5.17. Heterofermentative lactic fermentation or phosphoketolase pathway.

acetyl phosphate. Glyceraldehyde phosphate enters the other phases of the glycolysis pathway, in which it is converted to lactic acid, while acetyl phosphate is reduced to ethanol via acetyl-CoA and acetaldehyde.

The summarily expressed fermentation equation looks like:

$$C_6H_{12}O_6 \rightarrow CH_3CHOHCOOH + CH_3CH_2OH + CO_2.$$

Some heterofermentative lactic acid bacteria (*Lactobacillus brevis*, *L. buchneri*) do not use glucose during anaerobic growth, because they lack enzymes catalyzing the reduction of acetyl phosphate to ethanol necessary to maintain the total redox equilibrium. However, instead of glucose, they can ferment fructose, which they convert via mannitol dehydrogenase to mannitol. Lactic acid, acetic acid, and CO_2 are also produced according to the following equation:

$$3\ C_6H_{12}O_6 \xrightarrow{\ H_2O\ } CH_3CHOHCOOH + CH_3COOH + 2\ C_6H_{12}O_6 + CO_2.$$

The reduction of fructose to produce mannitol allows to maintain the necessary reaction equilibrium.

5.5.1.3. Pentose sugars fermentation

The phosphoketolase pathway is also used in the fermentation of pentoses, for example by *Lactobacillus plantarum*. This fermentation results in the formation of lactic acid and acetic acid (**Fig. 5.18**). The acetic acid is produced by converting acetyl phosphate in the presence of acetokinase with the formation of ATP:

$$CH_3CO \sim P \xrightarrow[\substack{ADP \qquad ATP}]{acetate\ kinase} CH_3COOH$$

The energy balance represents a net gain of two moles of ATP per mole of fermented hexose or pentose. A distinct mechanism of heterofermentative glucose fermentation using the phosphoketolase pathway was observed in *Lactobacillus bifidus* strains, which have a dominant position in the intestinal microbiome of infants. This bacterium lacks both aldolase and glucose 6-phosphate dehydrogenase but contains two active phosphoketolases. The former catalyze the cleavage of fructose 6-phosphate to erythrose 4-phosphate and acetyl phosphate; the latter allow the formation of glyceraldehyde 3-phosphate and acetyl phosphate from xylulose 5-phosphate. Glyceraldehyde is cleaved through the glycolysis pathway via pyruvic acid in the final phase to lactic acid, while acetyl phosphate forms acetic acid. The course of the process is schematically shown in **Fig. 5.19**.

It can be summarily expressed in the equation below:

$$3\ C_6H_{12}O_6 \rightarrow 2\ CH_3CHOHCOOH + CH_3COOH.$$

This process is more energy-efficient than glycolysis, because two molecules of hexose produce five molecules of ATP.

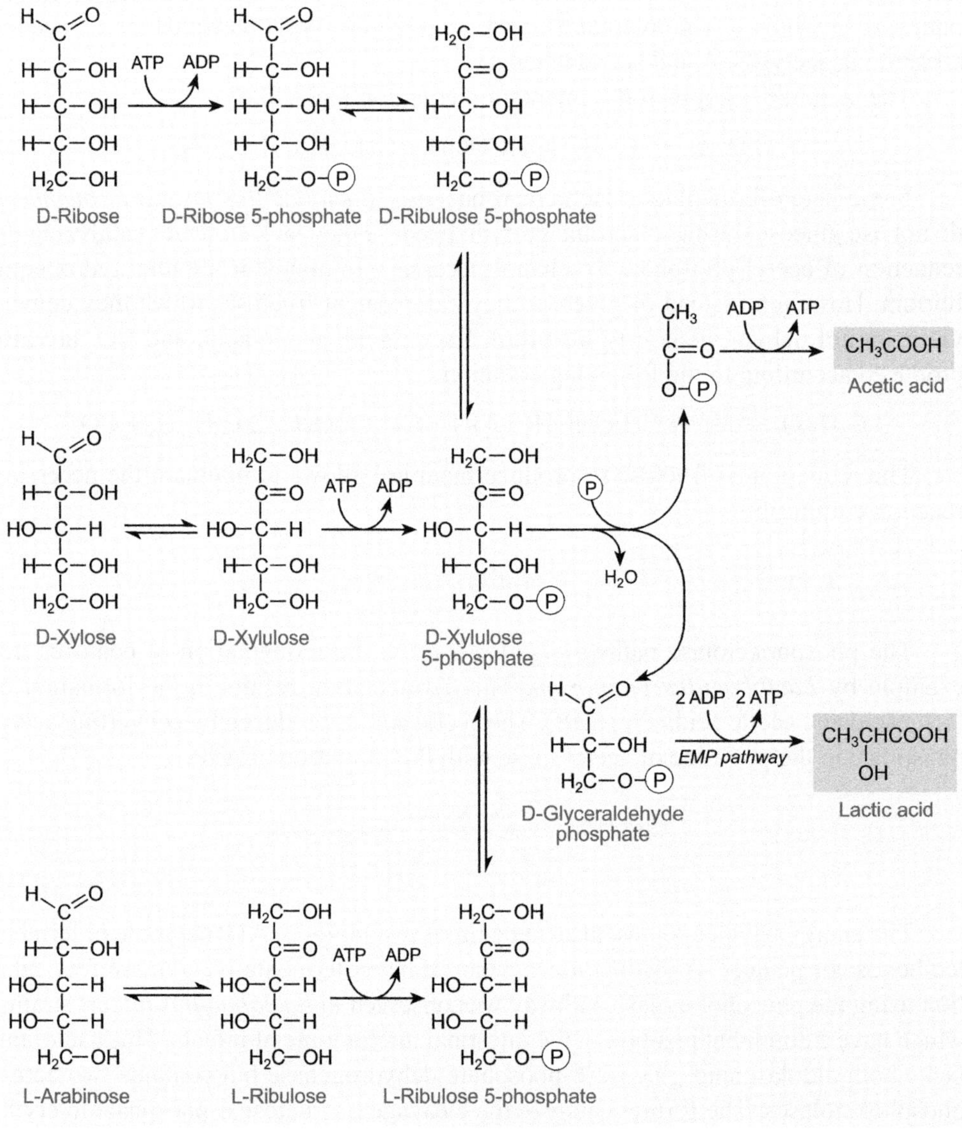

Fig. 5.18. Pentose fermentation.

Homofermentative and heterofermentative lactic fermentation has a wide use, especially in the food industry, where it is used to produce dairy products (sour milk, yoghurt, kefir, kumis, etc.), fermented cabbage and pickled gherkins. It is also important in agriculture, especially in the production of silage.

Thus, the central intermediate of pentose fermentation (e.g. ribose, xylose, and arabinose) is xylose 5-phosphate, from which acetyl phosphate, acetate, and glyceraldehyde phosphate are formed followed by the formation of lactic acid. In both pathways through acetate phosphate or glyceraldehyde phosphate, ATP is accumulated.

The process of heterofermentative fermentation of glucose in *Lactobacillus bifidus* occurs by the phosphoketolase pathway with formation of xylose 5-phosphate as a central intermediate (**Fig. 5.19**).

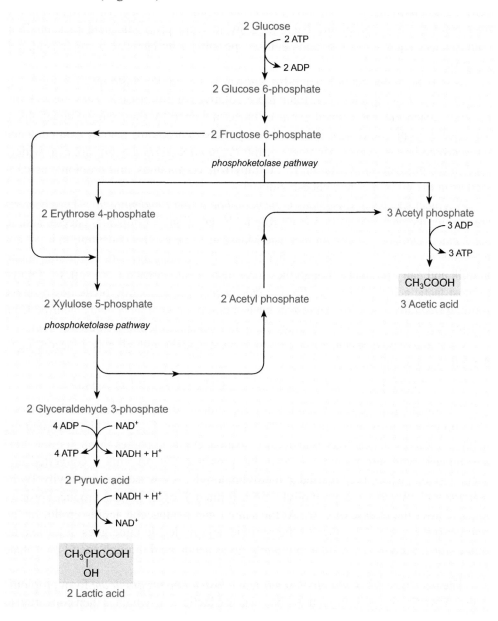

Fig. 5.19. Heterofermentative fermentation of glucose in *Lactobacillus bifidus* following the phosphoketolase pathway.

Thus, as a result of the heterofermentative fermentation of two moles of glucose, three moles of acetate and two moles of lactic acid are formed.

5.5.1.4. Propionic acid fermentation pathway

Propionic fermentation agents are mainly representatives of the *Propionibacterium* genus, *Veillonella alcalescens*, and *Clostridium propionicum*. Bacteria of the *Propionibacterium* genus or propionic bacteria grow to form propionic acid in a medium with glucose, sucrose, lactose and pentoses, possibly with lactic acid, malic acid, and glycerol.

This type of fermentation is also known as the ***methylmalonyl-CoA pathway***. *Propionibacterium* species is the main representatives of this process. They are a Gram-positive, anaerobic, rod-shaped genus of bacteria named for their unique metabolism. The species of propionibacteria are able to synthesize propionic acid by using unusual transcarboxylase enzymes. Members of the *Propionibacterium* genus are also widely used in the production of vitamin B_{12}, tetrapyrrole compounds, and propionic acid, as well as in the probiotics and cheese industries.

A typical species of this genus is *Propionibacterium freudenreichii*. These bacteria are non-motile and play an important role in the creation of Emmental cheese, and, to some extent, Jarlsberg, Leerdammer, and Maasdam cheeses. Its concentration in Swiss-type cheeses is higher than in any other cheese. Propionibacteria are commonly found in milk and dairy products, though they have also been extracted from soil. In contrast to most lactic acid bacteria, the *Propionibacterium* species mainly break down lipids, forming free fatty acids. Recent research has focused on possible benefits incurred from consuming *P. freudenreichii*, which are thought to cleanse the gastrointestinal tract.

Cleavage of hexoses takes place up to pyruvic acid through glycolysis according to the following equation:

$$3\ C_6H_{12}O_6 \rightarrow 4\ CH_3CH_2COOH + 2\ CH_3COOH + 2\ CO_2 + H_2O.$$

The mechanism of the formation of propionic acid consists in the carboxylation of pyruvic acid to oxaloacetic acid. This reaction is catalyzed by biotin-containing methylmalonyl-CoA carboxyltransferase. Oxaloacetic acid passes through malic acid and fumaric acid into succinic acid. In the presence of CoA, this is converted into succinyl-CoA, which is converted to methylmalonyl-CoA by isomerase activity. By its reaction with pyruvic acid, propionyl-CoA is formed. Propionyl-CoA produces propionic acid after the release of CoA. At the same time, oxaloacetic acid is produced and returns to the cycle. A portion of the pyruvic acid molecule undergoes decarboxylation to form acetyl-CoA, which ultimately gives acetic acid through acetyl phosphate (**Fig. 5.20**).

The bacteria *Veillonella alcalescens* and *Clostridium propionicum* form propionic acid differently. As a substrate, these bacteria use lactic acid, which is then converted to propionic acid via acrylic acid. Lactic acid is converted to lactyl-CoA by acetyl-CoA-transferase, which is dehydrated by the action of lactyl-CoA-dehydratase to acryloyl-CoA. Its conversion to propionyl-CoA is catalyzed by acetyl-CoA-dehydrogenase. From propionyl-CoA, propionic acid is again produced by acetyl-CoA-transferase. Acetyl-CoA is involved in this reaction (**Fig. 5.21**).

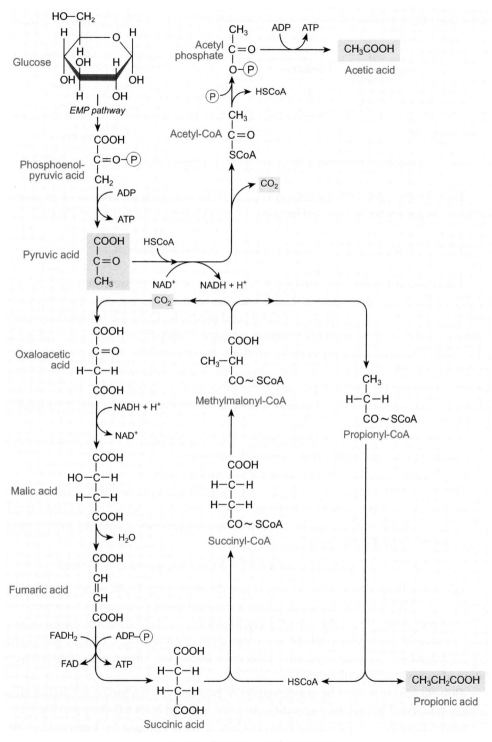

Fig. 5.20. Propionic acid fermentation.

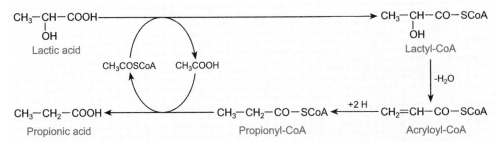

Fig. 5.21. Propionic acid fermentation by *Veillonella alcalescens* and *Clostridium propionicum*.

Thus, propionic acid fermentation is used in the industrial production of propionic acid and, in particular, in the food industry in the production and aging of cheeses.

5.5.1.5. Butyric acid fermentation and solvent formation

The origins of these fermentation processes are anaerobic bacteria, which mostly belong to the *Clostridium* genus. These bacteria ferment saccharides, some polysaccharides, and organic acids to form different end products. The most common are acids (butyric, acetic, lactic), alcohols (butanol, ethanol, isopropanol), acetone, and gases (H_2, CO_2). The representation of these products is governed by the generic properties of the fermentation agent and by the conditions of the fermentation process. In processes carried out by clostridiums, acylation processes are essential for the production of acetyl-CoA from pyruvic or acetic acid. The main phases of the whole procedure then include (**Fig. 5.22**):

- coupling of acetyl-CoA to acetoacetyl-CoA and further on butyryl-CoA to form acetic acid and butyric acid, optionally solvents;
- reduction of acetyl-CoA to acetaldehyde and ethanol.

Butyric acid fermentation is caused by *Clostridium butyricum*. The main products of this process, which follows the EMP pathway in hexoses, are butyric acid and acetic acid. The formation of these acids, usually bound in the form of calcium salts, can be expressed by the following equation:

$$4\ C_6H_{12}O_6 \rightarrow 3\ CH_3CH_2CH_2COOH + 2\ CH_3COOH + 8\ CO_2 + 10\ H_2.$$

Glycolically produced pyruvic acid is first decarboxylated with pyruvate dehydrogenase and ATP, from which two electrons are transferred to ferredoxin. The enzyme-TTP complex, under the influence of acetyltransferase with CoA, carries acetyl-CoA, which converts to acetyl phosphate in the presence of phosphate. Reduced ferredoxin is reoxidized by hydrogenase, forming molecular hydrogen. The process of decarboxylation of pyruvic acid is illustrated in **Fig. 5.23**.

Butyric acid is formed by condensation of two acetyl-CoA molecules in the presence of thiolase. The resulting acetyl-CoA is first reduced by NAD-β-hydroxybutyryl-CoA-dehydrogenase to β-hydroxybutyryl-CoA. This is reduced by butyryl-CoA-dehydrogenase to butyryl-CoA, resulting in butyric acid after CoA cleavage.

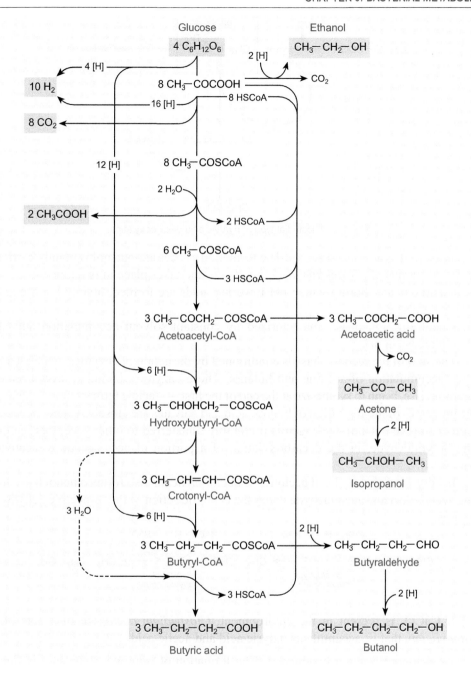

Fig. 5.22. Butyric acid fermentation.

Thus, butyric acid fermentation is a process, where various organic compounds and solvents are formed. The final products of this process are carbon dioxide, ethanol, acetate, acetone, isopropanol, butanol, and butyric acid.

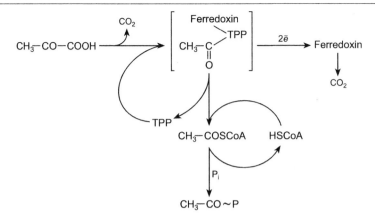

Fig. 5.23. Clostridial type of pyruvic acid decarboxylation.

Acetone-butanol production. The formation of acetone-butanol is mainly carried out on the strains of *Clostridium acetobutylicum* and takes place in two phases:

- acidic phase, during which pH-lowering acids are formed, mainly butyric and acetic acid;
- neutralization phase characterized by constant solvent accumulation and an increase in pH.

The start of the second phase is conditioned by the action of enzymes, which leads to the accumulation of acetone and butanol. These enzymes, including acetoacetate decarboxylase, begin to synthesize at the end of the first — acidic fermentation — phase. The processes of the formation of butyric acid and solvates are closely related, because in order to give butanol, hydrogen is preferably transferred to other acceptors, including butyraldehyde, in the decarboxylation of a portion of acetoacetate to acetone **(Fig. 5.24)**.

In addition to butanol and acetone, *C. acetobutylicum* also forms ethanol by reducing acetyl-CoA and acetaldehyde under the catalytic action of the respective dehydrogenases:

$$CH_3-COSCoA \xrightarrow[\text{acetaldehyde dehydrogenase}]{NADH + H^+ \quad NAD^+} CH_3-CHO \xrightarrow[\text{alcohol dehydrogenase}]{NADH + H^+ \quad NAD^+} CH_3-CH_2-OH$$

If $CaCO_3$ is present in the environment, *Clostridium acetobutylicum* acts like *C. butylicum*, that is, accumulates butyric acid and acetic acid.

Isopropanol-butanol production. The formation of isopropanol–butanol is similar to acetone–butanol fermentation, except that acetone is further reduced to isopropanol. This process is carried out by some strains of *Clostridium butyricum*. The formation of isopropanol occurs even when the substrate is acetone itself. Neutralization leads to preferential acid production and the suppression of solvent formation:

$$3\ C_6H_{12}O_6 \rightarrow 2\ CH_3(CH_2)_3OH + CH_3CHOHCH_3 + 7\ CO_2 + H_2O + 3\ O_2.$$

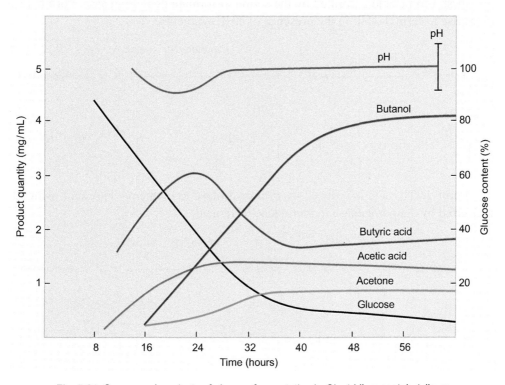

Fig. 5.24. Course and products of glucose fermentation in *Clostridium acetobutylicum*.

In these processes, the energy yield does not exceed two ATPs generated by the EMP path. Clostridial fermentation processes to produce solvents are of considerable significance, as these substances are widely used in the industrial manufacture of varnishes and other technological sectors.

Acetic acid production. The fermentation of glucose results in acetic acid, which is produced as the only product in an amount of three molecules per molecule of glucose. A more detailed study of this process has found that while two molecules of acetic acid are derived from pyruvic acid following the EMP pathway, the third molecule is formed by the conversion of CO_2 in the presence of tetrahydrofolic acid (THF) and corrinoid coenzyme (a vitamin B_{12} derivative).

THF is a cofactor in many reactions; especially the THF derivatives serve as donors of C_1 units in a variety of biosynthetic reactions involving amino acids (methionine, serine, and glycine), pyrimidines (thymine), vitamins (pantothenic acid), purine bases (inosinic acid), and the initiation of protein synthesis in bacteria. 10-Formyltetrahydrofolate acts as a donor of a group with one carbon atom. Tetrahydrofolate gets this extra carbon atom by sequestering formaldehyde produced in other processes. The principal source of the C_1 metabolite is 5,10-methylene-THF, normally derived from glucose via glycine, although it is also formed in a non-enzymatic reaction by condensation of formaldehyde with THF.

First, formic acid is formed by the action of formate dehydrogenase. Formic acid gives rise to formyl-THF in the presence of THF synthetase.

Formyl-THF acid is reduced by cyclohydrolase to methenyl-TFA acid, which is converted by dehydrogenase to methylene-THF acid.

Methylene-THF acid provides methyl-THF by further hydrogenation.

However, the methyl group of this acid is transferred to the reduced corrinoid to form the CO-methyl corrinoid protein complex, which is carboxylated to form carboxymethyl-corrinoid.

It should be noted that corrinoids are a group of compounds based on the skeleton of corrin, a cyclic system containing four pyrrole rings similar to porphyrins. These include compounds based on octadehydrocorrin, which has the trivial name corrole.

In the final phase, this complex is cleaved to form acetic acid. The whole process is illustrated in **Fig. 5.25.**

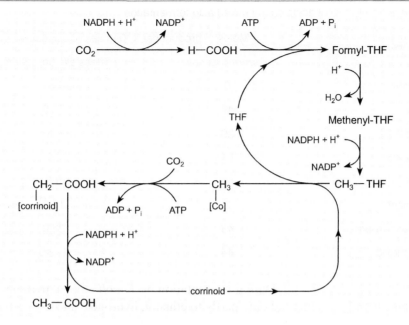

Fig. 5.25. Direct conversion of CO_2 to acetic acid in clostridia.

Thus, the acetate as a final product is synthesized from CO_2 via reactions catalyzed by the enzymes, and 5-methyltetrahydrofolate and a methylcorrinoid are intermediates in this synthesis.

5.5.1.6. Mixed acid fermentation pathway

Mixed acid fermentation is the process by which a six-carbon sugar, for example, glucose, is converted into a complex and variable mixture of acids. The most common are acetic acid, formic acid, lactic acid, succinic acid, acetoin, 2,3-butanediol, hydrogen, and CO_2. Therefore, these processes are referred to as mixed fermentation. It is characteristic for members of the *Enterobacteriaceae*, a large family of Gram-negative bacteria that includes *E. coli*, *Pseudomonas* species, and some *Bacillus*.

Metabolic pathways include primarily the Embden–Meyerhof–Parnas pathway. It is also known as the **h**exose **m**ono**p**hosphate (HMP) shunt or the phosphogluconate pathway. The pentose phosphate pathway is an alternative pathway to glycolysis and the TCA cycle for oxidation of glucose. The process of fermentation and the character of end products depend on the originator and the conditions of the process. **Table 5.4** demonstrates a different way of conversion of pyruvic acid in two typical representatives of *Enterobacteriaceae*.

The formation of these end products depends on the presence of certain key enzymes in the bacteria. The proportion, in which they are formed, varies between different bacterial species. The end products of mixed acid fermentation can have many useful applications in biotechnology and industry.

Table 5.4. Mixture of acids during mixed acid fermentation

Product	Moles of product per 100 moles of glucose	
	Escherichia coli	Klebsiella aerogenes
Butanediol	0	66.5
Ethanol	42	70.0
Succinic acid	29	0.0
Lactic acid	84	3.0
Acetic acid	44	0.5
Formic acid	2	18.0
Molecular hydrogen	43	36.0
Carbon dioxide	44	172.0

Fermentation by *E. coli*. During fermentation by *E. coli* strains, pyruvic acid is transformed mainly to acids and only partly to ethanol, hydrogen, and CO_2. Ethanol is formed from acetyl phosphate, because enterobacteria lack pyruvate carboxylase catalyzing decarboxylation of pyruvic acid to acetaldehyde. A part of the acetyl phosphate formed is converted to acetic acid. The cleavage of pyruvic acid to acetyl phosphate is accompanied by the formation of formic acid, which is further decomposed by the action of formate hydrogenase to hydrogen and CO_2 (**Fig. 5.26**).

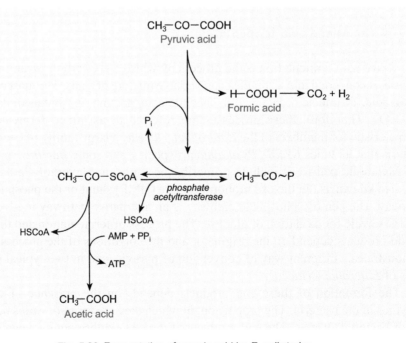

Fig. 5.26. Fermentation of pyruvic acid by *E. coli* strains.

Lactic acid is formed by the reduction of pyruvic acid. Succinic acid is formed by carboxylation of phosphoenolpyruvate to oxaloacetic acid, which is further reduced to succinic acid (**Fig. 5.27**). Predominant acid formation accompanied by a decrease in the pH of the environment is the essence of the diagnostic, the so-called Methyl Red test (MR-test).

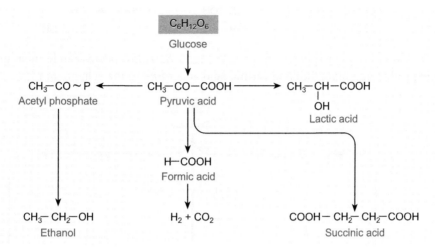

Fig. 5.27. Glucose fermentation products in *E. coli*.

Fermentation by *Klebsiella aerogenes*. Under the same conditions, *K. aerogenes* forms some organic acids, the main fermentation products acetoin and 2,3-butanediol acetoin being formed from two pyruvic acid molecules that are condensed to acetolactic acid by simultaneous decarboxylation. The formation of this acid involves active acetaldehyde (oxyethyl-TPP) and the corresponding synthetase. In the next step, the acetic acid is decarboxylated by the action of α-acetolactate decarboxylase to form acetoin. The reduction of acetoin with butanediol dehydrogenase produces 2,3-butanediol (2,3-butylene glycol):

Butanediol passes in the air to diacetyl, which gives a guanine derivative in the alkaline environment with a typical red coloration. The Voges–Proskauer test (VP test) is based on this reaction. Thus, bacteria *K. aerogenes* metabolize pyruvate to diacetyl through butanediol as an intermediate product.

Similarly, some other bacteria, such as certain *Pseudomonas* and *Bacillus* species, form acetoin. *K. aerogenes* and *Citrobacter freundii* bacteria can also use glycerol, which reduces to trimethylene glycol. This reaction takes place via β-hydroxypropionaldehyde:

$$
\begin{array}{ccccc}
CH_2-OH & \xrightarrow{\quad H_2O \quad} & CH_3 & \xrightarrow{\quad NADH+H^+ \quad NAD^+ \quad} & CH_2-OH \\
| & & | & & | \\
CH-OH & & CH-OH & & CH_2 \\
| & & | & & | \\
CH_2-OH & & CH-O & & CH_2-OH
\end{array}
$$

Fermentation by species of *Serratia* genus. *Serratia* species can ferment glucose, while producing butanediol and formic acid according to the scheme in **Fig. 5.28**.

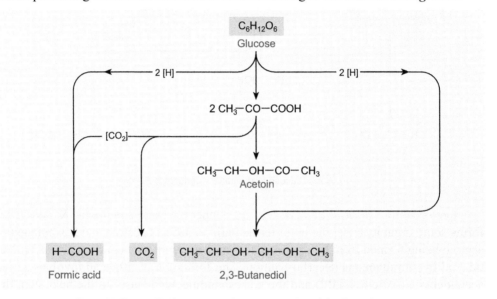

Fig. 5.28. Butanediol fermentation by some species of the *Serratia* genus.

Thus, the species of the *Serratia* genus ferment glucose to the following final products: 2,3-butanediol, formic acid, and CO_2.

5.5.1.7. Fermentation of sugars and polysaccharides

These substances are cleaved by the system of amylolytic enzymes produced by various bacteria, but mainly by representatives of the *Clostridium* and *Bacillus* genera. Most of these bacteria produce effective α-amylases. By their action, starch is gradually decomposed — the first generated oligosaccharides are ultimately broken down into maltose or isomaltose. This enzyme is hydrolyzed by maltase to glucose, which is fermented to end products by glycolysis, possibly by the HMP, phosphoketolase, or Entner–Doudoroff pathway.

Fermentation of cellulose. Cellulose is fermented with bacteria having the cellulase enzyme, which breaks down cellulose into a cellobiose disaccharide. It is further cleaved with cellobiase to glucose. Among the agents of anaerobic degradation of cellulose, there are predominantly *Clostridium* species, for example, *C. cellobioparum*, *C. thermo cellum*. The final products of anaerobic digestion of glucose are formic, acetic, butyric and lactic acids, ethanol, carbon dioxide, and hydrogen. In the presence of anaerobic methane bacteria, CO_2 can be reduced to methane.

Fermentation of pectin. Pectins are broken down by complexes of enzymes known as protopectinases. Their activity produces cellulose and protopectins, which are complexes with other polysaccharides (arabinans, galactans), which are further cleaved to pectin and components of the complex mentioned above. Pectin, which is a methylated polygalacturonic acid ester, provides pectin acids after methyl alcohol is cleaved by the pectinase enzyme. By the action of pectolase, galacturonic acid, arabinose, and xylose are produced as final products, which can be further fermented into butyric acid, acetic acid, carbon dioxide, and hydrogen. The originators of anaerobic degradation of pectin substances are *Clostridium felsineum* and *Bacillus macerans*.

5.5.2. Anaerobic respiration

Anaerobic respiration is a process in which an organic substrate is usually oxidized by transferring hydrogen and electrons to the oxygen bound in the molecule of an inorganic substance. The most common are salts of nitric acid, sulfur, and carbon dioxide. The hydrogen acceptor is oxygen bound in the inorganic compound, which is simultaneously being reduced; the respective processes are carried out under anaerobic conditions. Transmission of hydrogen and electrons occurs through the appropriate cytochrome reductase, eliminating the formation of ATP, which occurs during aerobic respiration in the last stage of terminal oxidation. Anaerobic respiration acquires energy in certain chemoorganotrophic and in isolated cases of chemolithotrophic bacteria. In *chemoorganotrophic bacteria*, an organic substrate serves as a hydrogen and electron donor and a carbon source. *Chemolithotrophic bacteria* use an inorganic substance as a hydrogen and electron donor and CO_2 as a carbon source.

Depending on the nature of the reduced substance, we recognize these metabolic processes in anaerobic respiration:

(1) nitrate reduction and nitrification;
(2) sulfate reduction (desulfurization);
(3) reduction of CO_2 to methane.

5.5.2.1. Nitrate reduction and denitrification

The dissimilatory reduction of nitrates can be accomplished in three ways:
- uncompleted reduction accompanied by accumulation of nitrites in the environment;
- complete reduction to ammonia accompanied by transient occurrence of nitrite;
- complete reduction accompanied by the production of gaseous products, especially N_2, or denitrification:

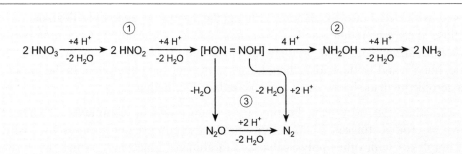

Bacteria that reduce nitrate to ammonia can use it as a source of nitrogen. Other nitrate-reducing bacteria require the supply of a nitrogen source either in the form of an ammonium salt or other reduced nitrogen compounds (e.g., amino acids).

Under anaerobic conditions, nitrate can be used as the ultimate hydrogen acceptor instead of oxygen by many chemoorganotrophic bacteria. The respiration mechanism of the respective reduction processes consists in the transfer of electrons, which pass to the nitrate via nitrate reductase to produce nitrite in the last stage. Thus, during the transport of electrons, only two ATP molecules are generated, as is evident from the following scheme:

Aerobic respiration

$$[H] \longrightarrow NAD^+ \underset{ATP}{\longrightarrow} FP \underset{ATP}{\longrightarrow} Cyt\ b \longrightarrow Cyt\ c \underset{ATP}{\longrightarrow} Cyt\ a \longrightarrow O_2$$
$$\longrightarrow NO_3^-$$

Anaerobic respiration

Reduction of nitrate to nitrite can be expressed by equation:

$$NO_3^- \xrightarrow[2\,H^+;\,-H_2O]{nitrate\ reductase} NO_2^-.$$

In the presence of acetate, the reaction is carried out according to the equation:

$$CH_3COOH + 4\ NO_3^- \rightarrow 2\ CO_2 + 2\ H_2O + 4\ NO_2^-.$$

Similarly, glucose is subject to change:

$$C_6H_{12}O_6 + 12\ NO_3^- \rightarrow 6\ CO_2 + 6\ H_2O + 12\ NO_2^-.$$

Nitrite is toxic to numerous bacteria and its accumulation suppresses growth. Therefore, it is often further reduced by nitrate reductase to nontoxic products.

The reduction of nitrites to nitrates is performed by facultative anaerobic bacteria, which can be divided into two groups. The first group includes *Proteus vulgaris* and *Staphylococcus aureus*, which can reduce in aerobic conditions. However, the oxygen present does not act as a terminal electron acceptor, but as an inducer of cytochrome biosynthesis mediating electron transfer to nitrate reductase. The prostatic group of cytochromes is represented by hemin. Under anaerobic conditions, nitrate reduction is inhibited. Other bacteria, especially *E. coli* and *Pseudomonas aeruginosa*, use nitrate as the ultimate acceptor of electrons in anaerobic and aerobic conditions. Their cytochrome systems contain metallo-flavoproteins.

Reduction of oxyacid salts of selenium, bismuth, and tellurium. In addition to nitrates, some bacteria can also reduce other oxyacid salts such as selenium, bismuth, and tellurium. This ability is used in diagnostics. For example, *Corynebacterium diphtheriae* reduces tellurites to metallic tellurium, which gives the mature colonies a metallic shine. Similarly, salmonellas reduce bismuth compounds.

Denitrification refers to an anaerobic respiration process that involves the conversion of nitrate to N_2, N_2O, or a mixture of both gases. In the first stage of denitrification, nitrate is reduced by the addition of two electrons to nitrite, which can serve as the starting substrate instead of the nitrate for all denitrification bacteria. However, for the other bacteria, the nitrite at a higher concentration is toxic. The ultimate product is above all molecular nitrogen. However, under certain conditions, reductions of NO_2 may occur, producing a substantial amount of N_2O, which is further reduced to N_2:

$$2\ NO_2^- + 4[H] \rightarrow N_2O + H_2O + 2\ OH^-;$$
$$N_2O + 2[H] \rightarrow N_2 + H_2O.$$

It is therefore possible that denitrifying bacteria can either form N_2O or N_2 directly from nitrite or N_2 from N_2O. If the carbon source is acetic acid, the reduction of the nitrate can take place in two ways:

$$CH_3COOH + 2\ NO_3^- \rightarrow 2\ CO_2 + N_2O + H_2O + 2\ OH^-$$

or

$$5\ CH_3COOH + 6\ NO_3^- \rightarrow 10\ CO_2 + 4\ N_2 + 6\ H_2O + 8\ OH^-.$$

If the substrate is glucose, the denitrification process can be expressed by:

$$C_6H_{12}O_6 + 4\ NO_3^- \rightarrow 6\ CO_2 + 6\ H_2O + 2\ N_2.$$

According to the current knowledge, the denitrification process is carried out by transferring an electron pair with simultaneous changes of the oxidized form of nitrogen to the reduced form. Individual reactions are catalyzed by appropriate reductases (nitrate, nitrite, hyponitrite, and hydroxylamine reductases). These enzyme systems play an important role in the nitrogen cycle in the nature.

The method of transferring electrons in denitrification is not clear yet. Some observations show that, like oxygen, it involves cytochrome systems. The nitrate reductase involved in the reduction of nitrate is branched out from cytochrome *b*, whereas nitrite reductase and other enzymes that catalyze other reduction processes are derived from cytochrome *c*. This method of electron transfer has been observed, for example, in *Micrococcus denitrificans* and *Pseudomonas denitrificans*:

In *Pseudomonas stutzeri*, NO_2^- and NO^- reductases have almost the same properties, which lead to the formation of N_2 according to the scheme:

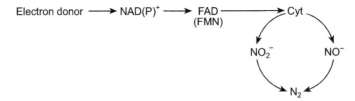

The question of linking electron transport with phosphorylation is also unclear. The formation of ATP during the transport of electrons has been found in the reduction of nitrate or nitrite, but not in the reduction of lower oxides of nitrogen (NO, N_2O) or hydroxylamine (NH_2OH). It follows that phosphorylation takes place only in the presence of an inorganic nitrogen as a final acceptor containing at least two oxygen atoms.

Denitrification by chemolithotrophic bacteria. The denitrification agents are predominantly chemoorganotrophic bacteria that require organic acids such as succinic acid, malic acid, and lactic acid as a carbon source. In addition, certain chemotrophic bacteria can also be denitrified. These include, in particular, *Thiobacillus denitrificans*, which oxidizes thiosulfate to sulfate under anaerobic conditions, while reducing the nitrate to molecular nitrogen according to the following equation:

$$2\,Na_2S_2O_3 + 8\,KNO_3 + 2\,NaHCO_3 \rightarrow 6\,Na_2SO_4 + 4\,K_2SO_4 + 2\,CO_2 + 4\,N_2 + H_2O.$$

Likewise, some hydrogen bacteria growing in a minimal environment (autotrophic) can carry out the reduction of nitrates to molecular nitrogen under anaerobic conditions. Nitrate reduction occurs with the use of molecular hydrogen:

$$5\,H_2 + 2\,NO_3^- \rightarrow N_2 + 4\,H_2O + 2\,OH^-.$$

Nitrate respiration is competitively inhibited by molecular oxygen. The competition between nitrate and oxygen in *Pseudomonas denitrificans* leads to the conclusion that nitrate reduction can only take place when the oxygen concentration is below the critical level so that enzymes requiring aerobic conditions are blocked. Depending on the relationship between nitrate reduction and dissolved oxygen tension, the reduction is not achieved if the oxygen concentration is greater than 0.2 ppm. Similarly, it is true for *Achromobacter genus*.

Denitrification plays an important role in metabolic processes in soil. This process takes place at a high speed, especially at that time. At this high speed, this process occurs especially if there is low oxygen tension. It has been found that approximately 1% of oxygen in soil suppresses denitrification by up to 12%. The main product of denitrification is the molecular nitrogen, which amounts to 83–95% of the nitrate present in the soil. NO occurs in smaller quantities (around 5%), and N_2O is the rest.

5.5.2.2. Sulfate reduction (desulfurization)

Some bacteria are characterized by the ability to use oxidized sulfur compounds as an energy source in anaerobic respiratory processes. The hydrogen donor is mostly an organic substrate, but it can also be hydrogen gas. In the first case, this is a chemoorganotrophic process, in the second case, chemolithotrophic. Sulfates, sulfites, thiosulfates, tetrathionates, etc. may be used as acceptors. The reduction of these substances to hydrogen sulfide is carried out by anaerobic bacteria, especially by representatives of *Desulfovibrio* and *Desulfotomaculum* genera.

Chemolithotrophic sulfate reduction takes place in the presence of hydrogen gas. The agents of this process are facultatively anaerobic strains of *Desulfovibrio desulfuricans*. The reaction proceeds according to the following equation:

$$4\,H_2 + SO_4^{2-} \rightarrow S^{2-} + 4\,H_2O.$$

Most strains contain cytochrome c and active hydrogenase. Energy is obtained by the oxidation of hydrogen with oxygen bound in a sulfate molecule. Sulfate reduction takes place in several stages. At first, the substrate is activated by ATP in the presence of adenyltransferase, while two molecules of inorganic phosphate are removed simultaneously. The activation is accompanied by the formation of energy-rich adenosine 5'-phosphosulfate (APS), resulting from the binding of SO_4^{2-} to AMP (**Fig. 5.29**).

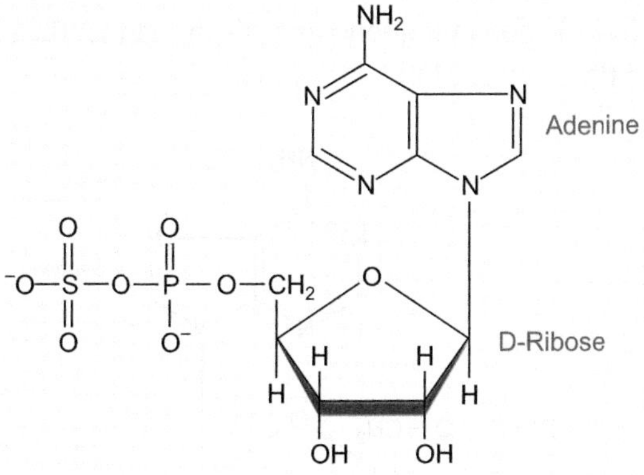

Fig. 5.29. Structure of adenosine 5'-phosphosulfate.

The reaction can be expressed by the equation:

$$SO_4^{2-} + ATP \rightarrow APS + 2\,H_2PO_4.$$

In the second step, molecular hydrogen is first oxidized with hydrogenase and cytochrome c representing low redox potential hemoprotein ($H_0 = -250$ mV). By reducing the cytochrome, sulfate is reduced to sulfite simultaneously:

$$APS + cyt\ c_{red} \rightarrow AMP + cyt\ c_{ox} + SO_3^{2-}.$$

The reaction is catalyzed by adenylyl-sulfate reductase.

The final reduction takes place in the presence of cytochrome c, hydrogenase, and sulfate reductase:

$$SO_3^{2-} + 3\,H_2 \rightarrow S^{2-} + 3\,H_2O.$$

The presence of cytochrome c appears to affect the low redox potential required for the growth of *Desulfovibrio desulfuricans*. The whole process can be expressed using a scheme comprising the reactions shown in **Fig. 5.30**.

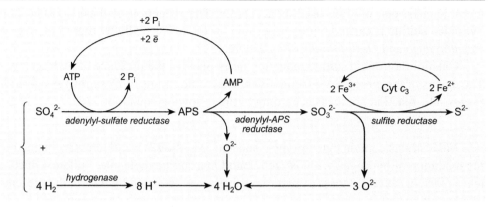

Fig. 5.30. Scheme of sulfate reduction process.

For some bacteria and yeasts, the reduction of sulfate is associated with the formation of energy-rich 3′-phosphoadenosine-5′-phosphosulfate (PAPS), which has the same function as APS (**Fig. 5.31**):

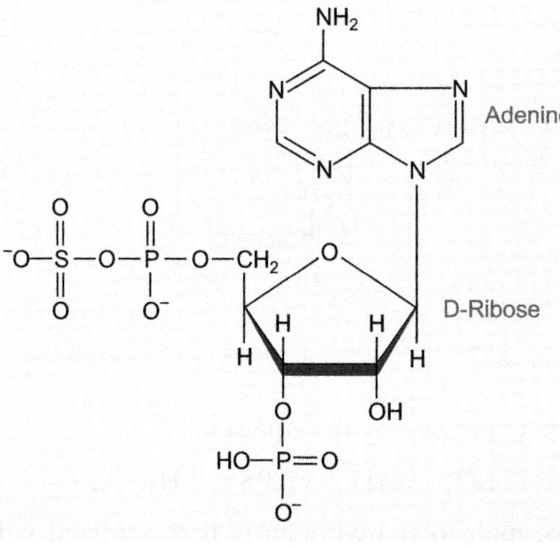

Fig. 5.31. Structure of 3′-phosphoadenosine-5′-phosphosulfate.

Unlike adenosine sulfate transferase, which catalyzes the first stage of the reaction, there is a need for another adenylyl-sulfate kinase enzyme catalyzing the transfer of the second phosphate group to APS.

$$ATP + SO_3^{2-} \rightarrow APS + 2\ P_i;$$
$$ATP + APS \rightarrow ADP + PAPS.$$

The final stage of sulfate reduction to sulfite takes place with the participation of NADPH and H^+:

$$NADPH + H^+ + PAPS \rightarrow NADP^+ + PAP + HSO_3^-.$$

This reaction is catalyzed by PAPS reductase. PAPS reduction probably occurs with dihydrolipoic acid. The expected reduction mechanism is as follows:

$$PAP-O-SO_3^- + Lip\begin{matrix} SH \\ SH \end{matrix} \rightleftharpoons PAP-OH + Lip\begin{matrix} SH \\ S-SO_3^- \end{matrix}$$

$$Lip\begin{matrix} SH \\ S-SO_3^- \end{matrix} \rightleftharpoons Lip\begin{matrix} S \\ | \\ S \end{matrix} + HSO_3^-$$

According to these two ways of sulfate reduction, it is possible to divide the originators into two groups. For the first group (*Desulfovibrio*, etc.), this process is predominantly of a dissimilatory character and serves mainly for the acquisition of energy. It is therefore carried out with the formation of hydrogen sulfide, the amount of which is proportional to the concentration of the substrate (hydrogen or organic matter).

On the other hand, the second group, including *E. coli*, *Salmonella enterica*, and yeast, uses sulfate to grow as a source of sulfur, and the products resulting from its reduction (sulfite, sulfide) are therefore incorporated into assimilation processes. Generated hydrogen sulfide here usually significantly exceeds the amount of reduced sulfate resulting from the concurrently occurring catalysis of sulfuric amino acids.

Thiosulfate reduction. In addition to sulfate, thiosulfate can also be used as an electron acceptor. Its reduction by the *Desulfovibrio* species, by the action of thioreductase and cytochrome *c*, produces sulfides and sulfites that can be further reduced to sulfides. *Thiobacillus denitrificans* anaerobically reduces thiosulfate in a different way. This is done through a specific enzymatic system of rhodanese (thiosulfate sulfur-transferase) with the participation of lipoic acid. Rhodanese, also known as rhodanase, thiosulfate sulfurtransferase, thiosulfate cyanide transsulfurase, and thiosulfate thiotransferase, is an enzyme that detoxifies cyanide (CN⁻) by converting it to thiocyanate (SCN⁻). This enzyme belongs to the family of transferases, specifically the sulfurtransferases, which transfer sulfur-containing groups. The two substrates of this enzyme are thiosulfate and cyanide, while its two products are sulfite and thiocyanate. The enzyme contains probably two pairs of spatially separated sulfur atoms, which can alternately pass between the oxidized and reduced forms. Upon reduction, one sulfur atom from the thiosulfate is transiently bound to rhodanese, from which it is transferred to lipoic acid to form lipoate persulfide. It is immediately cleaved to hydrogen sulfide and the oxidized form of lipoic acid. The whole process can be illustrated by the following scheme:

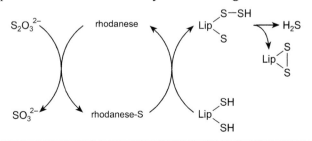

The carrier system is the same as for the reduction of sulfate. The relationship between the reduction of sulfate, thiosulfate, and sulfite can be expressed by the following scheme:

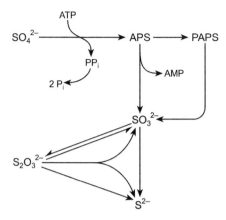

Chemoorganotrophic sulfate reduction. This process is also carried out by representatives of the *Desulfovibrio* family or some other anaerobic bacteria. It happens in the presence of a simple organic compound that also serves as a source of carbon. Lactic and pyruvic acids appear to be the most suitable hydrogen and electron donors. Less often, glucose is used, which is preferably dissimilated in the EMP or in the Entner–Doudoroff pathway under anaerobic conditions. Desulfurization in the presence of organic hydrogen and electron donors is a more frequent phenomenon than sulfate reduction with hydrogen gas. In the presence of pyruvate, its course can be expressed by the equation:

$$4\ CH_3COCOONa + 5\ MgSO_4 \rightarrow 5\ MgCO_3 + 2\ Na_2CO_3 + 5\ H_2S + 5\ CO_2 + H_2O.$$

Sometimes, the oxidation of the organic substance is not complete until it stops, but it stops on an incompletely oxidized intermediate, which then accumulates in the environment. This is the case, for example, in the dehydrogenation of lactic acid; in addition to CO_2 and H_2S, acetic acid is produced:

$$2\ CH_3CHOHCOOH + Na_2SO_4 \rightarrow 2\ CH_3COOH + Na_2S + 2\ CO_2 + 2\ H_2O.$$

During this process, pyruvate is first formed, which is converted in the presence of CoA and inorganic phosphate to acetyl phosphate, CO_2, and hydrogen. The cleavage of the phosphate by acetokinase results in the formation of acetate. At the same time, ATP used in the formation of APS is being regenerated. The different stages of the whole process are expressed by the following equations:

$$2\ CH_3–CHOH–COOH \rightarrow 2\ CH_3–CO–COOH + 4\ H^+$$

$$2\ CH_3–CO–COOH + 2\ P_i \rightarrow 2\ CH_3–CO–P + 2\ CO_2 + 4\ H^+$$

$$2\ CH_3–CO–P + AMP + 2\ H^+ \rightarrow 2\ CH_3–COOH + ATP$$

$$SO_4^{2-} + ATP + 8\ H^+ \rightarrow S^{2-} + 2\ H_2O + AMP + 2\ P_i + 2\ H^+$$

$$2\ CH_3–CH–COOH + SO_4^{2-} \rightarrow 2\ CH_3–COOH + 2\ CO_2 + 2\ H_2O + S^{2-}$$

The decarboxylation of pyruvic acid associated with the formation of acetyl phosphate, CO_2, and H_2O also involves ferredoxin, which at the same time stimulates the reduction of sulfite. Schematically, the reduction of sulfate in the presence of lactic acid is shown in **Fig. 5.32**.

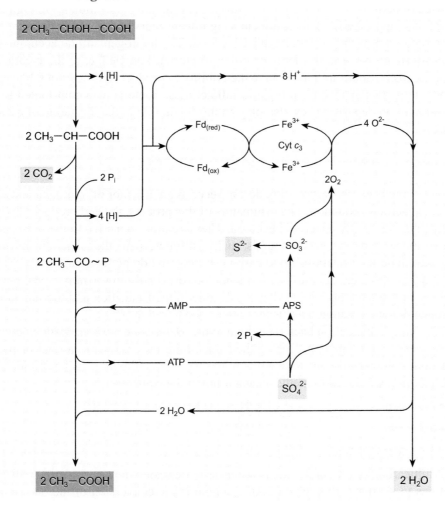

Fig. 5.32. Sulfate reduction in the presence of lactic acid by sulfate-reducing bacteria.

The energy balance here is the same as in the reduction of oxygenated sulfur compounds by molecular hydrogen. To reduce one mole of sulfate to sulfide, four electron pairs obtained by oxidizing two molecules of lactate and one mole of ATP to produce APS are used. The macroergic binding of ATP is restored at the expense of acetyl phosphate. The energy required for growth is apparently obtained by the oxidative phosphorylation associated with the transfer of electrons from lactate and pyruvate to sulfate.

If the lactate is replaced by pyruvate as the only source of energy, then *Desulfovibrio desulfuricans* grows in the absence of sulfate, while releasing hydrogen.

The transfer of electrons from pyruvate to hydrogen is done in the presence of ferredoxin, cytochrome c and hydrogenase.

Desulfurizing bacteria are widely distributed in nature both in the soil and the seas, but also in freshwater reservoirs. They are also found in hot springs and in the presence of petroleum. The resulting hydrogen sulfide often accumulates in high concentrations (up to 2 g/L). This results in a significant reduction in redox potential and thus a reduction of the growth of other bacteria. The hydrogen sulfide accumulation of desulfurizing bacteria has led, according to current knowledge, to the creation of sulfate deposits. Some authors assume their ability to produce hydrocarbons-like compounds, in particular fats, from organic sulfate; these bacteria also contributed to the formation of oil deposits. They also play an important role in the formation of mineral springs.

5.5.2.3. Carbon (IV) oxide reduction to methane

For some bacteria, carbon dioxide can serve as an acceptor of hydrogen and electrons in anaerobic oxidation. The originators of this process, the end product of which is methane, are strictly anaerobic bacteria. They are more sensitive to the presence of oxygen and certain oxygen compounds, such as nitrates, than other anaerobic bacteria. They have a definite relationship to carbon sources and do not use carbohydrates and amino acids. Products of cellulose fermentation, especially lower fatty acids, normal alcohols and isoalcohols, are used as a substrate in their nutrition and the function of hydrogen donors. Hydrogen gas is less often used to reduce CO_2.

Methane fermentation bacteria belong to the *Methanobacterium*, *Methanococcus*, and *Methanosarcina* genera. During their operation, they carry out a total or partial oxidation of the organic substrate while releasing carbon dioxide which is either completely or partly reduced to methane according to the equation:

Oxidation	$CH_3COOH + 2\,H_2O \rightarrow 2\,CO_2 + 8\,[H]$
Reduction	$CO_2 + 8\,[H] \rightarrow CH_4 + 2\,H_2O$
Molecular equation	$CH_3COOH \rightarrow CH_4 + CO_2$

Multicarbon (C_3–C_6) fatty acids are usually degraded during oxidative reduction reactions to shorter carbon chains. An example is the conversion of the butyric acid produced by *Methanococcus mazei* and the formation of acetic acid according to the equation:

$$2\,CH_3CH_2CH_2COOH + 2\,H_2O + CO_2 \rightarrow 4\,CH_3COOH + CH_4.$$

Some bacteria use lower alcohols (ethanol, methanol) as a substrate or reduce carbon dioxide to methane with hydrogen gas. *Methanobacterium omelianskii* was identified as the originator of this process. Later, however, it was found that it is not a pure culture, but a symbiosis of two anaerobic bacteria, the first of which (*Methanobacterium* sp.) produces methane from CO_2 and H_2, whereas the second, still unidentified ("S Organism") converts ethanol to acetic acid and hydrogen. The whole process can be represented by the following reactions:

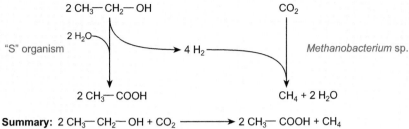

Summary: $2\ CH_3—CH_2—OH + CO_2 \longrightarrow 2\ CH_3—COOH + CH_4$

Similarly, methanol can be used as a substrate for methane formation. However, while the ethanol oxidation is fully dependent on the presence of CO_2 and stops when it is depleted, methanol is first converted to CO_2, which is reduced to methane:

$$4\ CH_3OH + 4\ H_2O \rightarrow 4\ CO_2 + 24\ [H]$$

$$3\ CO_2 + 24\ [H] \rightarrow 3\ CH_4 + 6\ H_2O$$

Molecular equation $4\ CH_3OH \rightarrow 3\ CH_4 + CO_2 + 2\ H_2O$

Carbon monoxide can also be used as hydrogen acceptor. For example, *Methanobacterium formicicum* converts CO to CO_2, which is further reduced to methane. The reaction also takes place in two stages:

$$CO + H_2O \rightarrow CO_2 + H_2$$

$$CO_2 + 4\ H_2 \rightarrow CH_4 + 2\ H_2O$$

Molecular equation $CO \rightarrow 3\ H_2 + CH_4 + H_2O$

The mechanism of processes leading to the formation of methane is not yet clear. Some reactions have shown that methane formation depends on the presence of CO_2 in certain substrates and different organisms. This is the case with the oxidation of butyric acid and ethanol. On the other hand, in other cases, methane is produced at the expense of CO_2, released during the oxidation of the organic substrate. Meanwhile, methane bacteria do not assimilate CO_2 through the Calvin cycle and lack an electron transport chain containing cytochrome. According to current ideas, an unknown carrier of single-carbon fragments (XH) is involved in methane formation. In the final stage, the methane is produced with the action of methyltransferase, which may be either methylcobalamin or **N5,N10-methylenetetrahydrofolic acid**. The reaction can be expressed in the scheme below. The energy required for the reduction process is probably provided by the ATP, one mole of which is required per every mole of methane produced. CoA is also a part of the reduction mechanism (**Fig. 5.33**).

Biological methane formation is a significant geochemical process that takes place everywhere, where organic matter is subject to anaerobic decomposition. The methane-forming bacteria thus represent final elements of the nutrient conversion chain, because they use fermentation products of other anaerobes. An important component of the overall microbial activity is also methanogenesis in the digestive tract of ruminants, where it is associated with digestion.

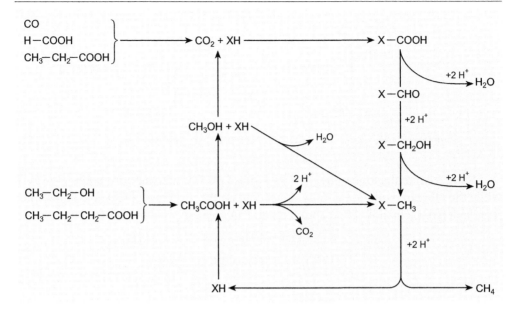

Fig. 5.33. Mechanisms of methane formation.

Thus, methyltransferase is one of the important enzymes involved in the methane formation process. Methyltransferases are a large group of enzymes that methylate their substrates but can be split into several subclasses based on their structural features.

5.5.3. Aerobic respiration in chemolithotrophic bacteria

The starting substrate, providing hydrogen and electrons in the aerobic respiration of chemolithotrophic bacteria, is an inorganic substance. The ultimate hydrogen and electron acceptor is molecular oxygen. As a carbon source, the bacteria use CO_2, which is reduced to carbon in the *Calvin cycle*.

The ability of the chemolithotrophic lifestyle of bacteria is not yet clear. According to present knowledge, the following factors can be individually or collectively involved in this phenomenon:

(1) decreased permeability for organic substances caused by the absence of specific permeases;

(2) inability to obtain energy by oxidation of organic substrates, regardless of the possibility of their entry into the cell;

(3) limited ability to synthesize compounds required for growth from substances other than CO_2. The cause may be the absence of certain enzymes (NADH + H-oxidase, α-ketoglutarate-dehydrogenase, etc.) in a number of chemolithotrophic bacteria, but also the nature of these enzymes, for example, different amino acid sequences for inorganic and organic oxidation catalysts;

(4) inhibition of growth induced by an excess of external organic nutrients in relation to CO_2 metabolism;

(5) inhibition caused by products of metabolism of organic compounds (e.g., pyruvic acid as a metabolite of glucose dissimilation strongly inhibits the growth of *Thiobacillus* genus);

(6) specific dependence of growth on intermediates of respiratory processes occurring on an inorganic substrate.

Depending on the nature of the oxidized substrate, aerobic respiration processes performed by chemolithotrophic bacteria include:

- oxidation of ammonia;
- oxidation of reduced sulfur compounds;
- oxidation of iron compounds;
- oxidation of hydrogen;
- oxidation of methane.

5.5.3.1. Oxidation of ammonia

The oxidation of ammonia is the process of so-called nitrifying bacteria. It takes two stages according to the equations:

$$2\,NH_3 + 3\,O_2 \rightarrow 2\,NO_2^- + 2\,H_2O + 2\,H^+;$$
$$2\,NO_2^- + O_2 \rightarrow 2\,NO_3^-.$$

The first stage reaction, during which ammonia is oxidized to nitrite, is carried out by the strains *Nitrosomonas* and *Nitrosococcus*. The intermediate of this reaction is hydroxylamine, which passes through nitroxyl (NOH) and nitric oxide (NO) to the nitrite according to the scheme:

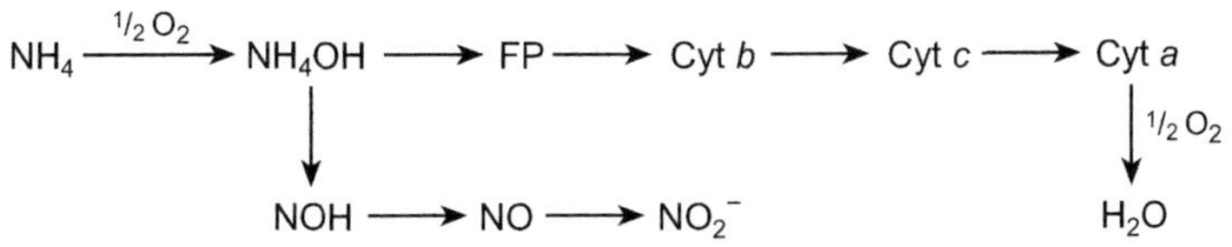

The oxidation of hydroxylamine takes place with the participation of the cytochrome oxidase system. FAD probably works as a hydrogen carrier. Oxidative phosphorylation occurs during oxidation. Due to the autotrophic metabolism, the reduction of CO_2 requires the presence of reduced NADP. Its production takes place by the back-flow of electrons from ferrocytochrome with the participation of ATP and the hypothetical energy-rich intermediate ($\sim$X) according to the equation:

$$ATP + X \rightarrow \sim X + ADP + P_i;$$
$$AH_2 + \sim X + NADP^+ \rightarrow A + NADPH + H^+ + X.$$

Compound AH_2 may be hydroxylamine. Electrons released by hydroxylamine are transmitted via the cytochrome system to oxygen. From cytochrome *c*, another branch can be separated, including the formation of reduced $NADPH + H^+$ to reduce carbon dioxide to the cell material. The required energy is provided by the ATP generated by electron transfer between cytochromes *a* and *c*.

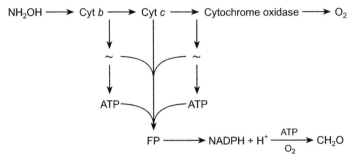

The electron backflow follows the scheme below:

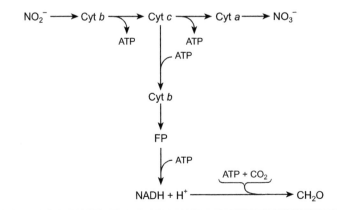

In the second stage, the nitrite is oxidized to nitrate. This process is carried out by representatives of the *Nitrobacter* genus. The oxidation is catalyzed by cytochromes and is associated with the formation of ATP:

$$NO_2 + \tfrac{1}{2} O_2 + n \, ADP + n \, P_i \rightarrow NO_3^- + n \, ATP.$$

According to today's knowledge, water provides oxygen needed to oxidize nitrite. It also serves as a hydrogen donor for NADP$^+$ reduction. The corresponding reactions are carried out according to the following equations:

$$NO_2^- + ADP + P_i + H_2O \xrightarrow{cytochromes} NO_3^- + ATP + 2 \, H^+$$

$$NADP^+ + 2 \, H^+ + 2 \, \bar{e} \xrightarrow{energy} NADPH + H^+$$

$$NADP + H^+ + ADP + P_i + \tfrac{1}{2} O_2 \rightarrow NADP^+ + ATP + H_2O$$

Summarized equation $NO_2 + 2 \, ADP + 2 \, P_i + \tfrac{1}{2} O_2 + 2 \, \bar{e} \rightarrow NO_3^- + 2 \, ATP$

The mechanism of this process is expressed in the scheme:

The released energy is partly used to reduce O_2 through $NADPH + H^+$. Enzymatic equipment shows that the energy metabolism of this organism is separate from carbon dioxide assimilation, so that the oxidation of nitrite takes place independently of the formation of reduction systems for assimilation of CO_2. In addition, the oxidation of nitrite to nitrate in this organism may be in both directions, as the cells also contain enzymes catalyzing the reduction of nitrate to nitrite. Nitrate reduction occurs after its accumulation, which results in the inhibition of oxygen consumption and thus inhibition of the further oxidation of nitrite. The resulting nitrite can be oxidized to nitrate in the recycle process according to the following scheme:

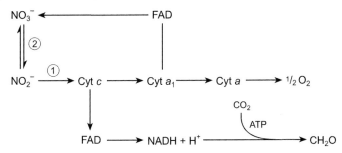

Electron transfer can be catalyzed by cytochrome c reductase (reaction 1), which is associated with phosphorylation with ATP, and nitrate reductase (reaction 2) that reduces nitrate to nitrite. Nitrifying bacteria are sensitive to the presence of organic substances in the environment, which inhibit their growth even in small concentrations. Isolation of these organisms, which are typical for soils, is therefore usually carried out in purely mineral environments, most commonly on silica gel plates.

5.5.3.2. Oxidation of reduced sulfur compounds

Many bacteria can oxidize hydrogen sulfide, thiosulfates, and other reduced sulfur compounds to sulfates. The originators of these processes are sulfuric bacteria, which mostly belong to the *Thiobacillus* genus. Some species of this genus are obligatory chemolithotrophic (*T. thiooxidans* and *T. neapolitanus*); others are facultatively chemolithotrophic or mixotrophic (*T. intermedius*, *T. novellus*). Oxidation of the reduced sulfur compounds to form sulfate as the final product can be expressed by the equations:

$$H_2S + 2\,O_2 \rightarrow H_2SO_4;$$
$$S + \tfrac{3}{2}\,O_2 + H_2O \rightarrow H_2SO_4;$$
$$Na_2S_2O_3 + 2\,O_2 + H_2O \rightarrow Na_2SO_4 + H_2SO_4.$$

Oxidation of thiosulfate occurs in some bacteria in two stages, either via tetrathionate or through elemental sulfur. Tetrathionate is formed by thiosulfate oxidation carried out by *T. thiooxidans*. The further oxidation of this intermediate is carried out by *T. neapolitanus*:

$$6\,Na_2S_2O_3 + 5\,O_2 \rightarrow 4\,Na_2SO_4 + 2\,Na_2S_4O_6;$$
$$2\,Na_2S_4O_6 + 7\,O_2 + 6\,H_2O \rightarrow 2\,Na_2SO_4 + 6\,H_2SO_4.$$

Sulfur as an intermediate is oxidized by thiosulfate (*T. thioparus*). Elemental sulfur is further converted to sulfate:

$$5\ Na_2S_2O_3 + 4\ O_2 + H_2O \rightarrow 5\ Na_2SO_4 + H_2SO_4 + 4\ S;$$

$$2\ S + 3\ O_2 + 2\ H_2O \rightarrow 2\ H_2SO_4.$$

Due to the existence of the permeation barrier, some processes of sulfur metabolism appear to be occurring on the cell surface. These processes result in the formation of thiosulfate, the transmission of which through the cytoplasmic membrane occurs through reduced glutathione. Within the cell, the thiosulfate is finally oxidized to sulfate. A key intermediate of the oxidation of thiosulfate is the sulfite, the conversion of which is carried out in two ways. The first of them is associated with phosphorylation for the formation of APS and ATP according to the equations:

$$2\ S_2O_3^{2-} + 4\ H^+ + 4\ \bar{e} \rightarrow 2\ SO_3^{2-} + 2\ H_2S$$
$$2\ H_2S + O_2 \rightarrow 2\ S + 2\ H_2O$$
$$2\ SO_3^{2-} + 2\ AMP \rightleftharpoons 2\ APS + 4\ \bar{e}$$
$$2\ APS + 2\ P_i \rightleftharpoons 2\ ADP + 2\ SO_4^{2-}$$
$$2\ ADP \rightleftharpoons ATP + AMP$$

Summarized equation $2\ S_2O_3^{2-} + AMP + O_2 + 2\ P_i + 4\ H^+ \rightarrow 2\ S + 2\ SO_4^{2-} + ATP + 2\ H_2O$

The cleavage of thiosulfate to hydrogen sulfide occurs with the effect of sulfur-transferase; oxidation of hydrogen sulfide to elemental sulfur takes place by the sulfide oxidase. The formation of APS is catalyzed by adenyl sulfate transferase. ATP is produced by phosphate transferase at the substrate level at one mole per mole of sulfate. Elemental sulfur formed by the oxidation of hydrogen sulfide is transformed to sulfate by the effect of oxygenase, the coenzyme of which is reduced glutathione oxidized to sulfite, as follows:

$$S + O_2 + H_2O \rightarrow H_2SO_3.$$

In the second pathway, the sulfite is directly converted to sulfite by sulfite oxidase. This direct oxidation step is associated with oxidative phosphorylation. Schematically, the conversion of thiosulfate to sulfate is illustrated in the scheme:

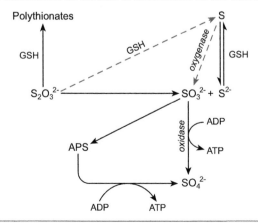

Overall, the metabolism of sulfur in nature can be represented by the following scheme:

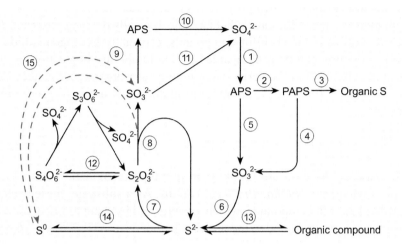

This scheme includes dissimilatory processes carried out by desulfurizing bacteria (*Desulfovibrio desulfuricans*) — reactions 1, 5, 6, processes of assimilation desulfurization — reactions 2, 3, 4 carried out by, for example, some enterobacteria, and, finally, oxidation of reduced sulfur compounds carried out by representatives of *Thiobacillus* genera — reactions 7–12, 13, and 14.

For autotrophic CO_2 fixation, thiobacilli use reduced pyridine nucleotides that are generated by the backflow of electrons, the same as other chemolithotrophic bacteria do. The required energy is supplied by ATP generated in the phosphorylation pathway. The formation of NAD(P)H and its introduction into the CO_2 reduction process can be illustrated by the following scheme:

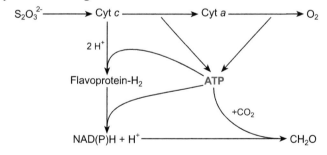

Some *Thiobacillus* species (*T. thioparus*) may also use organic compounds such as glucose, acetic acid, aspartic acid, and glutamic acid but do not grow in the absence of thiosulfate. Organic substrates are assimilated by these bacteria and used as a carbon source only in the chemolithotrophic way of obtaining energy, at the expense of thiosulfate oxidation.

Other species, such as *Thiobacillus novellus*, can also use organic substances in the absence of an inorganic hydrogen and electron donor. Both representatives are the so-called facultative chemolithotrophic organisms.

The ability of thiobacilli to withstand a very low pH environment (e.g., *Thiobacillus thiooxidans* cells can live in 1 N H SO solution) is used in the treatment of highly of highly alkaline soils, to which powdered sulfur is added.

In addition to thiobacilli, some other bacteria, typically the *Beggiatoa* and *Thiothrix* species, can oxidize reduced sulfur compounds, especially hydrogen sulfide. Hydrogen sulfide oxidation can be accompanied by the temporary deposition of elemental sulfur in the cells in these bacteria. They mainly use organic substances as carbon sources. The possibility of CO_2 autotrophic fixation has not yet been confirmed in these organisms.

5.5.3.3. Oxidation of iron compounds

Iron-oxidizing bacteria are involved in this type of oxidation, among the most important of which are *Thiobacillus ferrooxidans*. As a substrate, these bacteria most often use divalent iron compounds, which are oxidized by them to iron hydroxide. The released energy is partly bound to the macroergic bonds of ATP. *Thiobacillus ferrooxidans* emit energy by oxidizing ferric carbonate according to the equation:

$$4 \, Fe^{II}CO_3 + O_2 + 6 \, H_2O \rightarrow 4 \, Fe^{III}(OH)_3 + 4 \, CO_2.$$

Cytochromes of type *b*, *c*, and *a* have been found in the cells of these bacteria. The oxidation of iron is linked to an energy source that allows the formation of iron–oxygen complex $\left(\begin{smallmatrix} \text{Complex} \\ \diagdown \diagup \\ Fe-O- \end{smallmatrix} \right)$. This process takes place on the surface of the cell. During the reaction of the complex with Fe-oxidase, the electron transferred to the cell via flavoprotein is released. In the cell, the electron passes through cytochrome *c* and cytochromidase to oxygen. Sulfate can also act as the intermediate stage of the transfer. The transfer of electrons to cytochrome *c* is believed to be associated with phosphorylation, as in nitrifying bacteria. *Thiobacillus ferrooxidans* can also oxidize reduced sulfur compounds. However, this process is much slower than the oxidation of iron.

The formation of reducing equivalents needed to reduce CO_2 is allowed, as in the case with other chemolithotrophic bacteria, by reversing the flow of electrons.

More recent findings show that *T. ferrooxidans* can also use some glycides and amino acids as hydrogen and electron donors, after a short adaptation in glucose; therefore, they are rather facultative chemolithotrophs.

Some bacteria have the ability to oxidize iron and manganese compounds to form colloidal inclusions. Representatives of these bacteria are *Sphaerotilus natans* and *Leptothrix discophora*. The deposition of iron hydroxide on the surface of the fiber is characteristic of *Gallionella* representatives. The formation of $Fe(OH)_3$ can also occur in a nonspecific way in the presence of free trivalent iron ions. Sometimes, these ions can chelate with bacterial cell substances or bundles and thus induce their incrustation.

Thus, the process of oxidation of iron compounds occurs with the change of divalent iron to trivalent one with the formation of its hydroxide. This process takes place on the surface of the bacterial cell. Bacteria *T. ferrooxidans* are the main representative in this process.

5.5.3.4. Oxidation of hydrogen

Hydrogen-oxidizing bacteria were previously included in the genus *Hydrogeno-monas*, but later, with respect to the facultative chemoautotrophic way of life, they were differentiated into representatives of the *Alcaligenes*, *Paracoccus*, and *Nocardia* genera. These bacteria are obligatory aerobic, chemolithotrophic, and chemoorgano-trophic. Their cells contain hydrogenase enzyme, allowing the formation of reduced NAD. As to the composition, it is a metallo-flavoprotein hydrogenase. In addition, there is a cytochrome system in the cells allowing the transmission of electrons associated with oxidative phosphorylation.

In the cells of some hydrogen oxidizing bacteria, two hydrogenases with different functions were found. The first transmits hydrogen and electrons to the transport chain and through it to oxygen. This process is associated with phosphorylation and ATP production. The second enzyme functions as NAD dehydrogenase and catalyzes the transfer of electrons to oxygen directly through reduced FAD without ATP production. The transfer is accompanied by the formation of hydrogen peroxide or by equations:

$$NADH + H^+ + O_2 \rightarrow NAD^+ + H_2O_2;$$
$$NADH + H^+ + \tfrac{1}{2} O_2 \rightarrow NAD^+ + H_2O.$$

Hydrogen peroxide is decomposed by the catalase present.

The oxidation of hydrogen by hydrogen oxidizing bacteria can be expressed by the following equations:

$$H_2 \rightarrow 2 H^+ + 2 \bar{e};$$
$$2 H^+ + 2 \bar{e} + \tfrac{1}{2} O_2 \rightarrow H_2O.$$

In terms of energy, one mole of H_2 yields two ATPs. This energy is fully utilized to reduce O_2, since, due to the presence of hydrogen, there is no production of reduced NAD(P) by the backflow of electrons with the participation of ATP.

Most hydrogen-oxidizing bacteria belong to facultative chemolithotrophic organisms. They can therefore also use organic compounds, such as acetic acid, pyruvic acid, glucose, and fructose, as a carbon source. β-Hydroxybutyric acid is a common metabolic product under these conditions, and it can be further metabolized in the absence of a carbon source. This polymer can also be produced autotrophically from CO_2.

Some hydrogen-oxidizing bacteria have the ability to use uric acid and allantoin as the only source of nitrogen in heterotrophic growth. Similarly, some bases and amino acids can serve as a source of nitrogen in the presence of hydrogen.

Since hydrogen and oxygen are easily obtained by electrolysis, hydrogen can be used to regenerate air in enclosed spaces, for example spacecraft. Carbon dioxide contained in air and exhaled by human is assimilated by bacteria with the simultaneous oxidation of hydrogen and used for biosynthesis.

Thus, the process of hydrogen oxidation occurs with the participation of dehydrogenases allowing the formation of reduced NAD. The hydrogen-oxidizing bacteria are mainly facultative chemolithotrophic organisms.

5.5.3.5. Oxidation of methane

The species of the genus *Methylomonas* are considered chemolithotrophic, and representatives of the genus *Hyphomicrobium* live a heteroorganotrophic way of life. The oxidation of methane proceeds according to the equation:

$$CH_4 + 2\ O_2 \rightarrow CO_2 + H_2O.$$

The process consists of several intermediate steps. First, methane is oxidized to methanol and then formaldehyde. By oxidation, the latter provides formic acid, which is decarboxylated to CO_2.

$$CH_4 \xrightarrow{+\frac{1}{2}O_2} CH_3OH \xrightarrow{-2[H]} HCHO \xrightarrow{+\frac{1}{2}O_2} HCOOH \xrightarrow{-2[H]} CO_2.$$

Both chemolithotrophic and chemoorganotrophic methane-oxidizing bacteria carry out this oxidation up to the point of formaldehyde formation. Chemolithotrophs further incorporate formaldehyde into ribose 5-phosphate to form allulose 6-phosphate along a Calvin-like pathway. Optionally, methylotrophic bacteria bind acetaldehyde to glycine to form serine and then 2-phosphoglyceric acid via the so-called serine pathway.

Unlike species of the genus *Methylomonas*, the bacterium belonging to the genus *Hyphomicrobium* oxidizes formaldehyde heterotrophically to formic acid in the presence of tetrahydric acid. The course of this oxidation can be expressed by equations:

$$H{-}CHO + THF \rightarrow N5,N10\text{-methylenetetrahydrofolate};$$

$$N5,N10\text{-methylenetetrahydrofolate} + NADP^+ \rightarrow$$
$$N5,N10\text{-methenyltetrahydrofolate} + NADPH + H^+;$$

$$N5,N10\text{-methenyltetrahydrofolate} + H_2O \rightarrow N10\text{-formyltetrahydrofolate} + H^+;$$

$$N10\text{-formyltetrahydrofolate} + ADP + P_i \rightarrow \text{tetrahydrofolate} + HCOOH + ATP.$$

The formation of ATP is likely to occur at the level of cytochrome *c* at the expense of energy released in the reaction's intermediate step of methanol oxidation. As to the carbon source, the evidence suggests that preferably methane oxidizing bacteria assimilate reduced single-carbon intermediates by the oxidation of methane via CO_2. This type of metabolism shows that methane-oxidizing bacteria belong to chemoorganotrophic rather than chemolithotrophic kind.

Thus, the process of methane oxidation occurs through formaldehyde formation as a main intermediate with its oxidation to formic acid. The final product of this process is CO_2. Tetrahydrofolic acid is one of the important metabolic components in the methane oxidation process.

5.5.4. Aerobic respiration in chemoorganotrophic bacteria

Most bacteria acquire the necessary energy for life processes by oxidizing an organic substrate. The transmission of hydrogen and electrons takes place during this oxidation via a complete transport chain to oxygen, which combines with hydrogen to water. Phosphorylation reaction leading to the formation of ATP is a part of the oxidation, in which the organic substrate is oxidized to CO_2.

5.5.4.1. Incomplete oxidation of substrate

Some bacteria oxidize organic substrate incompletely, that is, not down to CO_2, but only to the formation of a particular metabolite that can accumulate in the environment. The originators of these processes are mainly the *Acetobacter* and *Gluconobacter* genera. Preferably, they utilize alcohols and carbohydrates.

Incomplete oxidation of alcohols. Primary alcohols, when incompletely oxidized, are converted to the corresponding fatty acids. The oxidation reaction is carried out through an aldehyde. The most notable process is acetic acid production carried out by acetic bacteria *Acetobacter aceti, A. suboxidans*, etc. in the presence of NAD dehydrogenase:

$$CH_3\text{—}CH_2\text{—}OH \xrightarrow{\quad NAD^+ \quad\quad NADH + H^+ \quad} CH_3\text{—}CHO$$

$$CH_3\text{—}CHO + {}^1/_2O_2 \longrightarrow CH_3\text{—}COOH$$

This process is used in the production of vinegar.

Secondary alcohols can be oxidized to ketones:

$$CH_2OH\text{–}CHOH\text{–}CH_2 \xrightarrow{\;-2\,H\;} CH_2OH\text{–}CO\text{–}CH_2.$$

Similarly, multivalent alcohols are oxidized to ketoses, for example, mannitol to fructose, sorbitol to sorbose.

Incomplete oxidation of glucose. In addition to alcohols, acetic bacteria can also use glucose as a substrate to form gluconic acid (**Fig. 5.34**). The *Gluconobacter* genus is well known for this ability. The oxidation of glucose takes place along the **h**exose **m**onophosphate (HMP) pathway. However, it can also be done without the formation of phosphates.

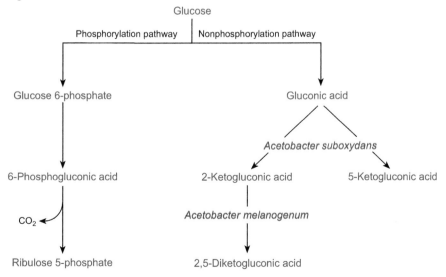

Fig. 5.34. Incomplete oxidation of glucose.

Gluconic acid can be further converted by acetic bacteria to ketogluconic acids. 2- or 5-ketogluconic acid is formed by *Acetobacter suboxide*; 2,5-diketogluconic acid is formed by *A. melanogenum*. Due to the fact that under certain conditions, such as intensive aeration, the complete oxidation of the substrate can occur, it is obvious that the bacteria of both genera, that is, *Acetobacter* and *Gluconobacter*, can form all enzymes needed for terminal oxidation, including cytochromes.

5.5.4.2. Complete oxidation of substrate

The processes of complete oxidation of the organic substrate are carried out by aerobic chemoorganotrophic bacteria generally equipped with enzymatic systems that allow the transfer of hydrogen and electrons to molecular oxygen to form water. The organic substrate, which undergoes decomposition to full oxidation, is mostly used not only as a source of energy but also as a carbon source.

Tricarboxylic acid cycle. The tricarboxylic acid (TCA) cycle is also known as the Krebs cycle. This cycle represents the major metabolic pathway in aerobic respiration processes. The starting compound entering this metabolic pathway is acetyl-CoA, so-called activated acetic acid, resulting from the oxidative decarboxylation of pyruvic acid with CoA, thiamine pyrophosphate (TPP), and lipoic acid (**Fig. 5.35**).

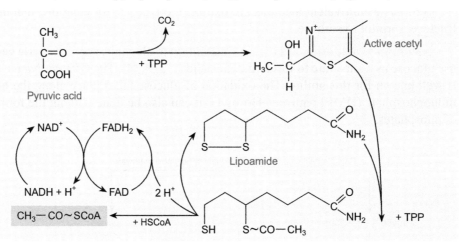

Fig. 5.35. Activation of acetic acid.

By pyruvic acid, the cycle can be linked to other metabolic pathways, in which the acid is formed (e.g., glycolysis, the HMP pathway, the ED pathway).

The Krebs cycle is initiated by the synthesis of citrate synthetase by the condensation of acetyl-CoA with oxaloacetic acid to citric acid. This is converted by isomerization to isocitric acid with the formation of *cis*-aconitic acid. The balance between these acids is maintained by the enzyme aconitase. Isocitric acid is dehydrogenated to oxalic acid which, when decarboxylated, provides α-ketoglutaric acid. Both reactions are catalyzed by isocitric acid dehydrogenase.

The chain of tricarboxylic acids ends with the decomposition of oxalosuccinic acid; dicarboxylic acids enter the cycle. The first one, α-ketoglutaric acid, is decarboxylated in the presence of LTPP (lipothiamine pyrophosphate) and CoA and, being dehydrogenated, converted to succinyl-CoA and further to succinic acid. The course of the reaction is the same as for the oxidative decarboxylation of pyruvic acid. The released energy is used for the phosphorylation of GDP to GTP. Succinic acid with flavoprotein-containing dehydrogenase is converted to fumaric acid, which is further hydrated to malic acid by the enzyme fumarase. By the catalytic effect of malate dehydrogenase, the coenzyme of which is NAD^+, malic acid is dehydrogenated to oxaloacetic acid, which returns to the cycle. Some bacteria, such as *Azotomonas* sp., *Micrococcus luteus*, *Serratia marcescens*, and *Pseudomonas fluorescens*, can carry out malic acid oxidation in the presence of enzyme systems other than NAD malate dehydrogenase. An overview of the reactions in the citric acid cycle is given in the diagram in **Fig. 5.36**.

The amount of energy released per one mole of acetyl-CoA is about 904.3 kJ. During the cycle, around 12 ATP type macroergic bonds containing about 351.7 kJ are generated. The formation of ATP occurs in the following reactions:

- oxidation of $NADH + H^+$ (dehydrogenation of isocitric,
  α-ketoglutaric, and malic acids) = 3 $NADH + H^+$ 9 ATP
- oxidation of $FADH_2$ (dehydrogenation of succinic acid)   2 ATP
- phosphorylation of GDP to GTP
  (cleavage of succinic acid to succinyl-CoA)   1 ATP

Thus, the **total yield** is   12 ATP

Because in the complete oxidation of acetic acid (acetyl-CoA), ΔG^0 = -904.3 kJ/mol, the amount of energy bound in macroergic bonds accounts for approximately 40% of the total amount of energy released in this metabolic pathway. When considering complete glucose oxidation, the number of ATP moles will correspond to:

- oxidation of 2 $NADH + H^+$ produced during glycolysis   6 ATP
- phosphorylation of ADP at substrate level   2 ATP
- oxidation of 2 $NADH + H^+$ resulting from decarboxylation
  of pyruvic acid   1 ATP
- oxidation of acetyl-CoA in the Krebs cycle   24 ATP

Total yield   **38 ATP**

Even in this case, the overall efficiency is 40% (in glucose oxidation, ΔG^0 = -2872 kJ/mol).

The meaning of the Krebs cycle is that, in addition to the complete oxidation of the bicarbonate metabolite in the system by the conversion of tricarboxylic and dicarboxylic acids to carbon dioxide and water, it provides building units for the biosynthesis of various products. This makes it an important link in the metabolism of vital substances such as saccharides, lipids, and amino acids. In addition, it provides a considerable amount of energy in the form of macroergic bonds to provide biosynthetic

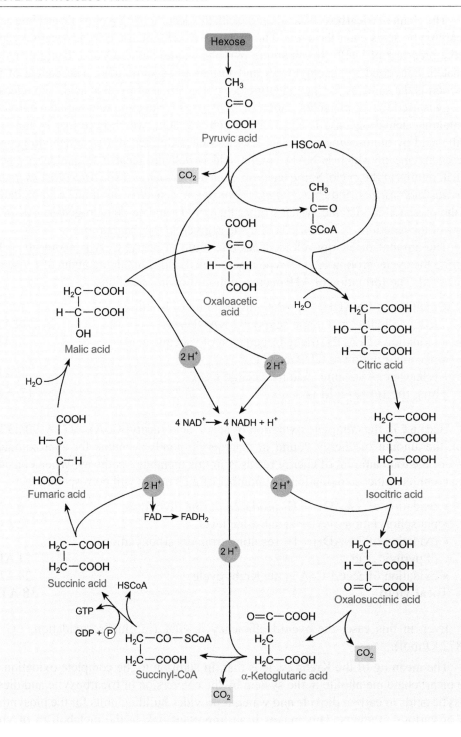

Fig. 5.36. Tricarboxylic acid cycle.

processes. This metabolic pathway has a general application in terminal oxidation processes in animal and plant organisms. It is also present in most heterotrophic aerobic bacteria, for example, *Azotobacter vinelandii*, *Micrococcus luteus*, *E. coli*, and *Klebsiella aerogenes*.

In addition to the cycle of tricarboxylic acids, alternative metabolic pathways can be applied to the substrate oxidation by bacteria. Most often, it is a cycle of dicarboxylic acids and glyoxalic acid.

Dicarboxylic acid cycle. The cycle of dicarboxylic acids represents a shortened metabolic pathway lacking the chain of tricarboxylic acids. It is also referred to as the Thunberg–Knoop cycle (**Fig. 5.37**).

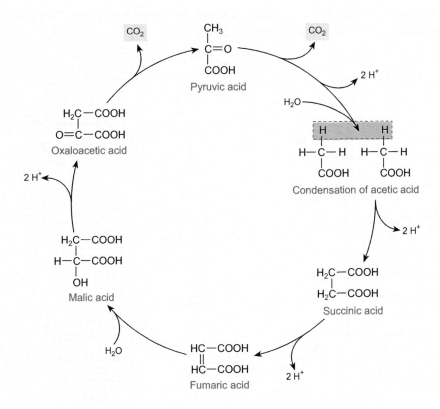

Fig. 5.37. Dicarboxylic acid cycle.

The starting compound is pyruvic acid. Its oxidative decarboxylation with CoA produces activated acetic acid (acetyl-CoA), which condenses to the succinic acid under the catalytic effect of NAD^+ dehydrogenase. The next reaction sequence until oxalic acid formation is the same as in the Krebs cycle. Oxalic acid is decarboxylated to pyruvic acid, which returns to the cycle.

The oxidation of pyruvic acid in the dicarboxylic acid cycle is carried out, for example, by *E. coli*, *Achromobacter globiformis* and *Klebsiella aerogenes* grown in the presence of acetate. The existence of this metabolic pathway does not exclude the ability of certain bacteria to oxidize pyruvic acid in the Krebs cycle.

Cycle of glyoxylic acid. This metabolic pathway is distinct from the citric acid cycle by isocitric acid not being converted to oxalosuccinic acid but cleaved by isocitrate synthase to succinic acid and glyoxalic acid. While the further conversion of succinic acid proceeds according to the Krebs cycle, glyoxalic acid is condensed with malate synthase and acetyl-CoA to malic acid. By dehydrogenation of this acid catalyzed by the malate dehydrogenase, oxaloacetic acid is produced and returns to the cycle.

The importance of this metabolic pathway lies in the fact that the increased production of malic and oxaloacetic acids provides enough substrate for further condensation reactions occurring during the biosynthesis of various metabolic products. The glyoxalic acid cycle has been described in some species of the genera *Pseudomonas* and *Xanthomonas*. Its course is shown in **Fig. 5.38**.

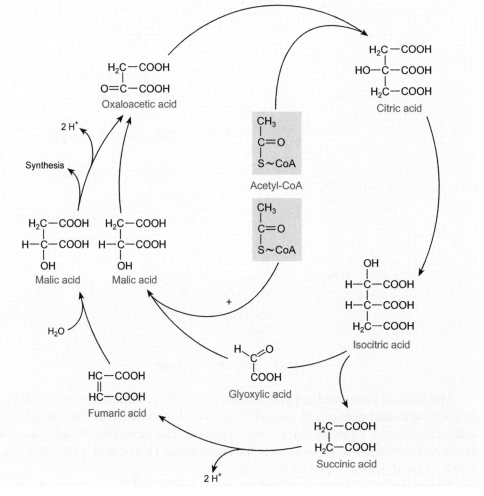

Fig. 5.38. Glyoxalic acid cycle.

5.5.4.3. Oxidation of saccharides and polysaccharides

Pentose phosphate pathway. In aerobic conditions, some bacteria can convert hexoses in the metabolic pathway referred to as the pentose phosphate cycle or the hexose monophosphate pathway (HMP). This type of metabolism is known especially in facultative anaerobic organisms such as *E. coli* and *K. aerogenes* (**Fig. 5.39**).

The pentose cycle is initiated by the phosphorylation of glucose to glucose 6-phosphate. It is further oxidized in the presence of glucose 6-phosphate dehydrogenase to 6-phosphogluconate, from which, after hydration with gluconolactonase, 6-phosphogluconic acid is formed. By the effect of phosphogluconate dehydrogenase, 6-phosphogluconic acid undergoes oxidative decarboxylation to form ribulose 5-phosphate, which is converted into two other pentoses: by the action of ribose 5-phosphate isomerase to xylulose 5-phosphate and by the action of ribose 5-phosphate isomerase to ribose 5-phosphate. The action of transketolase in the presence of TPP then converts the ketol group (active glycolaldehyde) from xylulose 5-phosphate to ribose 5-phosphate to form sedoheptulose 7-phosphate and glyceraldehyde 3-phosphate. In the next reaction, transketolase is involved in the transfer of the thiocarbon group from sedoheptulose to glyceraldehyde 3-phosphate to form erythrose 4-phosphate and fructose 6-phosphate. By converting fructose 6-phosphate to glucose 6-phosphate, the cycle closes. Erythrose 4-phosphate again receives a keto group and other xylulose 5-phosphate molecules through thus obtained transketolase, 3-β-glyceraldehyde, and fructose 6-phosphate return to the cycle. In summary, it is possible to express hexose conversion in this pathway by the equation:

$$C_6H_{12}O_6 + 12 \ NADP^+ + 6 \ H_2O + ATP \rightarrow 6 \ CO_2 + 12 \ NADPH + H^+ + P_i + ADP.$$

Aerobic bacteria can often oxidize saccharides in the incomplete HMP pathway by converting glyceraldehyde 3-phosphate to pyruvic acid by using the HMP pathway enzymes. Glucose oxidation is then carried out according to the equation:

$$3 \ \text{glucose 6-P} + 6 \ NADP^+ + ATP \rightarrow$$
$$2 \ \text{fructose 6-P} + 3\text{-P-glyceraldehyde} + 3 \ CO_2 + 6 \ NADPH + H^+ + ADP + P_i.$$

Pyruvic acid is then oxidized in the Krebs cycle. In this way, these bacteria acquire energy by producing ATP on the substrate level (2 ATP) and by active phosphorylation in the transport electron chain.

The importance of hexose monophosphate pathway is as follows:

(1) Unlike glycolysis, hydrogen transfer occurs in NADP containing dehydrogenases. The reduced form of these coenzymes is of no importance in the energy metabolism, as it does not enter the oxidation chain. However, it can be oxidized during the hydrogenating processes characteristic of biosynthesis.

(2) Important features of HMP are the formation of pentoses necessary for the synthesis of nucleic acids, coenzymes, and a number of other important compounds and the assimilation of CO_2 in photo- and chemolithotrophic bacteria incorporating this carbon source into ribulose 2-phosphate, which is created from ribulose 5-phosphate.

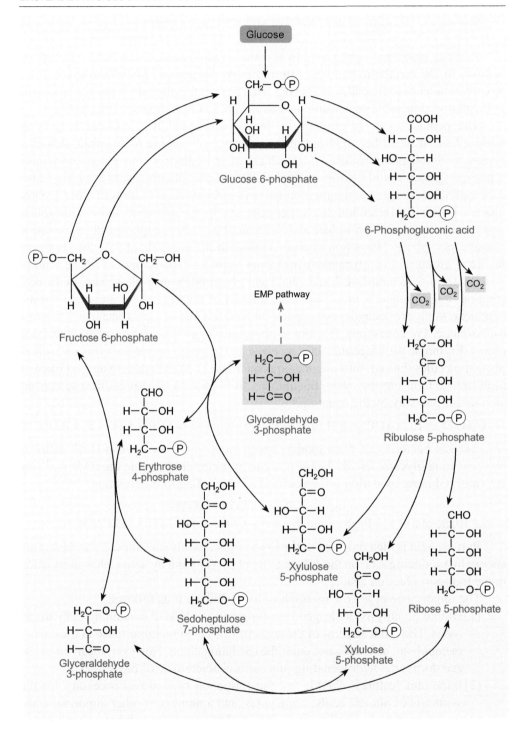

Fig. 5.39. Hexose monophosphate pathway.

(3) An important place in this metabolic pathway is finally taken by the formation of sedoheptulose, which is believed to function as a precursor of shikimic acid, which is the starting substrate for the biosynthesis of aromatic compounds.

Oxidation of polysaccharides. Under aerobic conditions, polysaccharides are oxidized in the presence of the respective enzymes to monomer units, which can be further oxidized in the pentose cycle or in the Entner–Doudoroff pathway to pyruvic acid entering the Krebs cycle.

Starch and *glycogen* are cleaved by amylases that are produced by a number of bacteria. The aerobic formation of α-amylase, catalyzing the cleavage of bonds 1–4 from the middle of the chain, has been demonstrated primarily in the *Bacillus* (*B. subtilis*, *B. cereus*) and *Pseudomonas* genera. β-Amylase enzyme, cleaving individual fragments (maltose with a reducing group) from the end of the polysaccharide chain, occurs less frequently in bacteria. The cleavage properties of both enzymes are present in the amylase of *Bacillus polymyxa* strains.

Cellulose is aerobically decomposed by the *Cytophaga* and *Cellulomonas* genera and some actinomycetes. Enzyme cellulose produced by these bacteria has an analogous effect to that of β-amylase, thus sequentially removing the cellobiose disaccharide from the end of the cellulose chain. The aerobic decomposition of cellulose by bacteria has a significant contribution to processes of degradation of plant residues in soil.

Xylans found in straw and phloem (15–20%) and in the wood of conifers and deciduous trees belong to hemicelluloses. Their aerobic decomposition to xylose and xylobiose or other fragments is induced by numerous bacteria via the enzyme xylanase (e.g., *Sporocytophaga myxococcoides*).

Pectins oxidize under aerobic oxidation conditions mainly by the action of the *Bacillus subtilis*, *Erwinia carotovora*, and *Pseudomonas fluorescens* strains. Like the anaerobic decomposers of pectin substances, these bacteria are also used to release phloem fibers during retting of flax.

5.5.4.4. Oxidation of lipids

This process is carried out by so-called lipolytic bacteria, including *Pseudomonas fluorescens*, *P. aeruginosa*, *Serratia marcescens*, *Corynebacterium bovis*, and some *Mycobacterium* species. All these bacteria form lipase enzyme, which breaks down fats into their constituents, that is, glycerol and fatty acids. The oxidation of glycerol may take place via pyruvic acid through the EMP pathway and through the Krebs cycle to CO_2 and water. Fatty acids are metabolized in processes known as β- and ω-oxidation. During the β-oxidation, a two-carbon fragment in the form of acetyl-CoA is sequentially cleaved out. At first, the acyl-CoA is formed with the involvement of ATP and CoA. The actual oxidation process is initiated by the dehydrogenation of acyl-CoA, which is converted into an unsaturated form by the formation of a double bond in the α and β positions. The reaction is carried out with FAD-containing acyl-CoA-dehydrogenase. The unsaturated form of acyl-CoA is then hydrated at the double bond sites by the appropriate hydratase to form the hydroxy group at the β position which is hydrogenated in the next reaction to a keto group. Acetyl-CoA is cleaved out by the effect of β-ketolase enzyme and the rest of the fatty acid starts a new cycle. This is illustrated in **Fig. 5.40**.

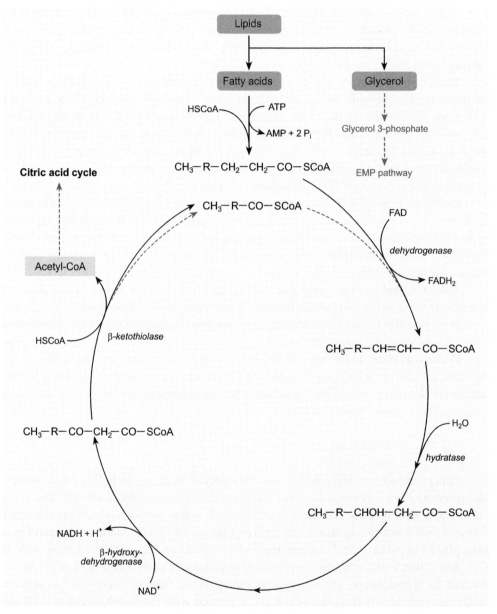

Fig. 5.40. β-Oxidation of fatty acids.

In the ω-oxidation of fatty acids, dicarboxylic acids are formed. During the process, ω-hydroxybutyric acids are formed as intermediates. The hydroxylation reaction is associated with NADPH + H⁺ in addition to the respective enzymes; the oxidation of ω-hydroxybutyric acid to dicarboxylic acid requires NAD^+. However, this process is of secondary importance.

Similarly, the fatty acids released during the degradation of bacterial waxes and phospholipids are oxidized. Most of these substances are resistant to bacterial action and can be only decomposed by specific bacterial species. The course of these oxidation processes is not fully understood.

5.5.4.5. Oxidation of hydrocarbons

Hydrocarbons are organic compounds consisting entirely of hydrogen and carbon. These compounds, from which one hydrogen atom was removed, have functional groups called hydrocarbyls. Because carbon has four electrons in its outermost shell (and because each covalent bond requires a donation of one electron per atom to the bond), carbon has exactly four bonds to make and is only stable if all the four bonds are used. The different types of hydrocarbons are aromatic hydrocarbons (arenes), alkanes, cycloalkanes, and alkyne-based compounds.

The oxidation of aliphatic and aromatic hydrocarbons is provided by species of the *Pseudomonas* genus (*P. fluorescens*) and some species of the *Mycobacterium* genus.

Alkanes (aliphatic hydrocarbons) are dissimilated in the process called **mono-terminal** or **diterminal oxidation**. Both processes are provided by species of the *Pseudomonas* genus. During monoterminal oxidation, first carbon is attacked and primary alcohol is formed and then transformed through aldehyde to a fatty acid. The fatty acid undergoes further α-oxidation with the formation of acetic acid. The reaction is catalyzed by NAD dependent dehydrogenase:

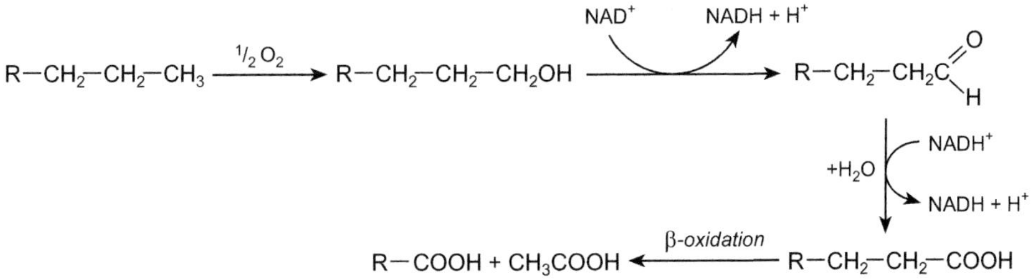

Diterminal oxidation is a process, in which any end of the chain can be oxidized with the formation of an acid mixture, for example, the oxidation of 2-methylhexane with the formation of 5-methylhexanoic and 2-methylhexanoic acids by *Pseudomonas aeruginosa*.

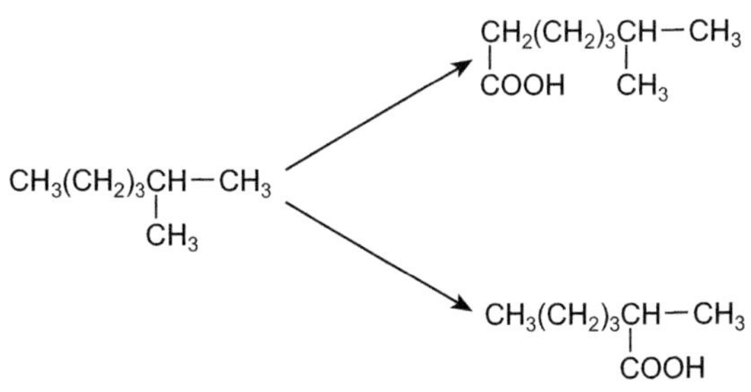

In general, the oxidation of alkanes can be presented by the following schema:

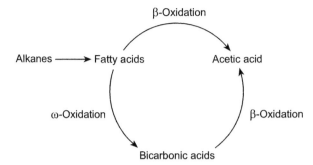

Aromatic hydrocarbons can also be oxidized by species of the *Pseudomonas* genus. The system of oxygenases, which is included to the oxidation process in bacteria, catalyzes incorporating one (monooxygenases) or two (dioxygenases) atoms of oxygen to the benzene core during bond cleavage between two adjacent carbon atoms.

The main intermediates of the oxidative dissimilation of aromatic compounds are catechol and protocatechuic acid. Both compounds are then cleaved in the ortho- or meta-position. Catechol, also known as pyrocatechol or 1,2-dihydroxybenzene, is an organic compound with the molecular formula $C_6H_4(OH)_2$. It is the ortho-isomer of three isomeric benzenediols.

The ***ortho-cleavage*** of catechol formed during the oxidation of various aromatic compounds (e.g., benzoic, mandelic, and anthranilic acids, phenols, naphthalene, phenanthrene) occurs between two hydroxyl C-atoms by catechol dioxygenase into *cis,cis*-muconic acid:

This acid is transformed by muconolactonase to muconolactone that is transformed to β-ketoadipic acid, which together with CoA provides the subsequent formation of succinic and acetic acids in the final phase. Succinic and acetic acids are subsequently oxidized in the tricarboxylic acid cycle. Protocatechuic acid has a similar way of ortho-cleavage and is formed by the conversion of cresol and *p*-hydroxybenzoic, phthalic, and vanillic acids. The intermediates of this dissimilation are β-carboxy-*cis,cis*-muconic acid and γ-carboxy-muconolactone.

The ***meta-cleavage*** of catechol, for example, protocatechuic acid, is carried out between the hydroxylated and the adjacent nonhydroxylated C-atoms. Catechol is also cleaved to semialdehyde of α-hydroxymuconic acid; protocatechuic acid is converted to semialdehyde of α-hydroxyl-γ-carboxymuconic acid. The end product is pyruvic acid together with acetic acid and CO_2. Catechol is oxidized in such a way, for example, by species of *Pseudomonas putida* and *P. arvilla*; protocatechuic acid is oxidized by *P. desmolytica* strains. Ortho- and meta-cleavage of catechol is demonstrated in **Fig. 5.41**.

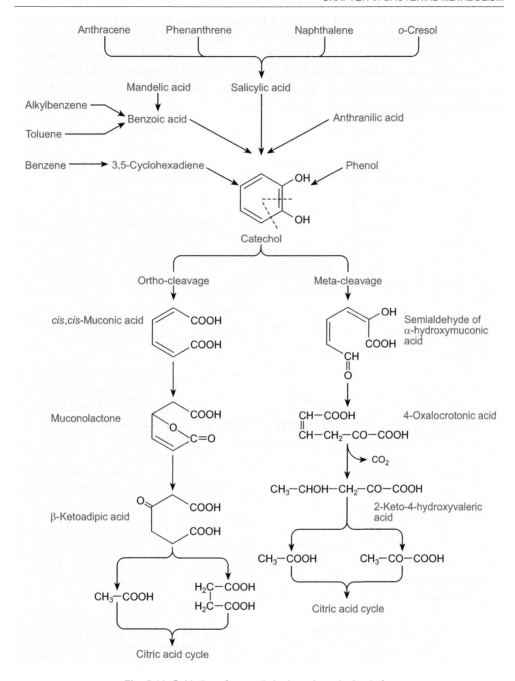

Fig. 5.41. Oxidation of aromatic hydrocarbons by bacteria.

Oxidation of aromatic hydrocarbons via gentisic acid. Some bacteria have the ability to oxidize aromatic compounds via gentisic acid. For example, *m*-hydroxybenzoic acid is transformed into gentisic acid by the hydroxylase of *Pseudomonas acidovorans* strains. Gentisic acid is then converted by dioxygenase to methyl pyruvic acid. Its hydrolysis produces pyruvic and malic acids. The strains of *P. testosteroni* in the presence of reduced glutathione (GSH) can transform maleylpyruvic acid by isomerization to fumarylpyruvic acid and then it hydrolyzed to fumaric and pyruvic acid. The species of *Nocardia* genus can oxidize via gentisic and anthranilic acids. The oxidation through gentisic acid is presented in **Fig. 5.42**.

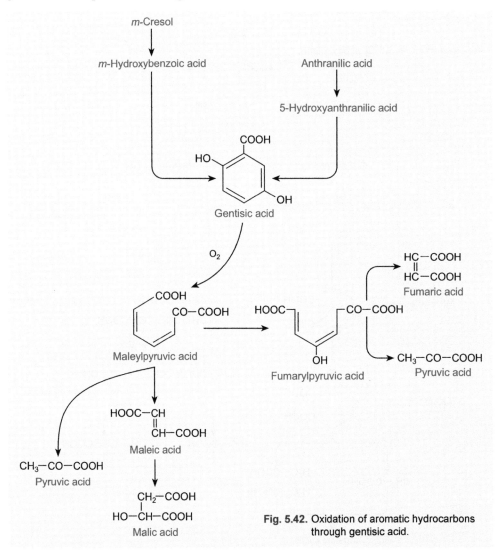

Fig. 5.42. Oxidation of aromatic hydrocarbons through gentisic acid.

Thus, the main final products in the process of oxidation of aromatic hydrocarbons via gentisic acid are pyruvic, fumaric, and malic acids. Gentisic acid is a main intermediate in the hydrocarbon oxidation metabolism in the species of *Pseudomonas* genus. The species of *Nocardia* genus can also oxidize anthranilic acid to gentisic acid with formation of the same intermediates and final products like the species of *Pseudomonas* genus.

5.5.5. Catabolism of nitrogenous compounds

Aside from inorganic compounds containing nitrogen, such as ammonia and nitrates, whose catabolism pathways were described in the previous chapter, also proteins and amino acids belong to the nitrogenous compounds.

5.5.5.1. Dissimilation of proteins and amino acids

Bacteria that produce proteolytic enzymes that are able to split a protein into monomers — amino acids are capable of this process. The production of these enzymes is typical for the bacterial genera *Bacillus*, *Clostridium*, and *Proteus*, but is not restricted only to those and can be found in other bacteria. Proteolytic enzymes are divided according to the target of their activity. Exopeptidases attack terminal peptide bonds, while endopeptidases break peptide bonds inside the peptide chain (see scheme):

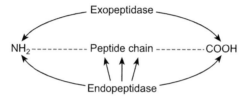

Exopeptidases consist of aminopeptidases and carboxypeptidases. Aminopeptidases require the chain end with a free amino group and the presence of metal ions, carboxypeptidases hydrolyze the peptide bond in the chain end with a free carboxy group. Due to the activity of these enzymes, the molecule of protein is cleaved to oligopeptides and single amino acids. Amino acids can be taken up by the bacterial cell and undergo further degradation by different metabolic pathways. Alongside the cleavage of proteins, the released energy is conserved in macroergic bonds of ATP and is used in biosynthetic processes.

Catabolism of amino acids is characteristic of anaerobic or facultative anaerobic bacteria, sporulating or non-sporulating. The main reactions of this process are deamination and decarboxylation, and the character of these reactions can be both reductive (anaerobic) or oxidative (aerobic).

5.5.5.1.1. Anaerobic degradation (amino acid fermentation)

If the degradation of amino acids takes place under anaerobic conditions, it is called fermentation. The main representatives of this process are *Peptococci* and *Clostridia*. The process of deamination is comprised of reductive reactions mediated by hydrogenases that lead to the formation of saturated fatty acids and ammonia release.

$$R-CH_2-\underset{\underset{NH_2}{|}}{CH}-COOH \xrightarrow{+2\,[H]} R-CH_2-CH_2-COOH + NH_3$$

Some amino acids can be deaminated with the formation of an unsaturated fatty acid instead of a saturated fatty acid. This is called deamination by desaturation.

$$R-CH_2-\underset{\underset{NH_2}{|}}{CH}-COOH \longrightarrow R-CH=CH-COOH + NH_3$$

Anaerobic dissimilation includes the catabolism of individual amino acids and coupled pairs of amino acids as well as reactions of amino acids with keto acids.

Fermentation of amino acids. Amino acids that undergo anaerobic dissimilation are aspartic acid, arginine, histidine, glutamic acid, glycine, serine, methionine, threonine, tryptophan, and lysine.

Aspartic acid is converted to fumaric acid by deamination. The reaction is catalyzed by the enzyme aspartase, produced, for instance, by *Clostridium cadaveris*:

$$COOH-CH_2-\underset{\underset{NH_2}{|}}{CH}-COOH \longrightarrow COOH-CH=CH-COOH + NH_3$$

Arginine is transformed by reductive deamination to citrulline, ornithine, carbamoyl phosphate, and ammonia by species of the genus *Mycoplasma*.

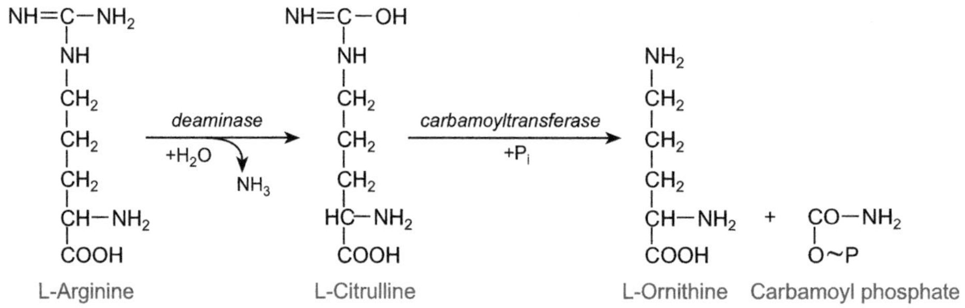

Carbamoyl phosphate is then degraded to NH_3 and CO_2 with the formation of ATP.

$$\underset{\underset{O\sim P}{|}}{CO-NH_2} + ADP \rightleftharpoons NH_3 + CO_2 + ATP$$

Some strains of *Clostridium botulinum* are capable of arginine degradation to ornithine the same way as mentioned before, but then the degradation goes through valeric acid to volatile fatty acids (acetic, butyric, and propionic acids). The reactions are not yet fully known.

Histidine is fermented by strains of *Clostridium tetanomorphum* through urocanic acid and mesaconic acid to formamide and glutamic acid, which is dissimilated to pyruvate, acetate, and ammonia (**Fig. 5.43**).

The desaturating deamination of histidine to urocanic acid is catalyzed by lyase. The following reaction leads to the creation of formiminoglutamic acid by the enzyme hydratase. The conversion to glutamic acid is proceeded by the enzyme containing cobalamin in the presence of THF. Pyruvate can be next converted to butyric acid.

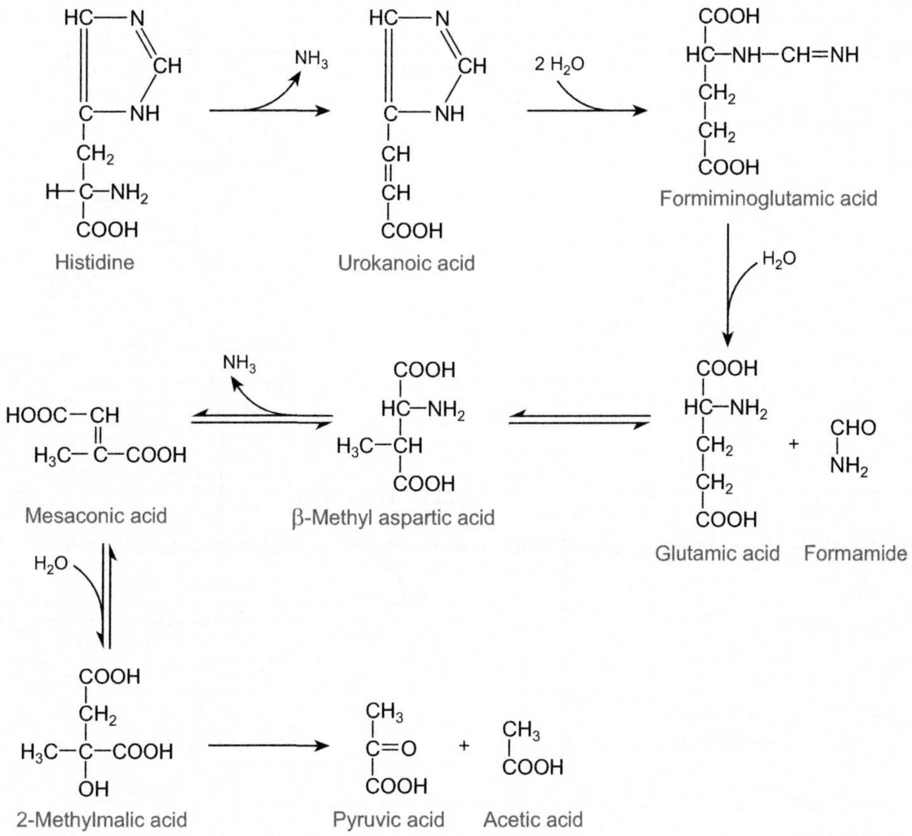

Fig. 5.43. Metabolism of histidine by *Clostridium tetanomorphum*.

Glycine serves as a substrate for the anaerobic bacterium *Peptococcus anaerobius*, and it is transformed to acetic acid, ammonia, and CO_2. The fermentation requires the presence of THF and pyridoxine; the reaction goes through intermediate products serine and pyruvate according to scheme in **Fig. 5.44**.

Methionine is fermented by *Clostridium sporogenes*. The products of the fermentation are α-ketoglutaric acid, methyl mercaptan (methanethiol), and ammonia:

$$CH_3S-CH_2-CH_2-\underset{\underset{NH_2}{|}}{CH}-COOH + H_2O \longrightarrow CH_3-CH_2-CO-COOH + CH_3SH + NH_3$$

Threonine. The anaerobic dissimilation of threonine can lead to different final products. The dehydratase of some strains of *Clostridium tetanomorphum* and *Veillonella alcalescens* deaminate threonine to α-ketoglutaric acid and ammonia:

$$CH_3-CHOH-\underset{\underset{NH_2}{|}}{CH}-COOH \xrightarrow[-H_2O]{+2 H^+} CH_3-CH_2-CO-COOH + NH_3$$

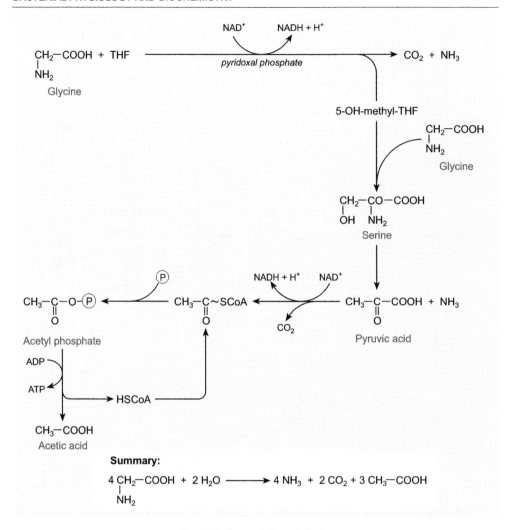

Fig. 5.44. Fermentation of glycine.

When *C. propionicum* is present, α-ketoglutaric acid is converted to propionic acid, CO_2, and hydrogen:

$$CH_3-CH_2-CO-COOH + H_2O \rightarrow CH_3-CH_2-COOH + CO_2 + H_2.$$

As a product of threonine fermentation by strains of *C. pasteurianum*, glycine and acetaldehyde are formed, but THF is required for the reaction. In the next step, acetaldehyde is reduced to ethanol:

Fermentation of tryptophan. This amino acid is reductively fermented by *Clostridium sporogenes* to 3-indolepropionic acid and ammonia. On the other side, facultatively anaerobic bacteria, such as *E. coli*, create indole, pyruvate, and ammonia through deamination (**Fig. 5.45**).

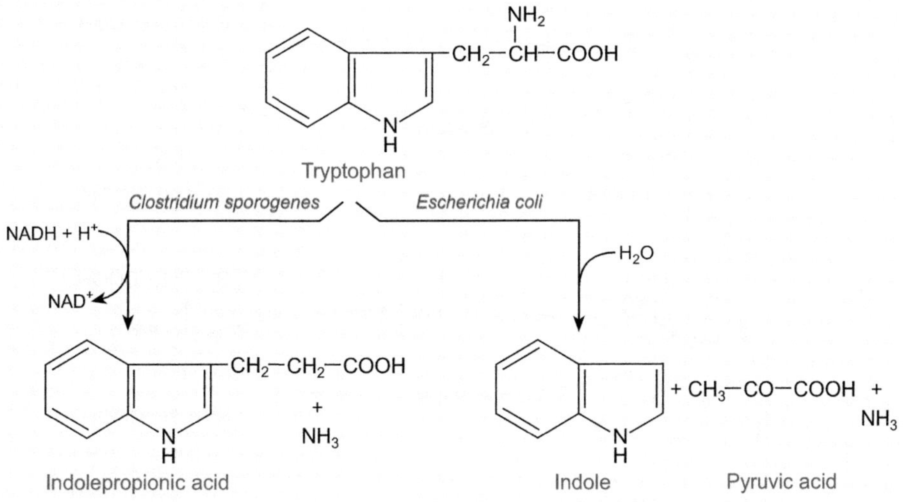

Fig. 5.45. Fermentation of tryptophan.

The formation of indole is an important feature in bacterial diagnostics. The presence of indole is indicated by dimethylbenzaldehyde, which changes color to red.

Fermentation of couple of amino acids (Stickland reactions). Anaerobic dissimilation of amino acids in the genus *Clostridia* often happens as coupled oxidation-reduction reactions between two amino acids. During the Stickland reactions, one of the amino acids serves as a donor, whereas the second one is an acceptor of electrons and hydrogen. Amino acids with the function of the donor are alanine, leucine, isoleucine, and valine, while the acceptors are glycine, proline, arginine, and tryptophan.

An example of the fermentation of two amino acids is the conversion of alanine and glycine by strains of *Clostridium propionicum* to acetic acid, ammonia, and CO_2:

$$CH_3-\underset{\underset{NH_2}{|}}{CH}-COOH + 2\,H_2O \longrightarrow CH_3-COOH + NH_3 + CO_2 + 4\,H^+$$

$$2\,\underset{\underset{NH_2}{|}}{CH_2}-COOH + 4\,H^+ \longrightarrow 2\,CH_3-COOH + 2\,NH_3$$

Summary:

$$CH_3-\underset{\underset{NH_2}{|}}{CH}-COOH + 2\,\underset{\underset{NH_2}{|}}{CH_2}-COOH + 2\,H_2O \longrightarrow 3\,CH_3-COOH + 3\,NH_3 + CO_2$$

According to the scheme (**Fig. 5.46**), in the Stickland reactions occur with the participation of dehydrogenases of respective amino acids (1) and dehydrogenases forming keto acids (2), which require LTPP, CoA, and phosphate acetyltransferases (3).

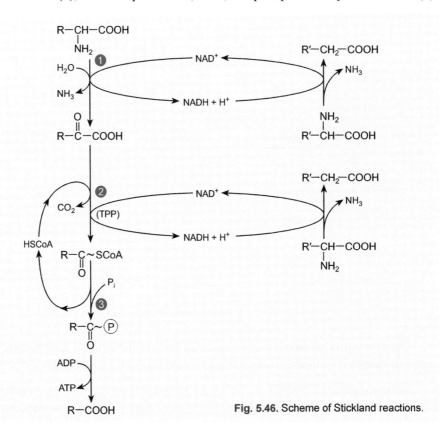

Fig. 5.46. Scheme of Stickland reactions.

Fermentation of amino acid in the presence of keto acid. Some amino acids could be deaminated simultaneously with the amination of keto acid, that serves as an acceptor of the amino group. This kind of fermentation can be performed by *Clostridium propionicum*. β-Alanine is converted to propionic acid in the presence of pyruvate. The whole process includes two phases. The amino group from β-alanine is transferred to pyruvate, from which α-alanine is formed. α-Alanine undergoes the second deamination, and in the presence of α-ketoglutaric acid, glutaric acid is formed. Catalysts for these reactions are respective aminotransferases and NAD glutamate dehydrogenase. The second phase leads to the formation of propionic acid. By the amination of acrylyl-CoA via aminases, β-alanine regenerates. The process is expressed by the scheme in **Fig. 5.47** and the equation:

$$
\underset{\substack{\text{β-Alanine}}}{\underset{\overset{|}{NH_2}}{CH_2-CH_2-COOH}} + \underset{\substack{\text{α-Ketoglutaric acid}}}{\overset{\overset{CO-COOH}{|}}{CH_2-CH_2-COOH}} \xrightarrow[\substack{4\,H^+}]{\text{transamination}} \underset{\substack{\text{Propionic acid}}}{CH_3-CH_2-COOH} + \underset{\substack{\text{Glutamic acid}}}{\overset{\overset{H_2N-CH-COOH}{|}}{CH_2-CH_2-COOH}}
$$

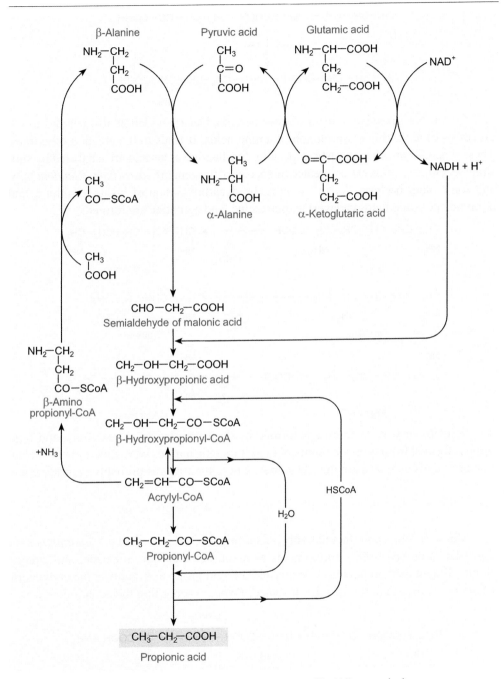

Fig. 5.47. Anaerobic dissimilation of β-alanine by *Clostridium propionicum.*

Fermentation of γ-aminobutyric acid by *Clostridium aminovalericum.* A similar pathway serves for *C. aminovalericum*; the fermentation of γ-aminobutyric acid in the presence of α-ketoglutaric acid leads to the formation of butyric and glutamic acids:

$$CH_2-CH_2-CH_2-COOH + COOH-CH_2-CH_2-CO-COOH$$

with NH_2 substituent on the first carbon

$$\Big\downarrow\ -H_2O \quad 4\,H^+$$

$$CH_3-CH_2-CH_2-COOH + COOH-CH_2-CH_2-CH-COOH$$

with NH_2 substituent

Anaerobic decarboxylation of amino acids. Decarboxylation also belongs to the processes of anaerobic dissimilation of amino acids. It leads to the creation of amines. These reactions are characteristic especially for diamino acids and are catalyzed by substrate specific decarboxylases, created by species of the genera *Clostridium*, *Pseudomonas*, and some other bacteria. As an example, by decarboxylation of lysine, ornithine, and arginine, the compounds cadaverine, putrescine, and agmatine are formed:

$$CH_2-CH_2-CH_2-CH_2-CH-COOH \longrightarrow CH_2-CH_2-CH_2-CH_2-CH_2$$

Lysine (NH_2, NH_2) $\;CO_2\;$ Cadaverine (NH_2, NH_2)

$$CH_2-CH_2-CH_2-CH-COOH \longrightarrow CH_2-CH_2-CH_2-CH_2$$

Ornithine (NH_2, NH_2) $\;CO_2\;$ Putrescine (NH_2, NH_2)

$$\overset{NH}{\underset{NH_2}{C}}-NH-CH_2-CH_2-CH_2-CH-COOH \longrightarrow \overset{NH}{\underset{NH_2}{C}}-NH-CH_2-CH_2-CH_2$$

Arginine (NH_2) $\;CO_2\;$ Agmatine (NH_2)

In a similar way, histamine is formed by decarboxylation of histamine, and tyramine is formed from tyrosine. Some of these biogenic amines were found in the nuclear matter of cells as well as in the ribosomes, where their role is probably neutralization.

5.5.5.1.2. Aerobic (oxidative) catabolism of amino acids

Many aerobic and facultatively anaerobic bacteria, such as *Pseudomonas* or enterobacteria, can utilize amino acids as a source of carbon, nitrogen, and energy. Some of them contain oxidase systems and a flavin group and catalyze the conversion of amino acids into peroxides (for instance, *Proteus*) according the equation:

$$R-\underset{NH_2}{CH}-COOH + O_2 \xrightarrow{\;H_2O_2\;} R-\underset{NH}{C}-COOH \xrightarrow{\;H_2O\;} R-\underset{O}{C}-COOH + NH_3$$

This way is typical for the catabolism of D-isomers of amino acids.

L-isomer is oxidized by oxidative deamination, which leads to the formation of keto acid and ammonia:

$$R-\underset{NH_2}{CH}-COOH + \tfrac{1}{2}O_2 \longrightarrow R-CO-COOH + NH_3$$

Most amino acids undergo the process of oxidative deamination. Deamination can be often connected with the decarboxylation of amino acids.

Thus, deamination is the removal of an amino group from a molecule. Enzymes that catalyze this reaction are called deaminases. The amine group is removed from the amino acid and converted to ammonia.

Oxidation of cysteine. Due to the effect of the enzyme called cysteine sulfur hydratase, cysteine is degraded to pyruvate, ammonia, and hydrogen sulfide:

$$SH-CH_2-CH-COOH + H_2O \longrightarrow CH_3-CO-COOH + NH_3 + H_2S$$
$$\overset{|}{NH_2}$$

This enzyme was found in *Proteus vulgaris, P. morganii, E. coli, B. subtilis*, etc.

Oxidation of lysine. L-lysine is degraded by *Pseudomonas putida* into CO_2 and H_2O. Oxygenase catalyzes the oxidation of L-lysine into aminovaleramide. Next, amidase oxidizes aminovaleramide into 5-aminovaleric acid. The acid is converted by transaminase into glutaric semialdehyde and then dehydrogenated by dehydrogenase to glutaric acid, which is oxidized to acetyl-CoA, CO_2, and H_2O (**Fig. 5.48**).

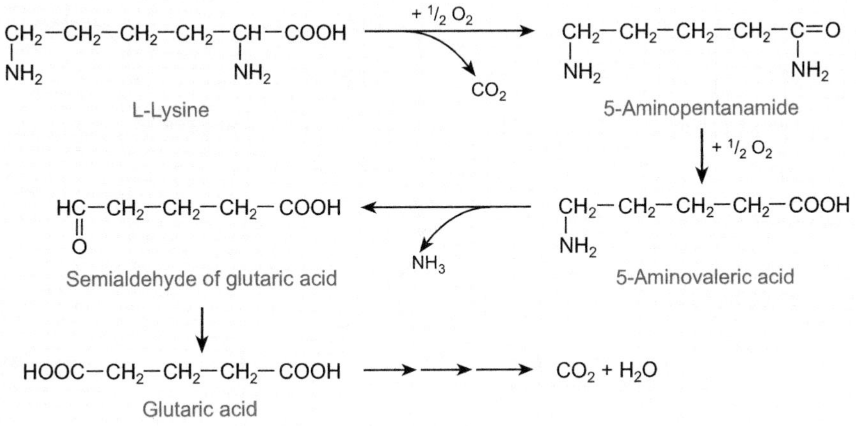

Fig. 5.48. Lysine metabolism by *Pseudomonas putida.*

Oxidation of valine, leucine, and isoleucine. These amino acids are oxidized mostly by different species of *Pseudomonas*. It was discovered that L-isomers are initially converted to α-keto acids by transaminases and then undergo decarboxylation in the presence of CoA. The corresponding acyl-CoA is formed. D-isomers are firstly oxidatively deaminated by dehydrogenases to keto acids. Acyl-CoA is further metabolized. The propionic acid is created from valine, acetic and acetoacetic acids from leucine and acetic and propionic acids from isoleucine (**Fig. 5.49**). The conversion of propionic acid (propionyl-CoA) can provide succinic acid. An interesting reaction is CO_2 fixation by carboxylase in the metabolism of leucine, which leads to the conversion of β-methylcrotonyl-CoA to β-methylglutaconyl-CoA. The enzyme for this reaction is dependent on biotin and was found in some bacteria of genera *Mycobacterium* and *Alcaligenes*.

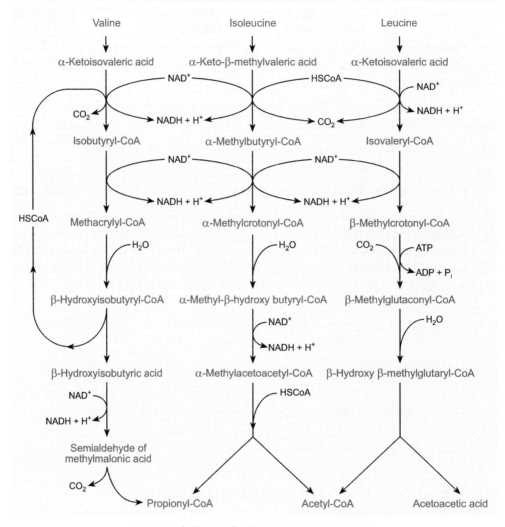

Fig. 5.49. Oxidation of valine, leucine, and isoleucine.

Oxidation of proline and hydroxyproline. Proline is firstly converted to pyrroline-5-carboxylic acid and then by NAD dehydrogenase to glutamic acid. Some strains of *Bacillus subtilis* are capable of this conversion. Glutamic acid can enter transamination reactions and the fermentation process.

Hydroxyproline serves as a substrate for *Pseudomonas fluorescens*. In the process of catabolism, it is converted through pyrroline-4-hydroxy-2-carboxylic acid according to the scheme:

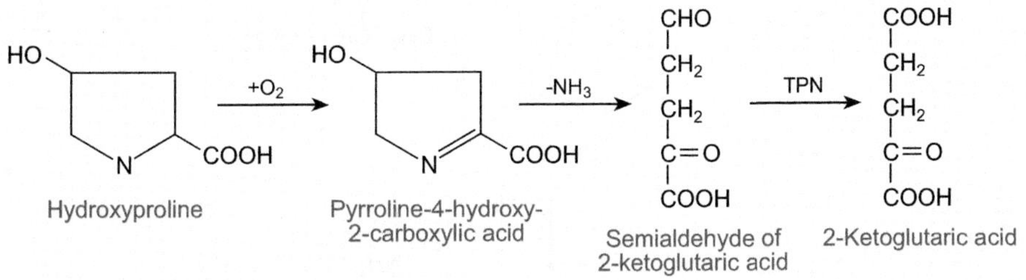

Hydroxyproline → Pyrroline-4-hydroxy-2-carboxylic acid → Semialdehyde of 2-ketoglutaric acid → 2-Ketoglutaric acid

Oxidation of phenylalanine and tyrosine. Phenylalanine and tyrosine are degraded by *Pseudomonas aeruginosa* to fumaric acid and acetoacetic acid. The process starts with the conversion of phenylalanine into tyrosine by the enzyme hydroxylase. The deamination of tyrosine leads to homogentisic acid, and, as the reaction goes further, final products are formed through fumarylacetoacetate according to the scheme in **Fig. 5.50**. The opening of homogentisic acid is catalyzed by oxygenase.

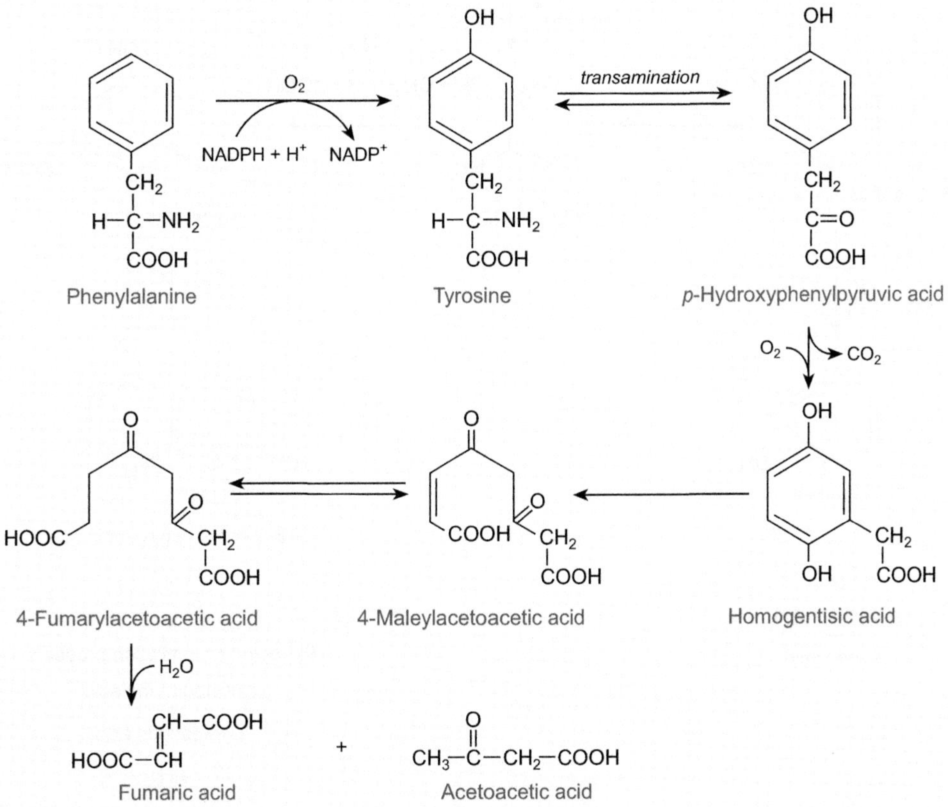

Fig. 5.50. Oxidation of phenylalanine and tyrosine.

243

Oxidation of tryptophan. The tryptophan is oxidized to kynurenine by various species of *Pseudomonas* genus (**Fig. 5.51**).

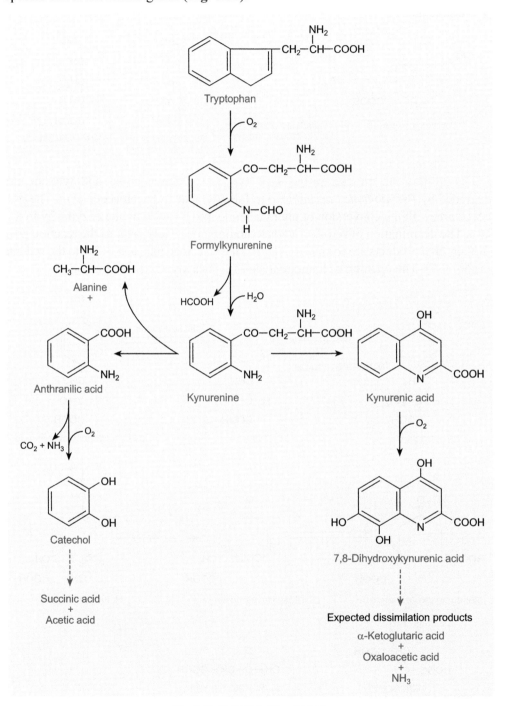

Fig. 5.51. Oxidation of tryptophan.

Kynurenine is broken down into alanine and anthranilic acid, which is further metabolized through catechol. The opening of the pyrrole cycle of tryptophan is catalyzed by dioxygenase; the formed formylkynurenine is transformed to kynurenine through the cleavage of formic acid due to the activity of formamidase. The products of oxidation of anthranilic acid are succinyl acid and acetic acid (**Fig. 5.51**). Few species, such as *Pseudomonas acidovorans*, are capable of kynurenine conversion to kynurenic acid instead of anthranilic acid. The final products are probably α-ketoglutarate, oxaloacetate, and NH_3.

Oxidation of spermidine. Aside of amino acids, amines and polyamines can be also oxidized, mostly by bacteria of the *Pseudomonas* genus. One of the examples is the oxidation of spermidine to succinic acid and β-alanine according to the scheme in **Fig. 5.52**. Individual reactions are catalyzed by corresponding NAD dehydrogenases and transferases. Spermidine is a polyamine compound ($C_7H_{19}N_3$) found in ribosomes and having various metabolic functions within some species of bacterial cells. Polyamines, such as spermidine, are polycationic aliphatic amines and are multifunctional. They play vital roles in cell survival.

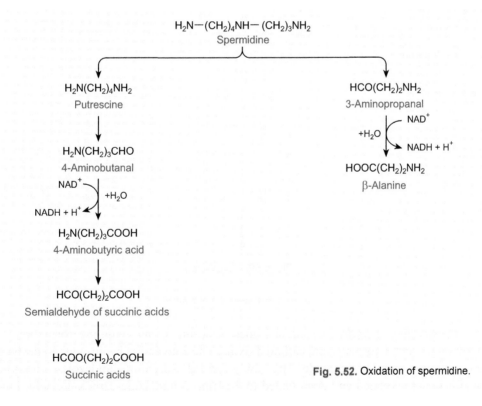

Fig. 5.52. Oxidation of spermidine.

Thus, final products of spermidine oxidation by bacterial species of the *Pseudomonas* genus are β-alanine and succinic acids, which can be further used in their metabolic processes.

5.5.6. Catabolism of heterocyclic compounds

The degradation of heterocyclic compounds can be as in anaerobic or aerobic conditions. In the first case, it is called **fermentation**, in the second — **oxidative dissimilation**.

5.5.6.1. Fermentation of heterocyclic compounds

Heterocyclic compounds, mainly purine, pyrimidine, and pyridine derivatives, can serve as a source of energy, carbon, and nitrogen. These anaerobic dissimilation processes are realized by clostridia and various anaerobic bacteria.

Fermentation of purines. Purines are degraded through fermentation processes by *Clostridium cylindrosporum* and *Veillonella alcalescens*. These derivatives are firstly converted to xanthine, which afterwards continues the fermentation process by the pathway in **Fig. 5.53**.

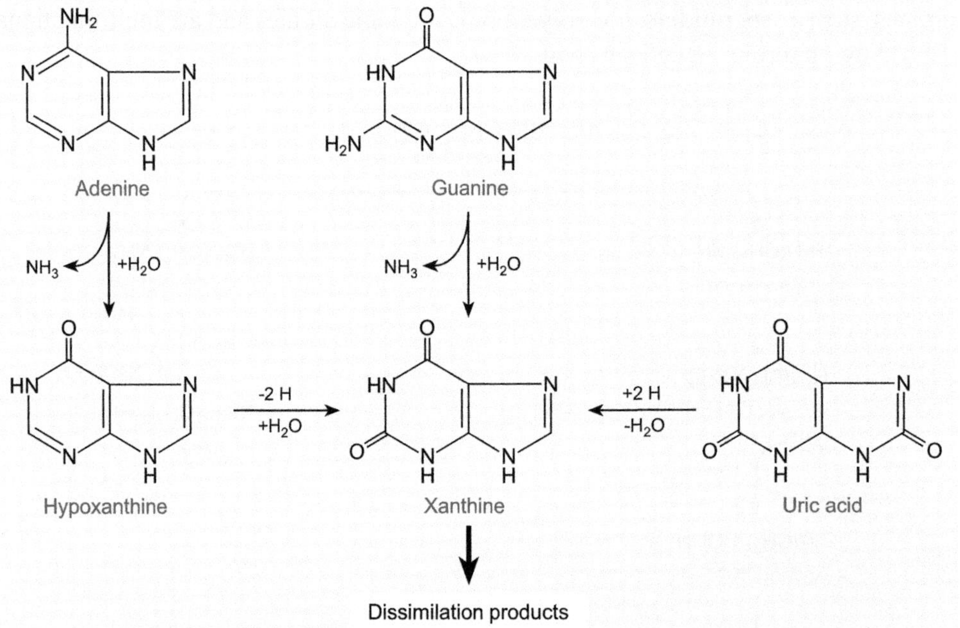

Fig. 5.53. Fermentation of purines.

Clostridium cylindrosporum degrades xanthine along with the formation of ammonia, acetate, formate, and carbon dioxide. The conversion begins with hydrolytic cleavage and opening the cycle. The newly formed 4-ureido-5-imidazole-carboxylic acid is decarboxylated and deaminated to 4-amino-5-imidazole-carboxylic acid. Further reactions lead to the creation of formiminoglycine. This compound splits into glycine, which is deaminated to acetic acid, and a formimine group. For the conversion of the formimine group into formic acid and ammonia, the presence of and interaction with tetrahydrofolic acid are required (**Fig. 5.54**).

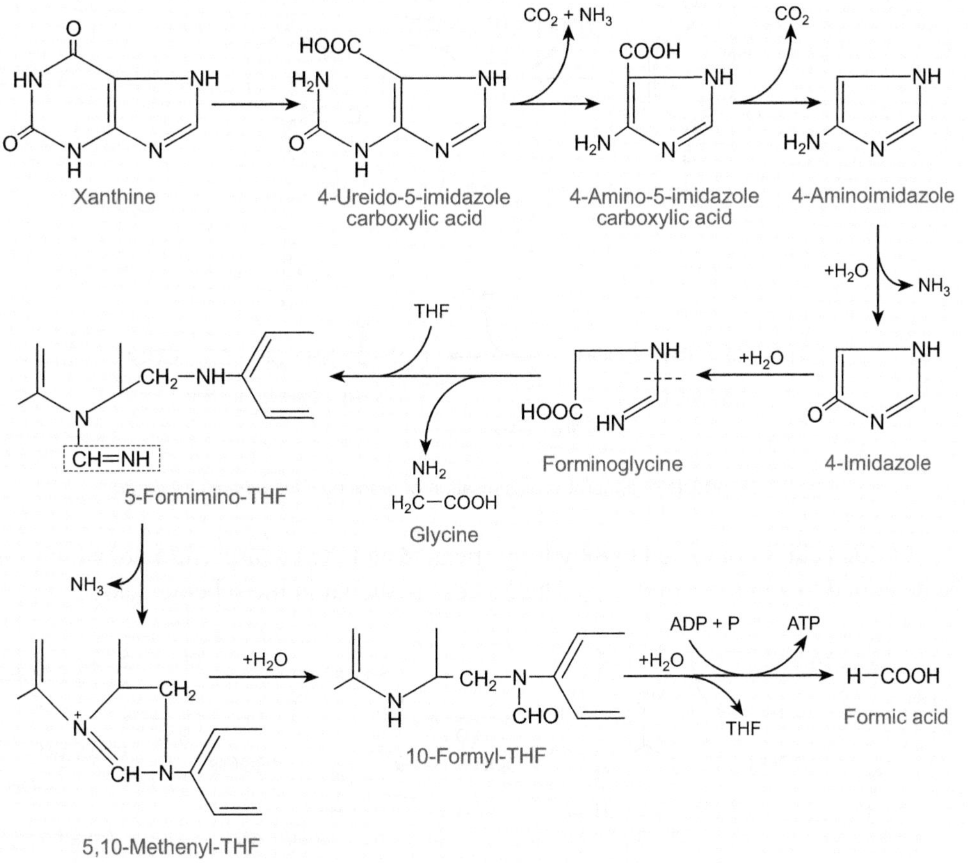

Fig. 5.54. Products of anaerobic dissimilation by *Clostridium cylindrosporum*.

The fermentation of xanthine by *C. acidiurici* leads to the formation of acetic acid, ammonia, and CO_2 according to the equation:

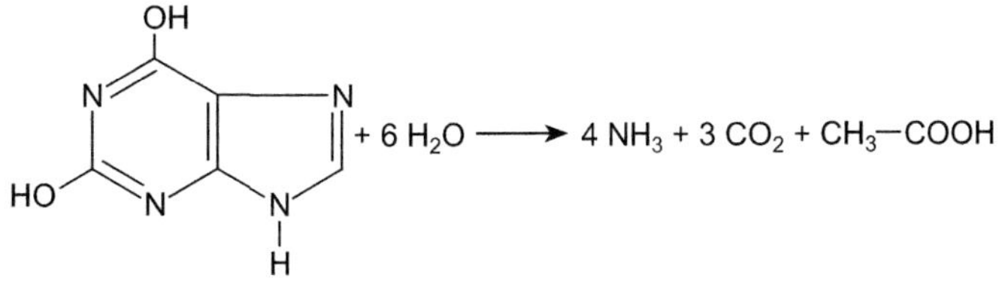

Fermentation of pyrimidine. The anaerobic degradation of pyrimidine can be found by *Clostridium uracilicum*, using uracil as a substrate, and *C. oroticum* that uses orotic acid as a substrate.

Uracil is first reduced to dihydrouracil and then undergoes hydrolytic cleavage to β-ureidopropionic acid. In the next phase, β-alanine, ammonia, and carbon dioxide are formed according to the scheme in **Fig. 5.55**.

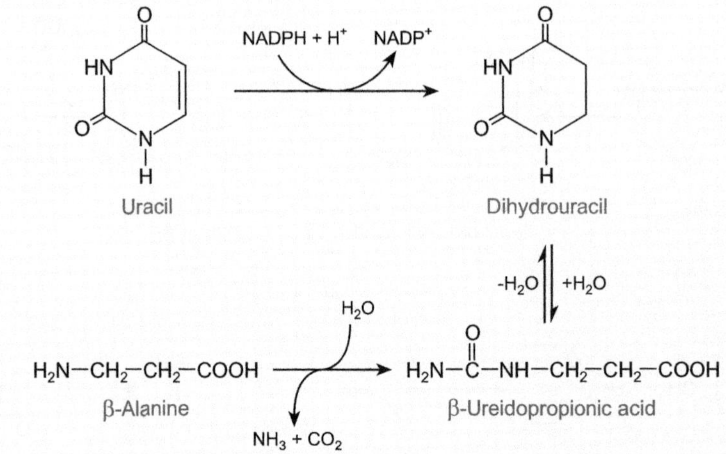

Fig. 5.55. Products of anaerobic dissimilation of uracil by *Clostridium uracilicum*.

Orotic acid is converted to dihydroorotic acid and then to aspartic acid, which can be degraded to ammonia and CO_2. The process is shown in the scheme below:

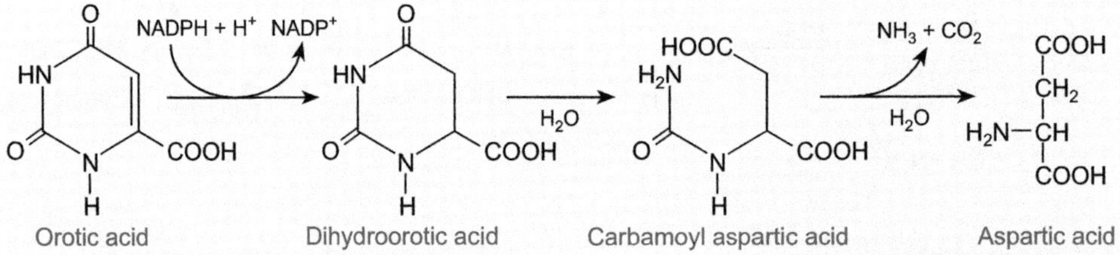

Thus, the final product of the fermentation of orotic acid is aspartic acid, which can be used later in the bacterial metabolism.

Fermentation of pyridine. Pyridine is a basic heterocyclic organic compound with the chemical formula C_5H_5N. It is structurally related to benzene, with one methine group ($=CH-$) replaced by a nitrogen atom. It is a highly flammable, weakly alkaline, water-soluble liquid with a distinctive, unpleasant fish-like smell. The fermentation of pyridine has not been yet fully studied. Partial discoveries gained from studies of nicotine acid catabolism by some clostridia species show that the final products of this reaction are propionic acid, acetic acid, and CO_2 according to the equation:

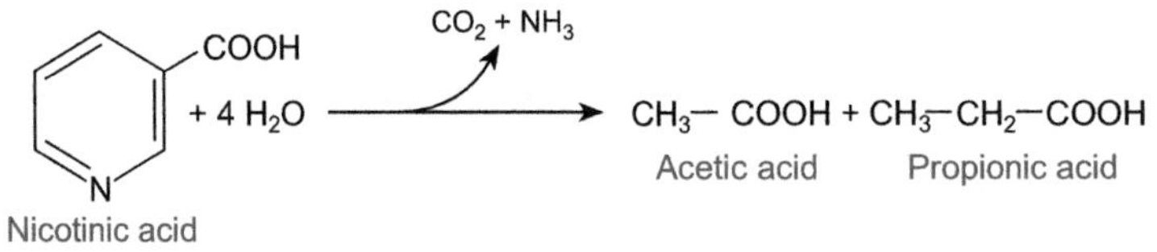

Thus, two final products of nicotinic acid fermentation are acetic and propionic acids.

5.5.6.2. Oxidation of heterocyclic compounds

The studies of oxidative deamination of heterocyclic compounds were mainly focused on allantoin and uric acid. These compounds are oxidized mostly by the genus *Pseudomonas*, *E. coli* strains, and hydrogen-oxidizing bacteria (*Alcaligenes eutrophus*). Under autotrophic conditions, these compounds serve as a source of nitrogen (the source of carbon is CO_2), whereas under heterotrophic conditions, they are used as a source of nitrogen as well as a source of carbon (**Fig. 5.56**).

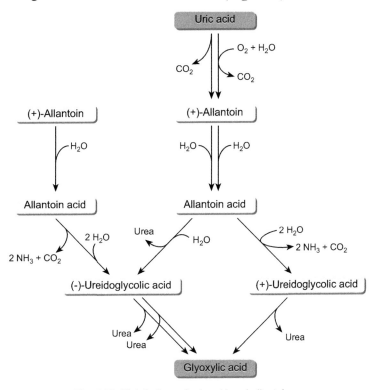

Fig. 5.56. Metabolism of uric acid and allantoin.

Strains of *Pseudomonas aeruginosa* and *P. fluorescens* oxidize uric acid to allantoin by the corresponding oxidase. Allantoin is converted by aminohydrolase to allantoic acid:

Uric acid   Allantoin   Allantoin acid

In the next reaction, allantoic acid is broken down by amidinohydrolase to (-) ureidoglycolic acid and urea:

$$
\begin{array}{ccc}
\underset{\text{Allantoic acid}}{
\begin{array}{c}
H_2N \\
| \\
O=C \\
| \\
HN\!-\!\!\!-\!\!\!-\!\!\!C\!-\!\!\!-\!\!\!-\!NH \\
\quad\quad | \\
\quad\quad H
\end{array}
\begin{array}{c}
\\
COOH \\
\end{array}
\begin{array}{c}
NH_2 \\
| \\
C=O \\
\end{array}
} & \longrightarrow &
\underset{\text{Urea}}{
\begin{array}{c}
H_2N \\
| \\
O=C \\
| \\
H_2N
\end{array}} + \underset{\text{(-)-Ureidoglycolic acid}}{
\begin{array}{c}
\\
COOH \\
| \\
HO\!-\!C\!-\!\!\!-\!\!\!-\!NH \\
\;| \\
\;H
\end{array}
\begin{array}{c}
NH_2 \\
| \\
C=O
\end{array}}
\end{array}
$$

By the effect of lyase, another molecule of urea and glyoxylic acid are formed.

The strains of *Pseudomonas acidovorans* lack amidinohydrolase. They possess amidohydrolase instead, that converts allantoic acid to (+) ureidoglycolic acid, ammonia, and CO_2. An intermediate product of this reaction is ureidoglycine. The process goes according to this scheme:

$$
\begin{array}{ccccc}
\underset{\text{Allantoic acid}}{
\begin{array}{c}
H_2N \\
| \\
O=C \\
| \\
HN\!-\!\!\!-\!\!\!-\!C\!-\!\!\!-\!\!\!-\!NH \\
\quad | \\
\quad H
\end{array}
\begin{array}{c}
\\
COOH
\end{array}
\begin{array}{c}
NH_2 \\
| \\
C=O
\end{array}
} & \longrightarrow &
\underset{\text{Ureido glycine}}{
\begin{array}{c}
H_2N \\
| \\
O=C \\
| \\
HN\!-\!C\!-\!NH_2 \\
\;\;| \\
\;\;H
\end{array}
\begin{array}{c}
\\
COOH + NH_3 + CO_2
\end{array}
} & \longrightarrow &
\underset{\text{(+)-Ureidoglycolic acid}}{
\begin{array}{c}
H_2N \\
| \\
O=C \\
| \\
HN\!-\!C\!-\!OH \\
\;\;| \\
\;\;H
\end{array}
\begin{array}{c}
\\
COOH + NH_3
\end{array}
}
\end{array}
$$

Allantoic acid is utilized by *E. coli* strains as well and converted to (-)-ureidoglycolic acid, two molecules of ammonia, and CO_2. Ureidoglycolic acid is further metabolized by ureidoglycolate lyase to urea and glyoxylic acid.

Glyoxylic acid or oxoacetic acid is an organic compound that is both an aldehyde and a carboxylic acid. It is an intermediate of the glyoxylate cycle, which enables certain organisms to convert fatty acids into carbohydrates. The glyoxylate cycle is initiated through the activity of isocitrate lyase, which converts isocitrate into glyoxylate and succinate. In the presence of water and sunlight, glyoxylic acid can undergo photochemical oxidation. Glyoxylic acid is converted through glyceric acid to pyruvate, which can enter the Krebs cycle, and urea, which is degraded to CO_2 and NH_3. The conversion is assisted by carboxylase, TPP, and dehydrogenase, and an intermediate product is tartronate semialdehyde:

$$
\begin{array}{c}
H\!-\!C=O \\
| \\
COOH
\end{array}
\xrightarrow[\text{TPP}]{CO_2}
\begin{array}{c}
CHO \\
| \\
CHOH \\
| \\
COOH
\end{array}
\xrightarrow{NADH + H^+ \quad NAD^+}
\underset{\text{Glyceric acid}}{
\begin{array}{c}
CH_2OH \\
| \\
CHOH \\
| \\
COOH
\end{array}}
\dashrightarrow
\begin{array}{c}
COOH \\
| \\
C=O \\
| \\
CH_3
\end{array}
\dashrightarrow TCA
$$

As a nitrogen source for hydrogen-oxidizing bacteria, thymine, cytosine, and uracil can be used. Hydrogen serves as a source of energy. The reactions are not fully known in detail.

Thus, the main example of the oxidation of heterocyclic compounds is uric acid, which through allantoin and allantoin acid is oxidized to glyoxylic acid. Glyoxalic acid can be converted through glyceric acid to pyruvate, which is involved in the citric acid cycle.

5.6. Processes of anabolism (biosynthesis)

Anabolism or biosynthesis is a wide group of biochemical pathways resulting in the creation of various ingredients and compounds, which are crucial for the cellular function, growth, and multiplication. These processes are dissimilatory, reducing the cellular pool of substrates by their structural and chemical conversion in general much bigger products. It is crucial to state that the smaller the substrate is the more chemical reactions have to occur in order to create complex cellular structures. Thus, the nature of the substrate has a direct impact on the synthesis.

The biosynthesis of compounds with structural and biocatalytic functions and reserve substances such as saccharides, lipids, amino acids, nucleic acids, and proteins has the decisive role in the cellular growth and facilitation of the vital function.

5.6.1. Biosynthesis of saccharides

Biosynthesis of saccharides is a complex anabolic pathway, which consists of the synthesis of monosaccharides as primary building blocks and energy source and the production of oligo- and polysaccharides which may later serve as parts of envelopes and cellular structures.

In autotrophic prokaryotes (photo- and chemolithotrophic), carbon dioxide serves as a sole source of a carbon atom. Its reduction is realized through the Calvin cycle (described in detail on page 313, **Fig. 5.100**), where part of the molecules of 3-phosphoglyceric acid undergoes the reversed Embden–Meyerhof–Parnas pathway through 1,3-diphosphoglyceric acid and P-trioses (glyceraldehyde 3-phosphate and dihydroxyacetone phosphate), resulting in the condensation of given molecules into hexoses. The energy needed for the condensation is facilitated through ATP, and a proton is provided by an external donor such as hydrogen sulfide or thiosulfate, eventually by molecular hydrogen.

In some photoautotrophic bacteria, the biosynthesis of saccharides may compete for carbon, which is assimilated in the reductive cycle of carboxylic acids during photosynthesis. The mechanism is described in detail on page 314, **Fig. 5.101**.

The biosynthesis of *saccharides* in *heterotrophs* (chemo- and organotrophic) is facilitated through the EMP pathway just like in autotrophs. The key molecule in this process is pyruvic acid, which is an intermediate of both anaerobic and aerobic dissimilatory reactions, for instance, glycolysis, Entner–Doudoroff pathway, and tricarboxylic acid cycle. The reversion of pyruvic acid to phosphoenolpyruvate does not go straight due to unfavorable energetic conditions. Rather, it undergoes shift through oxaloacetic acid by the effect of pyruvate carboxylase (EC 6.4.1.1) in the presence of HCO_3^- and ATP. Oxaloacetic acid is then decarboxylated by phosphoenolpyruvate carboxylase with GTP to phosphoenolpyruvic acid which later condenses with P-trioses into hexose.

Phosphoenolpyruvic acid also serves as a default intermediate in the biosynthesis of saccharides from the acetate. Given reactions are common for the *Pseudomonas* genus and some strains of *Escherichia coli*. The transformation of the acetate occurs through the TCA pathway to oxaloacetic acid, which is then decarboxylated by previously described mechanisms to phosphoenolpyruvate. Interconnections between the given pathways are then clearly visible in **Fig. 5.57**.

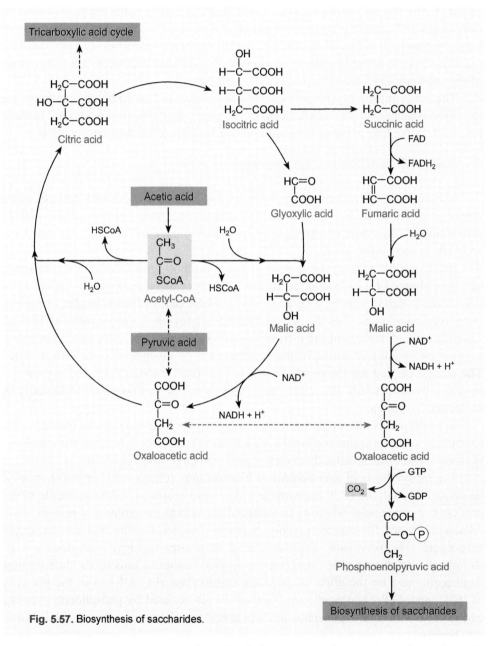

Fig. 5.57. Biosynthesis of saccharides.

Methylotrophic bacteria, such as *Methylomonas methanica* (syn. *P. methanica*), are able to utilize various oxidized derivatives of methane which are intermediates in the methane oxidation metabolism, mainly formaldehyde. The conversion into saccharides may go by two different pathways, allulose and serine pathways.

The **allulose pathway** is present in obligate methylotrophic bacteria and similar to the Calvin cycle. The key reaction is the condensation of formaldehyde and ribose 5-phosphate into phosphorylated ketohexose — allulose 6-phosphate, which later epimerizes to fructose 6-phosphate according to the scheme:

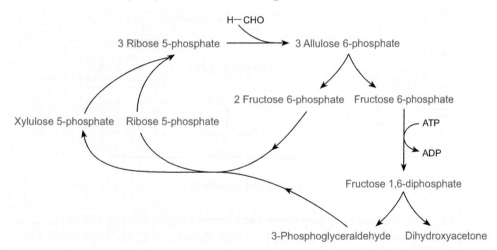

The condensation reaction occurs in the presence of thiamine pyrophosphate (TPP). Unlike in the Calvin cycle, in the allulose pathway, the epimerization prevents the creation of phosphoglyceric acid because fructose 6-phosphate is directly transformed into trioses glyceraldehyde 3-phosphate and dihydroxyacetone phosphate through fructose 1,6-phosphate as shown in **Fig. 5.58**.

Fig. 5.58. Allulose pathway.

The **serine pathway** occurs in the metabolism of facultative methylotrophs and comprises many different reactions. First, the formaldehyde is transferred to glycine by tetrahydrofolic acid (THF):

$$HC\underset{H}{\overset{O}{\big<}} + THF \longrightarrow 5,10\text{-methylene-THF} + \underset{NH_2}{CH_2}{-}COOH \longrightarrow \underset{OH\ \ NH_2}{CH_2{-}CH{-}COOH} + THF$$

Serine is then converted by various reactions to 3-phosphoglyceric acid, which is then partially transformed into saccharides and partially serves for the regeneration of glycine as a primary C_1-acceptor. The process comprises the transformation of 2-phosphoglyceric acid through phosphoenolpyruvate to malic acid together with the simultaneous fixation of CO_2. Malic acid is split into glyoxalic acid and acetyl-CoA, which enables the synthesis of another compounds through the glyoxylate cycle. The scheme of the serine pathway is shown in **Fig. 5.59.**

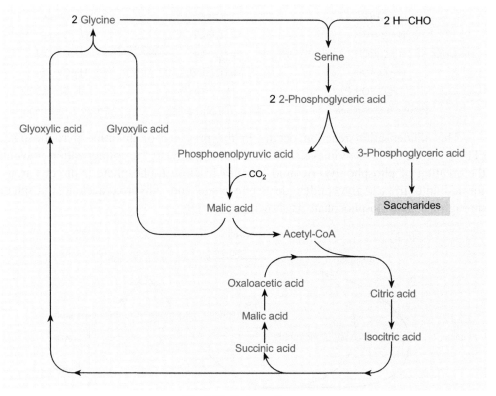

Fig. 5.59. Serine pathway.

Thus, the serine pathway comprises the transformation of 2-phosphoglyceric acid through phosphoenolpyruvate to malic acid with the simultaneous fixation of CO_2.

Biosynthesis of oligo- and polysaccharides — homoglycans. Biosynthesis processes leading to the creation of oligo- and polysaccharides have a common beginning. These processes are usually started by the activation of a hexose through ATP, hexokinase (EC 2.7.1.1), and phosphoglucomutase (EC 5.4.2.2) into glucose 1-phosphate. This intermediate is later converted to uridine diphosphate glucose in the presence of UTP **(Fig. 5.60)**. The activated glucose enables the connection of the reacting monosaccharides by glycosylic bond, resulting in the creation of oligo- and, later, polysaccharides. For instance, the biosynthesis of saccharose in some representatives of *Leuconostoc* genus goes according to the scheme in **Fig. 5.61.**

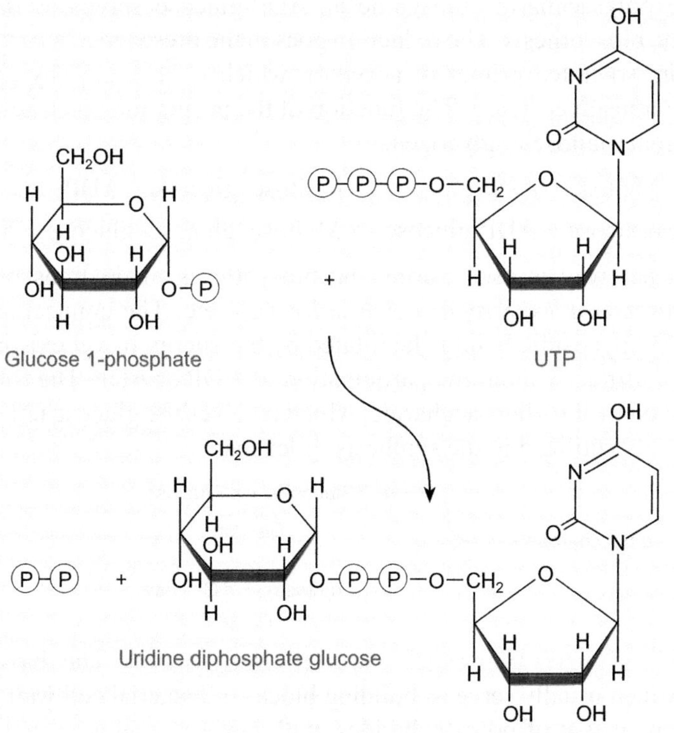

Fig. 5.60. Formation of uridine diphosphate glucose.

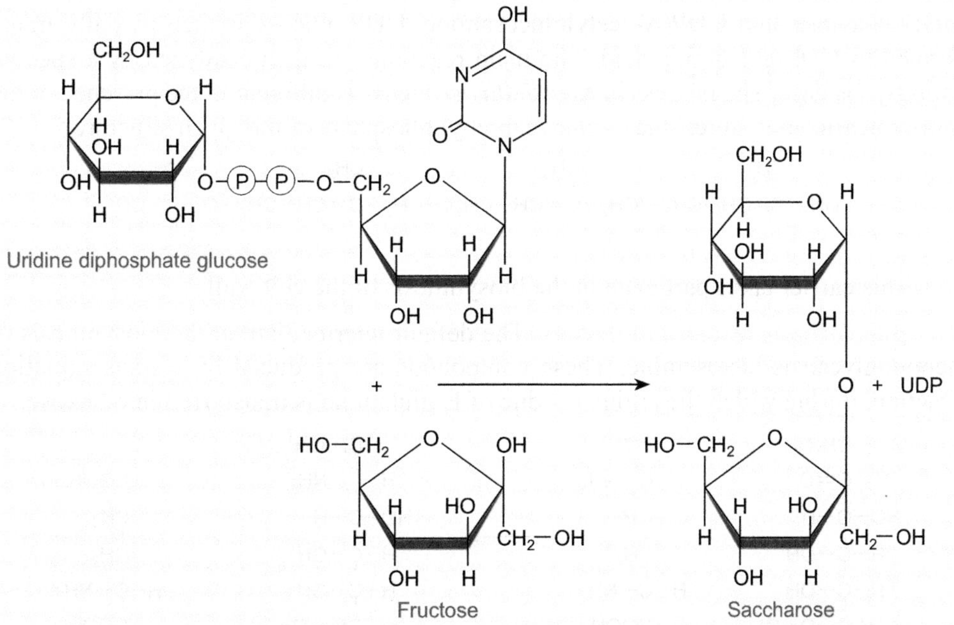

Fig. 5.61. Biosynthesis of saccharose in some species of *Leuconostoc* genus.

During the polysaccharide biosynthesis, ADP-glucose serves as the activated component of the biosynthesis. The reduction goes in the presence of a partly complete saccharide chain, so-called primer or acceptor, which is elongated by glucose residues by specific transglycosidase. The function of the primer may be taken by various maltodextrins from maltose to glycogen.

Maltose + ADP-glucose → Maltose–glucose + ADP;

Maltose–glucose + ADP-glucose → Maltose–glucose–glucose + ADP.

A different pathway is used during the biosynthesis of dextrans by strains of *Leuconostoc mesenteroides*, *Streptococcus salivarius*, etc. The connection of monomeric subunits by glycosidic bond is facilitated by the energy of yet existing bonds in an appropriate substrate without the participation of UDP or ADP. The role of primer is in this case taken by the oligosaccharide, which receives the glucose units. The same mechanisms are also utilized in the synthesis of levans:

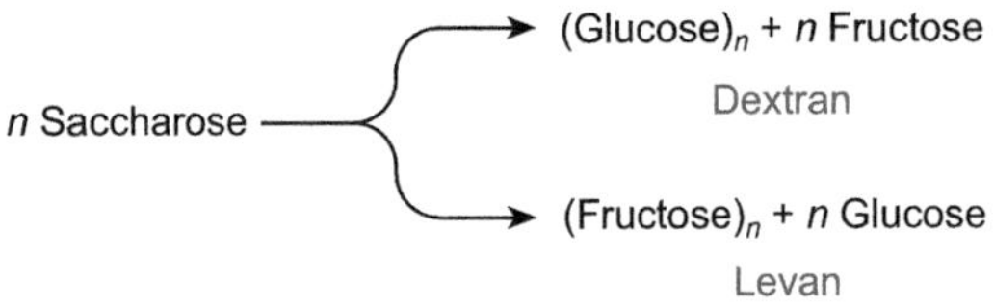

Biosynthesis of heteropolysaccharides (heteroglycans). The biosynthesis of heteroglycans, which usually serve as building blocks of bacterial cell walls, undergoes a similar pathway as that of polysaccharides, with hexose as an activated compound. There are some variations among genera; for instance, in *Streptococcus pneumoniae* and some *Neisseria* species, the activated compound may take the form of UDP-glucose, UDP-galactose, and UDP-*N*-acetylglucosamine. UDP-glucose is then partially transformed to UDP-glucuronic acid. The main reactions are then catalyzed by a specific transglycosidase. The process in *Klebsiella aerogenes* is different in the presence of the carrier bearing a lipid residue — undecaprenyl phosphate of the given structure:

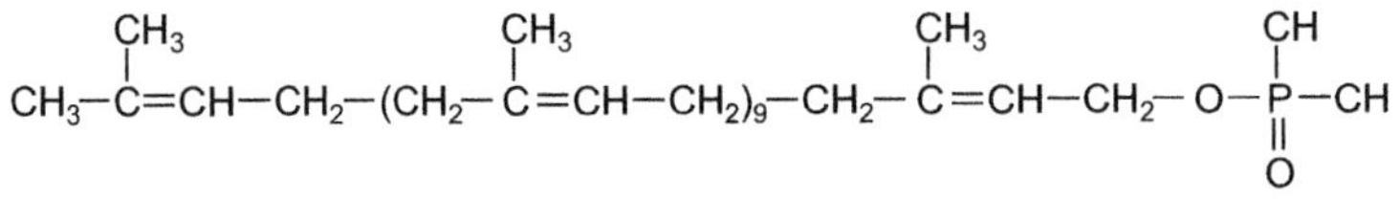

This carrier takes part even in the biosynthesis of the cell wall.

Biosynthesis of peptidoglycans. The default intermediate in the biosynthesis of peptidoglycan is hexosamine. These compounds are produced by a transamination reaction, during which the amine residue of L-glutamine is transferred into hexose:

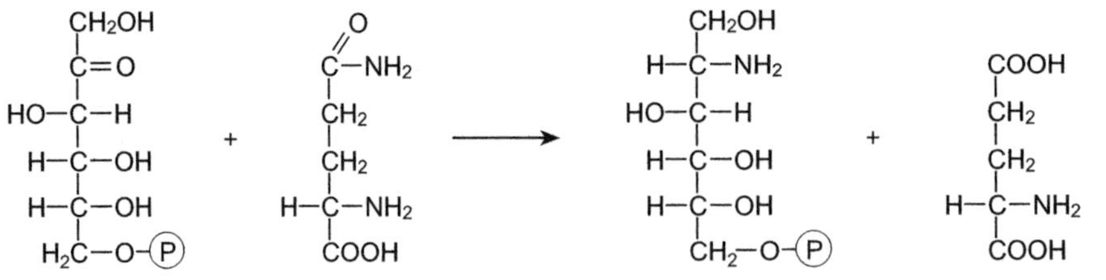

Fructose 6-phosphate Glutamine Glucosamine 6-phosphate Glutamic acid

Glucosamine 6-phosphate is transformed to glucosamine 1-phosphate, which is, in the presence of acetyl-CoA, further transformed into *N*-acetylglucosamine 1-phosphate. Peptidoglycans are synthesized from UDP-*N*-acetylglucosamine and phosphoenolpyruvate, resulting in the muramic acid. Muramic acid serves as a primer, to which individual amino acids of the peptide chain are connected. The creation of the peptide bond requires the energy of one high-energy bond in ATP. UDP–muramic acid pentapeptide is then assimilated into the newly synthesized cell wall, bonding with phospholipids. The reaction is accompanied by the release of UMP (**Fig. 5.62**).

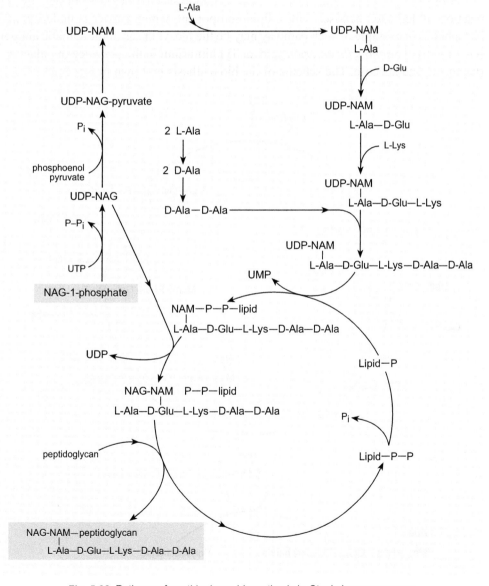

Fig. 5.62. Pathway of peptidoglycan biosynthesis in *Staphylococcus aureus.*

In the next stage, *N*-acetylglucosamine is transferred to UDP-*N*-acetylmuramyl-pentapeptide resulting in the formation of a disaccharide–pentapeptide complex and simultaneous UDP release. In the final phase of the cell wall synthesis, the peptide chains are transversely connected by the transpeptidase, so the penultimate molecule of D-alanine of muropeptide is binding to the amine residue of diaminopimelic acid or lysine in the neighboring chain, and the final molecule of D-alanine is simultaneously released from the complex. In some strains of *Staphylococcus aureus*, D-alanine of one strand and L-lysine of another strand are interconnected by a pentaglycine bridge.

Biosynthesis of lipopolysaccharides. Lipopolysaccharides are formed by the reaction of activated hexoses with a lipid component, which carrier is undecaprenyl phosphate. An example of such pathway may be the biosynthesis of the somatic antigen (*O*-antigen) of *Salmonella enterica* serovar Typhimurium in the presence of galactose, rhamnose, and abecose. The scheme of the biosynthesis is shown in **Fig. 5.63**.

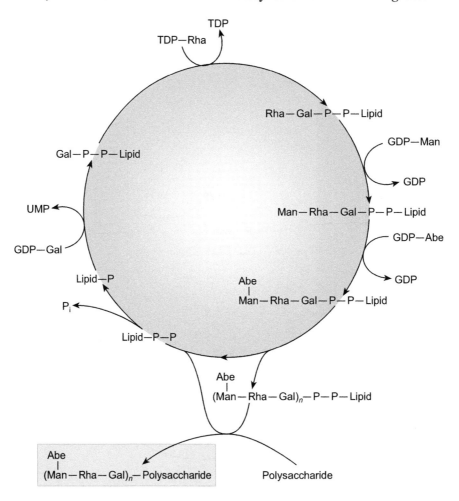

Fig. 5.63. Pathway of lipopolysaccharide biosynthesis in *Salmonella enterica* serovar Typhimurium.

All of the given glycans enter the reaction in the form of nucleoside diphosphate. After mutual connection, these saccharides polymerize into a chain, which is attached to the core oligosaccharide. The site of the lipopolysaccharide biosynthesis is a plasmatic membrane, from which individual components are transferred and further assimilated into the structure of the cell wall.

Biosynthesis of teichoic acids. Teichoic acids are integral parts of the cell walls of Gram-positive bacteria. The default building blocks for the synthesis are ribitol (adonitol) and glycerol. The activation of substrates occurs in the presence of cytidine-triphosphate (CTP) which transfer and assimilation into the polymer are catalyzed by glycosyltransferase according to the formula:

$$\text{Glycerol phosphate} + \text{CTP} \rightarrow \text{CDP-glycerol} + \text{P--P}_i$$

$$\text{CTP-glycerol} + (\text{Glycerol phosphate})_n \rightarrow (\text{Glycerol phosphate})_{n+1} + \text{P--P}_i$$

$$\text{Ribitol 5-phosphate} + \text{CTP} \rightarrow \text{CDP-ribitol} + \text{P--P}_i$$

$$n\,\text{CDP-ribitol} + (\text{Ribitol phosphate})_n \rightarrow \text{Ribitol-(ribitol phosphate)}_n + \text{CMP}_n$$

Afterward, the substituents form bonds with a newly synthesized polyol phosphate frame; D-alanine, glucose, galactose, and some amino sugars may form a bond in the presence of P-lipid. Ribitol teichoic acid finally creates bonds with the muropeptide, creating a final complex.

5.6.2. Biosynthesis of lipids

Biosynthesis of lipids is comprised of glycerol synthesis and the formation of fatty acids. The glycerol synthesis is provided by the dissimilation of saccharides. One way may be the conversion of alcohol fermentation during the fixation of aldehyde by sulfite according to these steps: dihydroxyacetone → glycerol phosphate → glycerol. Another way stems from the Entner–Doudoroff pathway, both in oxidative and reductive pentose cycles. The synthesis of fatty acids is facilitated by the presence of CoA, which plays a key role in the lipid synthesis process. The default substrate seems to be acetyl-CoA, the carboxylation of which results in malonyl-CoA and is catalyzed by biotin containing carboxylase according to this reaction:

$$\text{CH}_3\text{--CO--S--CoA} + \text{CO}_2 + \text{ATP} + \text{H}_2\text{O} \rightarrow \text{COOH--CH}_2\text{--CO--S--CoA} + \text{ADP} + \text{P}_i.$$

By the effect of the synthetase of fatty acid and the reaction of malonyl-CoA and acetyl-CoA, the carbon chain is elongated to the final phase, where acyl-CoA transforms to the fatty acid (**Fig. 5.64**). The transfer of acyl groups is mediated by a protein carrier (ACP, acyl protein carrier) containing phosphopantetheine. The thiol group of this compound is vital during the formation of a bond with an acyl residue, which goes the same way as the acyl — acetyl-CoA reaction:

$$-\text{Ser}-\text{O}-\overset{\overset{\text{O}}{\parallel}}{\underset{\underset{\text{OH}}{|}}{\text{P}}}-\text{O}-\text{CH}_2-\overset{\overset{\text{CH}_3}{|}}{\underset{\underset{\text{CH}_3}{|}}{\text{C}}}-\text{CHOH}-\text{CO}-\text{NH}-\text{CH}_2-\text{CH}_2-\text{CO}-\text{NH}-\text{CH}_2-\text{CH}_2-\text{S}-\text{CO}-\text{CH}_2-\text{COOH}$$

4-Phosphopantetheine

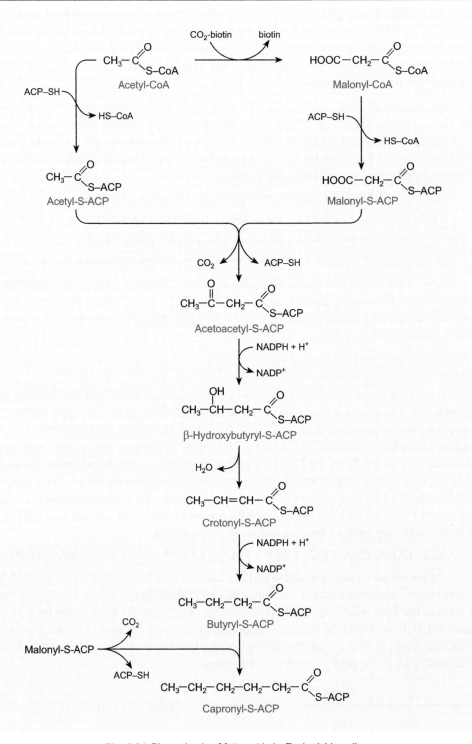

Fig. 5.64. Biosynthesis of fatty acids in *Escherichia coli.*

Fatty acids contain cyclopropane as in the case of lactobacillic acid in *Lactobacillus arabinosus* synthesized with *S*-adenosyl methionine, the methyl group of which is affiliated to a given carbon atom according to the scheme in **Fig. 5.65**.

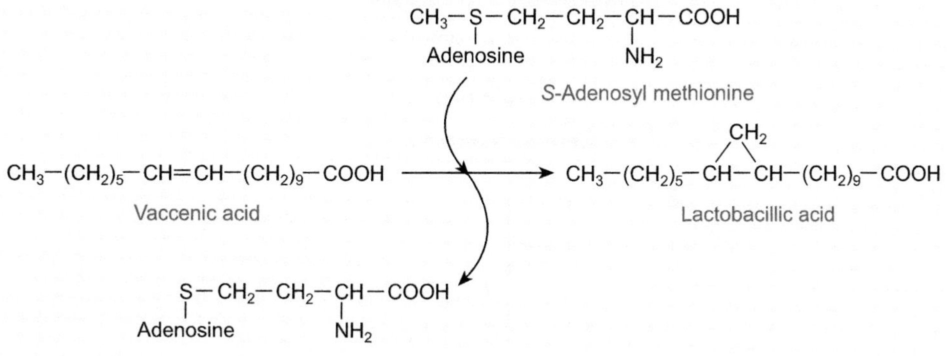

Fig. 5.65. Lactobacillic acid formation.

Biosynthesis of glycolipids and phospholipids. Glycolipids created by bonding give saccharides in the form of nucleoside diphosphate to substituted glycerol (diglyceride). According to this mechanism, compounds such as diaminosyl diglyceride of *Micrococcus luteus* are synthesized by the scheme in **Fig. 5.66**.

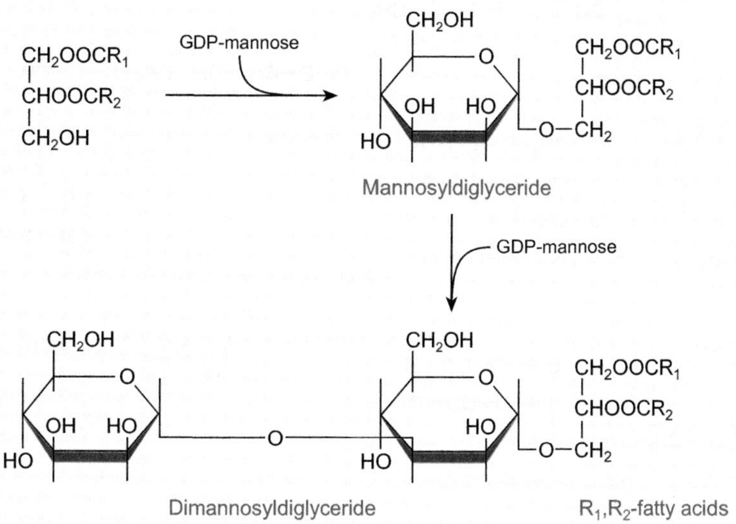

Fig. 5.66. Synthesis of glycolipids.

Similarly, the rhamnolipid of *Pseudomonas aeruginosa* strains is synthesized from 3-hydroxydecanoyl-3-hydroxydecanic acid (HDHD):

Thymidine diphosphate (TDP) + rhamnose + HDHD →
Rhamnosyl-HDHD + TDP-L-rhamnose → Dirhamnosyl-HDHD

Apart from glyceridic component, phospholipids contain nitrogen or phosphate part. The synthesis of phospholipids was studied mainly in *Escherichia coli* (**Fig. 5.67**).

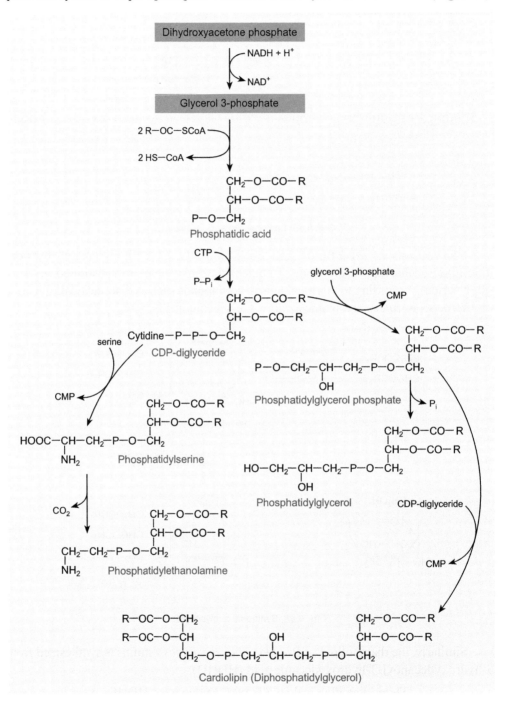

Fig. 5.67. Biosynthesis of phospholipids in *Escherichia coli.*

An important role in their synthesis is played by phosphatidic acid, which originates from the esterification of glycerol 3-phosphate in the presence of acyl-CoA and acyl protein carrier. The key intermediate is CDP-diglyceride, synthesized from phosphatidic acid and CTP. In combination with glycerol, this compound gives origin to phosphatidylglycerol and, in combination with serine, phosphatidylserine, which is further decarboxylated to phosphatidylethanolamine. The reaction between CDP-diglyceride and 3-phosphatidylglycerol phosphate results in cardiolipin (syn. phosphatidylglycerol).

Biosynthesis of carotenoids. Carotenoids are a large group of natural pigments, which have from red through orange to yellow colors. These compounds are synthesized by some microorganisms (e.g., microalgae, fungi, and bacteria). The light harvesting is one of the important roles of carotenoids in physiological functions. Apocarotenoids are carotenoid-derived compounds. They play important roles in various biological activities. Many carotenoids and apocarotenoids have high economic value in feed, food, and pharmaceutical industries, supplements, and cosmetics.

Furthermore, microorganisms usually produce mixtures of these molecules with very similar physical and chemical properties (such as α- and β-carotenes). The chemical synthesis of carotenoids is expensive due to structural complexity (e.g., astaxanthin has many unsaturated bonds and two chiral regions). Biotechnology via the rapidly advancing metabolic engineering and synthetic biology approaches has led to alternative ways to attain several carotenoids and apocarotenoids at relatively high titers and yields using fast-growing microorganisms. Carotenoids can be produced from fats and other basic organic metabolic building blocks by all these organisms.

There are over 1,100 known carotenoids, which can be further categorized into two classes, xanthophylls (which contain oxygen) and carotenes (which are pure hydrocarbons and contain no oxygen). All are derivatives of tetraterpenes, meaning that they are produced from 8 isoprene molecules and contain 40 carbon atoms. In general, carotenoids absorb wavelengths ranging from 400 to 550 nm (violet to green light). This causes the compounds to be deeply colored yellow, orange, or red.

Carotenoids belong among derivatives of isoprene, which are synthesized exclusively from acetyl-CoA. During the biosynthesis, acetyl-CoA condensates to acetoacetyl-CoA and then to 3-hydroxy-3-methylglutaryl-CoA (HMG-CoA), which is ultimately reduced to mevalonic acid. Mevalonic acid undergoes phosphorylation and decarboxylation to isopentenyl pyrophosphate, which after another set of reactions provides geranylgeranyl pyrophosphate. This final compound is later converted into carotenoids by the effect of specific enzymes. Isoprenyl pyrophosphate is the first substrate in the biosynthesis of farnesol and phytol, which are intrinsic parts of chlorophyll and also sterols and quinones. The scheme of carotenoid biosynthesis is shown in **Fig. 5.68**.

Carotenoids play two key roles in cyanobacteria: they absorb light energy for use in photosynthesis and protect chlorophyll from photodamage. Carotenoids that contain unsubstituted β-ionone rings (including β-carotene, α-carotene, β-cryptoxanthin, and γ-carotene) have vitamin A activity (meaning that they can be converted to retinol).

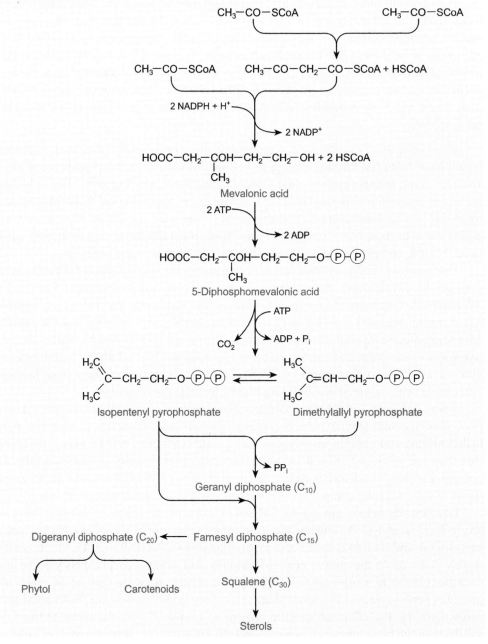

Fig. 5.68. Biosynthesis of carotenoids.

Biosynthesis of porphyrins. Primary substrates for the biosynthesis of the porphyrin rings, which are vital parts of *heme* enzymes and chlorophylls, are succinic acid and glycine. Their condensation by the effect of a specific synthetase is giving δ-aminolevulinic acid as a product. Two molecules of this acid are further condensed to substituted porphobilinogen, four molecules of which create a tetrapyrrolic ring of uroporphyrinogen III. The following oxidation of this compound results in protoporphyrin. The incorporation of a Fe atom during the chelating process results in the creation of hem. If the atom is Mg^{2+}, the final molecule is chlorophyll. Uroporphyrinogen III is also supposed to be part of the synthesis of cobamide. The process of porphyrin biosynthesis is illustrated in **Fig. 5.69**.

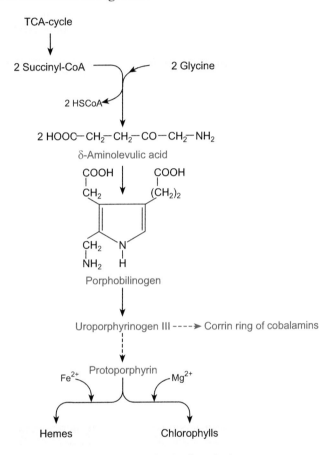

Fig. 5.69. Biosynthesis of porphyrins.

Thus, succinic acid and glycine are primary substrates for the biosynthesis of porphyrin rings. From protoporphyrin as an intermediate product, chlorophylls or hemes are formed.

5.6.3. Consumption of CO_2 by heterotrophic bacteria

The utilization of carbon dioxide and its connection to biosynthesis is not an exclusive trait of autotrophic microorganisms; however, it can be found even in heterotrophic bacteria. However, assimilated CO_2 does not represent the main source of carbon but is rather an additional option. The backward fixation of CO_2 in heterotrophs is mainly utilized in processes of carboxylation of keto acids. In some cases, the carboxylation may go simultaneously with reduction or may be accompanied by CoA. Depending on various energy sources, which may be utilized during carboxylation reactions, the reactions can be divided into three groups.

Reactions of the *first group* exploit energy during the oxidation of reduced pyrimidine coenzymes — given processes occur in the presence of malic acid, gluconic acid, and isocitric acid (**Fig. 5.70**). Malate dehydrogenase (EC 1.1.1.37) was found in *Lactobacillus arabinosus*; dehydrogenase of gluconic acid is present in *Bacillus subtilis*, *B. megaterium*, *Pseudomonas fluorescens*, and *Leuconostoc mesenteroides*.

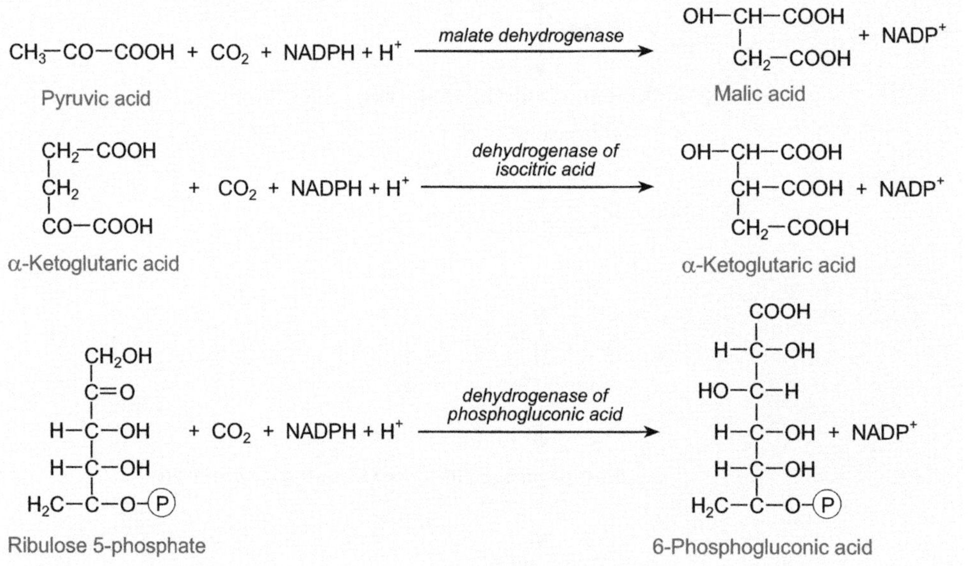

Fig. 5.70. Reactions of the first group.

Reactions of the *second group* depend on ATP as an energy source — carboxylation enzymes of such reactions were described in various bacterial genera. Propionyl carboxylase (EC 6.4.1.3) is synthesized in genera such as *Rhodospirillum rubrum* and *Mycobacterium smegmatis*. Methylcrotonyl carboxylase (EC 6.4.1.4) and acetyl carboxylase (EC 6.4.1.2) are described in mycobacteria and pyruvate carboxylase (EC 6.4.1.1) in *Pseudomonas citronellolis*. Fixation reactions of CO_2 that are dependent on ATP are mediated by biotin, which forms a peptide bond with CO_2 and thus enables the fixation of carbon. The process of carboxylation goes by the scheme in **Fig. 5.71**.

Carboxylation is a chemical reaction, in which a carboxylic acid group is produced by treating a substrate with carbon dioxide. The opposite reaction is decarboxylation. The term "carbonation" is sometimes used synonymously with carboxylation, especially when applied to the reaction of carbanionic reagents with CO_2. More generally, carbonation usually describes the production of carbonates.

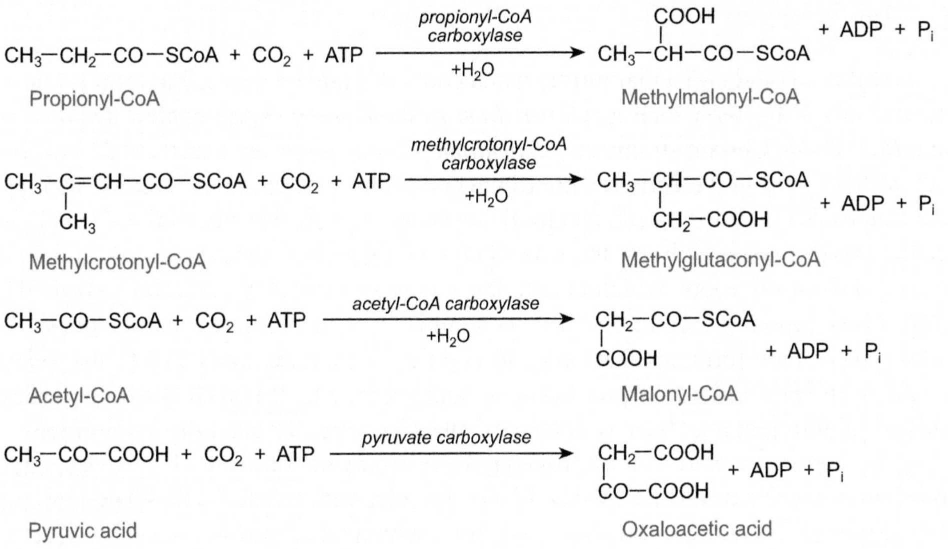

$$CH_3-CH_2-CO-SCoA + CO_2 + ATP \xrightarrow[\text{+H}_2\text{O}]{\text{propionyl-CoA carboxylase}} \begin{array}{l} COOH \\ | \\ CH_3-CH-CO-SCoA \end{array} + ADP + P_i$$

Propionyl-CoA   Methylmalonyl-CoA

$$\begin{array}{l} CH_3-C=CH-CO-SCoA \\ \quad\quad | \\ \quad\quad CH_3 \end{array} + CO_2 + ATP \xrightarrow[\text{+H}_2\text{O}]{\text{methylcrotonyl-CoA carboxylase}} \begin{array}{l} CH_3-CH-CO-SCoA \\ \quad\quad | \\ \quad\quad CH_2-COOH \end{array} + ADP + P_i$$

Methylcrotonyl-CoA   Methylglutaconyl-CoA

$$CH_3-CO-SCoA + CO_2 + ATP \xrightarrow[\text{+H}_2\text{O}]{\text{acetyl-CoA carboxylase}} \begin{array}{l} CH_2-CO-SCoA \\ | \\ COOH \end{array} + ADP + P_i$$

Acetyl-CoA   Malonyl-CoA

$$CH_3-CO-COOH + CO_2 + ATP \xrightarrow{\text{pyruvate carboxylase}} \begin{array}{l} CH_2-COOH \\ | \\ CO-COOH \end{array} + ADP + P_i$$

Pyruvic acid   Oxaloacetic acid

Fig. 5.71. Reactions of the second group.

Reactions of the ***third group*** are independent on the external sources of energy — these reactions utilize activated substrate, mostly phosphoenolpyruvic acid, and ribulose 2-phosphate. Specific enzymes are found both in autotrophic and heterotrophic microorganisms, for instance, *Thiobacillus thiooxidans* and *Pseudomonas oxalaticus*, and in propionic bacteria. Carboxylation reactions are summarized in the given scheme:

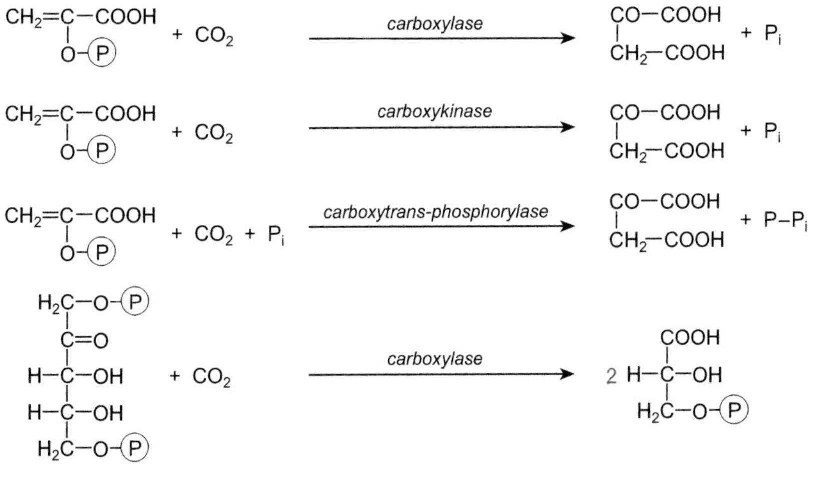

$$\begin{array}{l} CH_2=C-COOH \\ \quad\quad | \\ \quad\quad O-(P) \end{array} + CO_2 \xrightarrow{\text{carboxylase}} \begin{array}{l} CO-COOH \\ | \\ CH_2-COOH \end{array} + P_i$$

$$\begin{array}{l} CH_2=C-COOH \\ \quad\quad | \\ \quad\quad O-(P) \end{array} + CO_2 \xrightarrow{\text{carboxykinase}} \begin{array}{l} CO-COOH \\ | \\ CH_2-COOH \end{array} + P_i$$

$$\begin{array}{l} CH_2=C-COOH \\ \quad\quad | \\ \quad\quad O-(P) \end{array} + CO_2 + P_i \xrightarrow{\text{carboxytrans-phosphorylase}} \begin{array}{l} CO-COOH \\ | \\ CH_2-COOH \end{array} + P-P_i$$

$$\begin{array}{l} H_2C-O-(P) \\ | \\ C=O \\ | \\ H-C-OH \\ | \\ H-C-OH \\ | \\ H_2C-O-(P) \end{array} + CO_2 \xrightarrow{\text{carboxylase}} 2\begin{array}{l} COOH \\ | \\ H-C-OH \\ | \\ H_2C-O-(P) \end{array}$$

The meaning of the heterotrophic fixation of CO_2 lies in the fact that an additional source of carbon may enable the synthesis of an organic acid, which then serves as an input substrate for the synthesis of amino acids, lipids, and other compounds. Apart from this, they help maintain carboxylation processes in balance and harmonize the concentration of metabolites, which arise from various metabolic pathways.

5.6.4. Fixation of molecular nitrogen

Ability to bind molecular nitrogen and convert it into an accessible form is known mostly in soil bacteria such as *Clostridium pasteurianum* or the genera *Azotobacter* and *Rhizobium*. Certain strains of *Clostridium pasteurianum* are strictly anaerobic and very often found in the complex microbial community, where other bacteria facilitate the anaerobic condition, while metabolizing spare oxygen. Species of the *Azotobacter* genus are focused mainly on the rhizosphere of the soil, rich in organic compounds excreted from the roots, which enable the steady growth of the microorganism. The *Rhizobium* genus is well known for its symbiosis with *Fabaceae* family plants, on whose roots they form small tubers and nodules, which are connected to the xylem, enabling the molecular transport between both symbionts. It is well known nowadays that aerial nitrogen may be assimilated by other bacteria, for example, phototrophs.

The donor of electrons for nitrogen reduction may be formic acid, pyruvic acid, or molecular hydrogen (mostly in anaerobes). The electron transfer is facilitated through ferredoxin or flavodoxin in the reduced form, which takes part in a catalytic reaction with nitrogenases according to the scheme in **Fig. 5.72**.

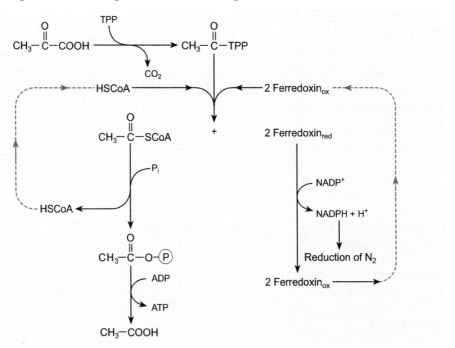

Fig. 5.72. Catalytic reaction of ferredoxin with nitrogenases.

Mechanisms of nitrogen fixation are not fully understood. Existing data suggest that the process is facilitated through a multienzyme complex with nitrogenase effect. This enzymatic complex probably comprised two subunits. The first is somewhat heavy; it has a molecular mass of about 200–300 kDa and contains non-heme-bound iron, molybdenum, and SH-groups. The second one is smaller (50 kDa) with a lesser amount of iron and sulfur and completely without molybdenum. Oxygen irreversibly inhibits the smaller subunit. In *Azotobacter* sp., the mechanism for the electron transfer is a bit different because of the existing TCA cycle. During the reduction of nitrogen, ATP is demanded in the amount of four moles per electron pair. Intermediates of the nitrogen reduction are probably compounds like diamine or hydrazine, which are converted to ammonia in the final stage.

$$N_2 \longrightarrow N{\equiv}N \xrightarrow[\substack{2\,[H]}]{\substack{n\text{ATP}}} HN{=}NH \xrightarrow[\substack{2\,[H]}]{\substack{n\text{ATP}}} H_2N{-}NH_2 \xrightarrow[\substack{2\,[H]}]{\substack{n\text{ATP}}} 2\,NH_3$$

$$\text{Diamide} \qquad \text{Hydrazine}$$

Thus, the reduction of nitrogen to ammonia is a complex process with the formation of diamine and hydrazine as intermediate products.

5.6.5. Biosynthesis of amino acids

The biosynthesis of amino acids takes place either by direct amination or transamination of a relatively small number of starting compounds. The source of nitrogen for **amino** and **imino groups** is nitrogen supplied to autotrophic bacteria in the form of ammonium salts, resulting from the denitrification of nitrates. Other bacteria may also use ammonia released during the decomposition of organic nitrogenous substances. As indicated previously, for some bacteria, the characteristic amino acid formation is caused by ammonia arising from atmospheric nitrogen fixation. The starting substrates of amino acid biosynthesis include:

(1) α-ketoglutaric acid from which glutamic acid is derived, which then provides glutamine, arginine, and proline;
(2) oxaloacetic acid producing aspartic acid, which then provides asparagine, methionine, threonine, lysine, and isoleucine;
(3) pyruvic acid, which is converted to alanine, valine, and leucine;
(4) phosphoglyceric acid, which produces serine, glycine, and cysteine;
(5) phosphoenolpyruvic acid and erythrose 4-phosphate, which are precursors to the biosynthesis of phenylalanine, tyrosine, and tryptophan;
(6) imidazole glycerol phosphate, from which histidine is formed.

Glutamic acid group. The chemical synthesis of glutamic acid was supplanted by the aerobic fermentation of sugars and ammonia in the 1950s, with the organism *Corynebacterium glutamicum* (also known as *Brevibacterium flavum*) being the most widely used for production. Glutamic acid is formed by the direct amination of α-ketoglutaric acid. The reaction occurs with the presence of glutamate dehydrogenase found in *E. coli* and *K. aerogenes*:

$$\begin{array}{c} CH_2{-}CH_2{-}COOH \\ | \\ CO{-}COOH \end{array} + NH_3 \xrightarrow[-H_2O]{\substack{NADH + H^+ \qquad NAD^+}} \begin{array}{c} CH_2{-}CH_2{-}COOH \\ | \\ CH{-}NH_2{-}COOH \end{array}$$

Glutamine is formed by the amination of glutamic acid in the presence of ATP:

$$H_2N-CH-COOH \atop CH_2-CH_2-COOH \quad + NH_3 \quad \xrightarrow[\;P_i\;]{ATP \quad ADP} \quad H_2N-CH-COOH \atop CH_2-CH_2-CO-NH_2$$

Glutamine is a growth factor for, for example, lactic acid bacteria but also plays an important role in amination processes in the biosynthesis of amino sugars and purines. Arginine is formed from glutamic acid via ornithine and citrulline (**Fig. 5.73**).

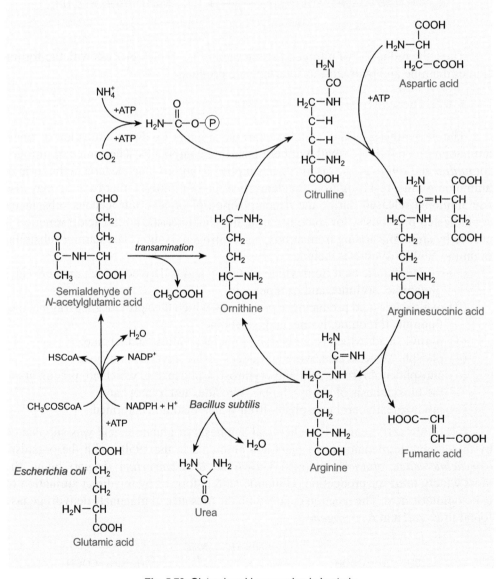

Fig. 5.73. Glutamic acid conversion in bacteria.

In the first stage of the process carried out by *Micrococcus glutamicus* and *E. coli* in the presence of acetyl-CoA and ATP, *N*-acetylglutamic acid is first formed and then converted by dehydrogenase to semialdehyde. The latter, after transamination via another glutamic acid molecule, provides ornithine. The conversion of ornithine to citrulline happens in the presence of carbamoyl phosphate. The reaction is catalyzed by transcarboxylase. In the next reaction, glutamic acid and citrulline are formed. Citrulline is converted to argininosuccinic acid, which converts to fumaric acid and arginine. In the final step, the cleavage of urea produces ornithine, which then returns to the cycle. Arginine conversion is performed by *Bacillus subtilis* strains.

Proline is synthesized from glutamic acid through its semialdehyde. The formation of the semialdehyde is associated with the phosphorylation of glutamic acid with the participation of ATP. By spontaneous reaction, the semialdehyde produces pyrrolidine-3-carboxylic acid, which is converted by dehydrogenase to proline. The reaction was described in *E. coli* and *K. aerogenes* (**Fig. 5.74**).

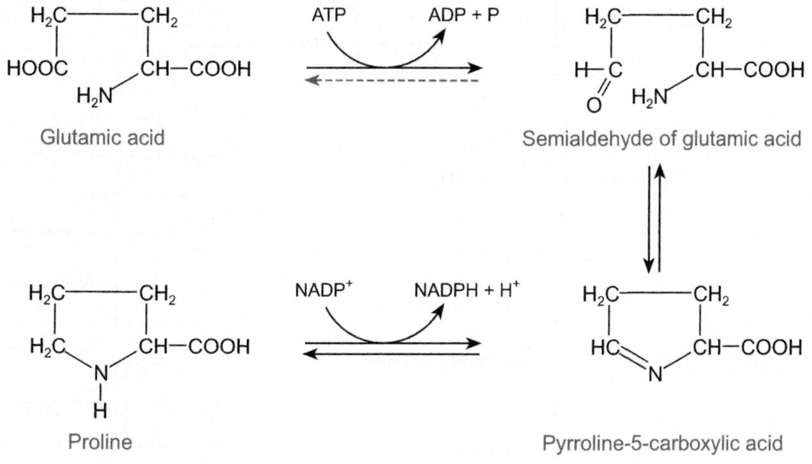

Fig. 5.74. Biosynthesis of proline.

Aspartic acid group. The formation of aspartic acid is effected by transamination from oxaloacetic acid with the participation of glutamic acid. The transmission of an amino group is catalyzed by a transaminase, the coenzyme of which is pyridoxal phosphate:

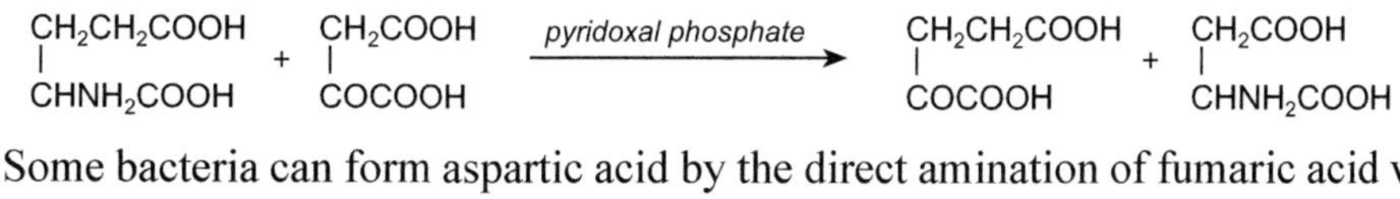

Some bacteria can form aspartic acid by the direct amination of fumaric acid with the participation of asparaginase:

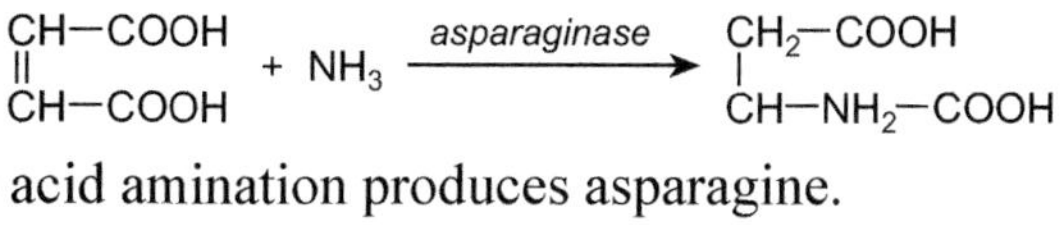

Further aspartic acid amination produces asparagine.

Threonine, *isoleucine*, and *methionine* are synthesized from aspartic acid semial-dehyde via homoserine (**Fig. 5.75**). After the formation of homoserine by reduction of aspartic acid semialdehyde, the process is divided into two branches. One leads through the phosphorylation of homoserine via ATP to form threonine and isoleucine.

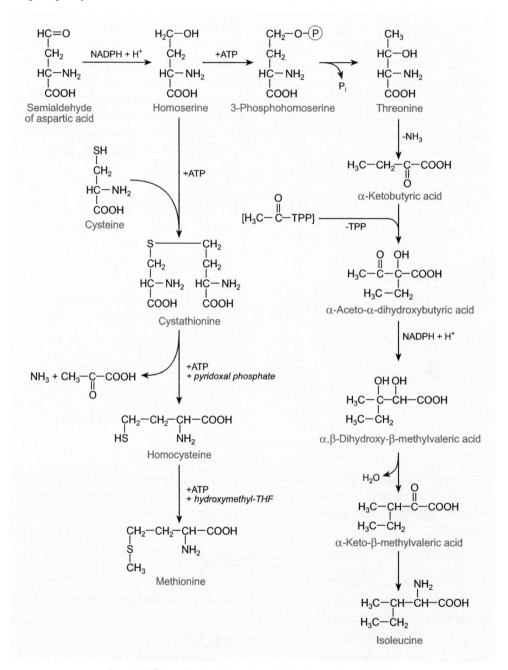

Fig. 5.75. Biosynthesis of threonine, isoleucine, and methionine.

In the second branch, homoserine condenses with cysteine to form cystathionine, which is dissimilated to homocysteine, pyruvic acid, and ammonia. ATP and pyridoxal phosphate participate in the reaction. Methionine is formed by the methylation of homocysteine mediated by carboxymethyl-THF in the presence of ATP, vitamin B_{12}, and decarboxylase. The course of individual reactions in *E. coli* and *Neurospora crassa* is shown in **Fig. 5.75**.

Isoleucine biosynthesis is derived from threonine, the deamination of which produces α-ketobutyric acid, which condenses acetyl-TPP to α-aceto-α-hydroxybutyric acid. The reaction is catalyzed by the same enzyme as the condensation reaction in acetobutyric acid formation. Acetohydroxybutyric acid is converted by reduction and dehydration to α-keto-β-methylvaleric acid, providing isoleucine after transamination mediated by glutamic acid **(Fig. 5.75)**.

Lysine is formed from aspartic acid semialdehyde via dihydropicolinic acid and diaminopimelic acid **(Fig. 5.76)**.

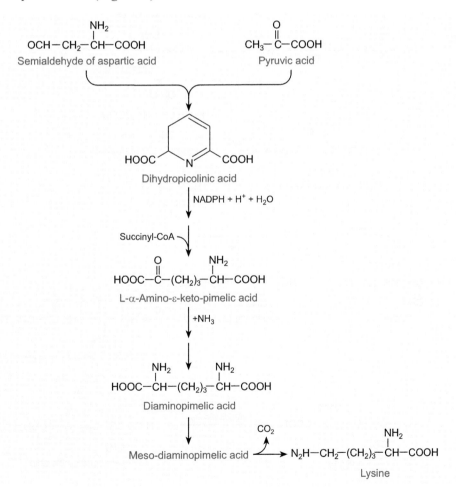

Fig. 5.76. Biosynthesis of lysine.

Group of alanine, leucine, and valine. Alanine is produced by members of the *Bacillus* genus from pyruvic acid in the presence of alanine dehydrogenase:

$$CH_3{-}CO{-}COOH + NH_3 \xrightarrow[-H_2O]{NADH + H^+ \quad NAD^+} CH_3{-}CH{-}NH_2{-}COOH$$

Leucine and valine are formed from pyruvate via α-ketoisovaleric acid (**Fig. 5.77**). Valine is produced by the transamination of α-ketoisovaleric acid via glutamic acid.

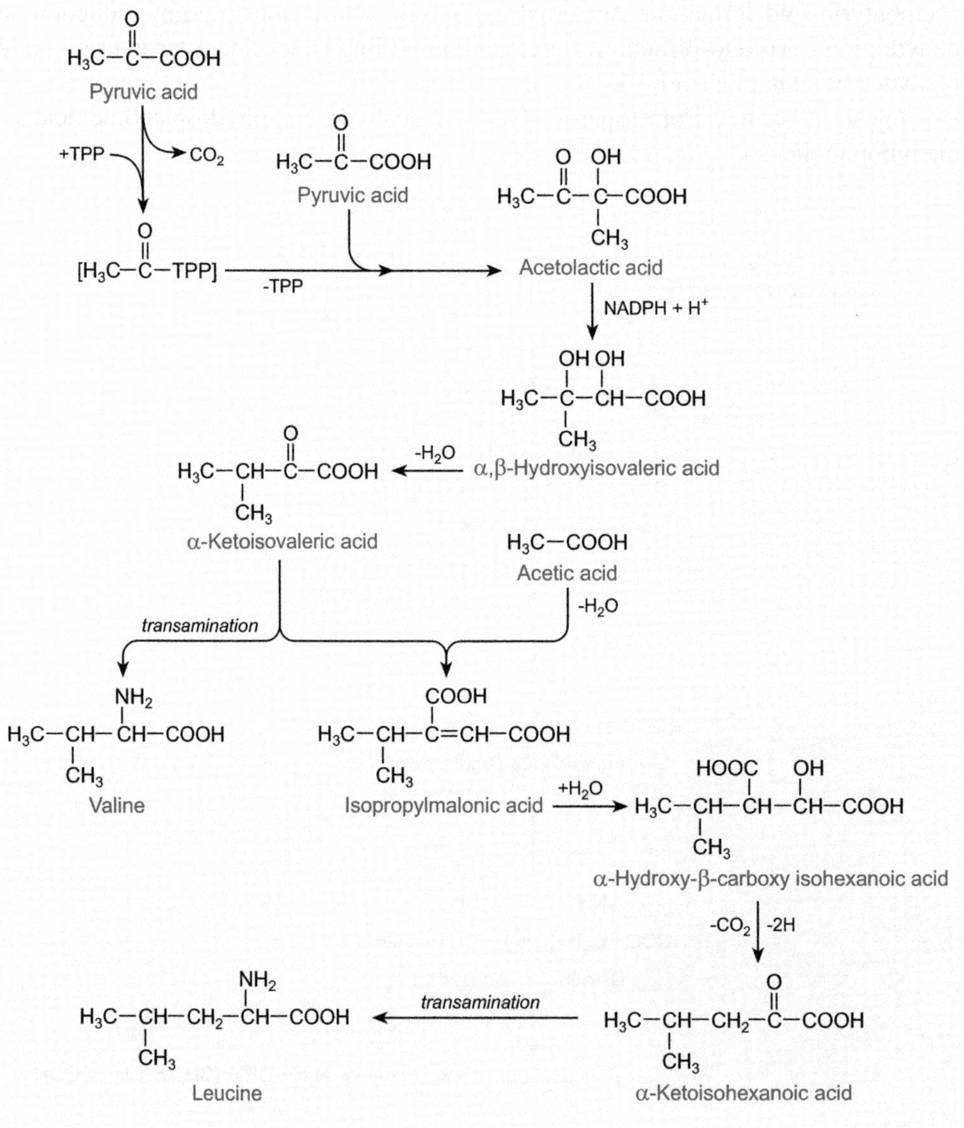

Fig. 5.77. Biosynthesis of leucine and valine.

The biosynthesis of leucine initially results in the condensation of α-ketoisovaleric acid and acetic acid to β-carboxy-β-hydroxyisocaproic acid, which is converted to isopropylmalic acid. Its hydration results in α-hydroxy-β-carboxyisocaproic acid, which is decarboxylated with the simultaneous reduction to α-ketoisocaproic acid. The transamination of this acid with the participation of glutamic acid again results in leucine.

Group of serine, glycine, and cysteine. These amino acids are derived from 3-phosphoglyceric acid, which converts to 3-phosphohydroxypyruvic acid, which then converts through transamination into serine in the presence of glutamic acid. In subsequent reactions, serine gives glycine after the cleavage of formaldehyde. In the presence of acetyl-CoA, serine may convert to acetylserine, which is converted to the cysteine by the conversion of the hydroxyl group to the sulfhydryl group and by the cleavage of the phosphate (**Fig. 5.78**).

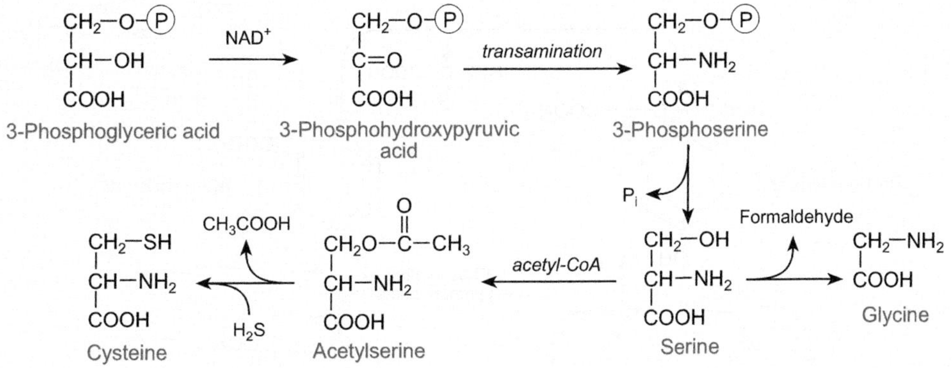

Fig. 5.78. Biosynthesis of serine, glycine, and cysteine.

Group of aromatic amino acids. Phenylalanine, tyrosine, and tryptophan are formed from aromatic compounds, whose precursor is shikimic acid (**Fig. 5.79**). This acid is formed via dehydroquinic acid, which is formed by the condensation of erythrose 4-phosphate and phosphoenolpyruvate. The phosphorylation of shikimic acid in the presence of ATP and condensation with phosphoenolpyruvic acid form prephenic or anthranilic acid via chorismic acid. Prephenic acid produces phenylpyruvic acid or hydroxyphenylpyruvic acid. The first one provides phenylalanine through transamination in the presence of glutamic acid, the second tyrosine. Anthranilic acid reacts with phosphoribosyl pyrophosphate to form *N*-(5′-phosphoribosyl) anthranilic acid, which is decarboxylated to indole glycerophosphate. Its amination in the presence of serine forms tryptophan. The presence of enzymes catalyzing aromatic amino acid biosynthesis was found in cells of *E. coli*, *K. aerogenes*, and *S. enterica*.

Histidine is formed from imidazole glycerol phosphate, which is an intermediate of the conversion of phosphoribosyl-ATP produced by the condensation reaction between ATP and phosphoribosyl pyrophosphate (**Fig. 5.80**).

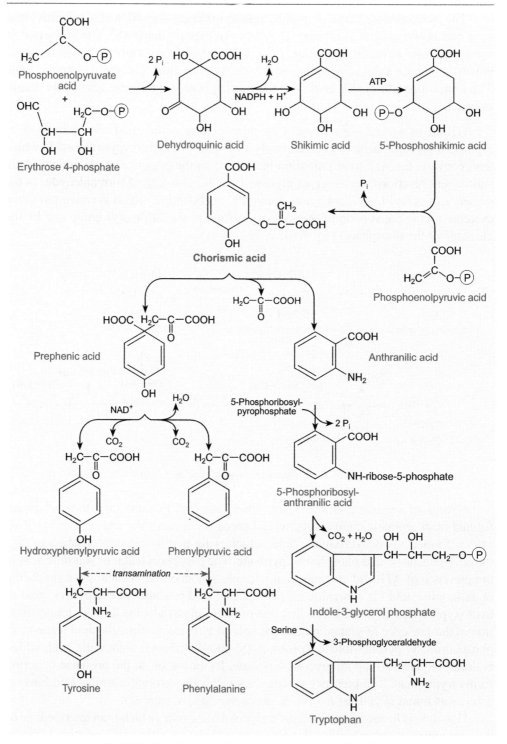

Fig. 5.79. Biosynthesis of phenylalanine, tyrosine, and tryptophan.

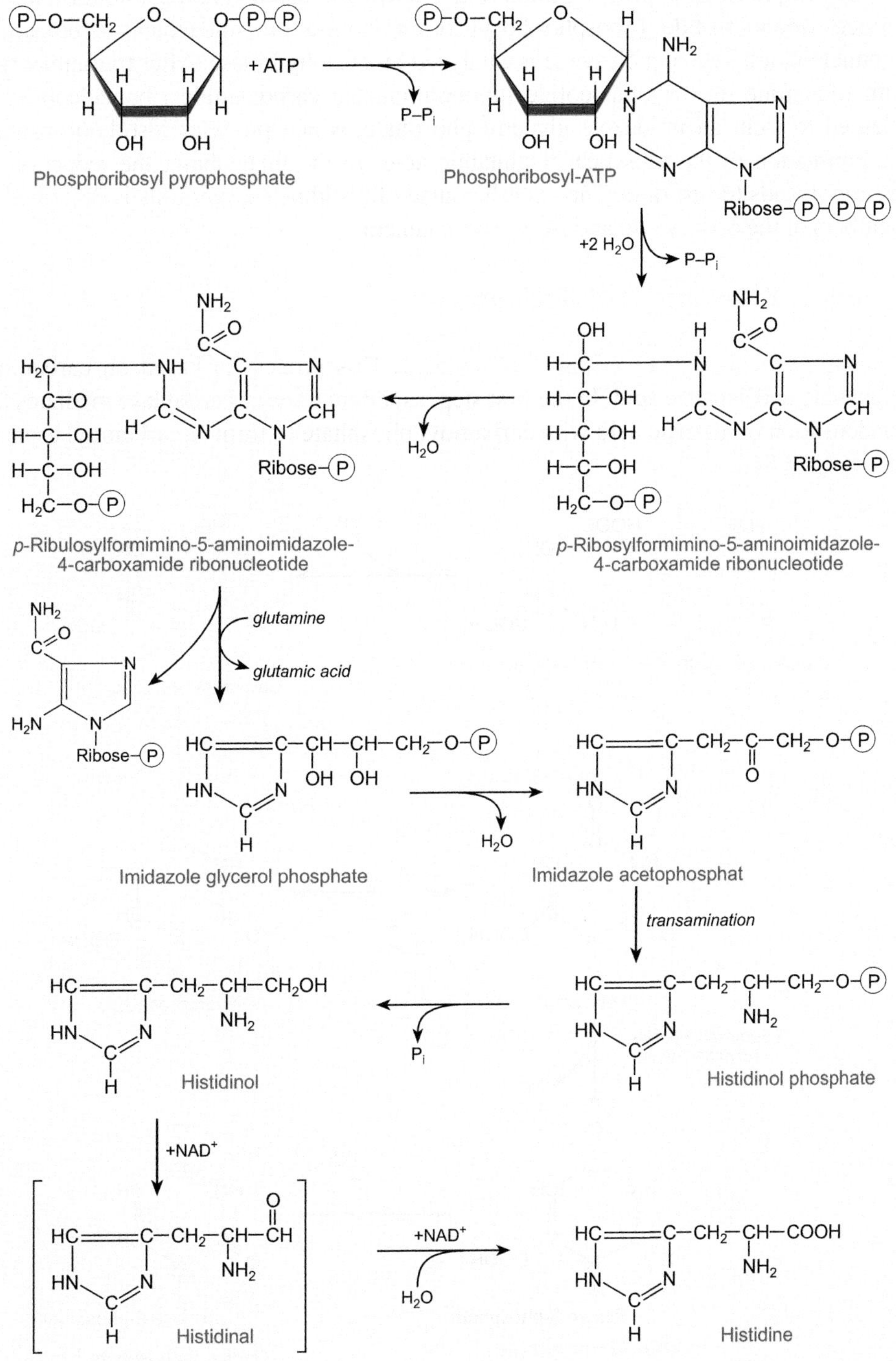

Fig. 5.80. Biosynthesis of histidine.

In phosphoribosyl-AMP, a purine ring is opened in the next reaction to form a substituted ribonucleotide (phosphoribosyl-formimino-5-aminoimidazole-4-carboxamide ribonucleotide). The ring cleavage is catalyzed by cyclohydrolase. After transamination with glutamine, the ribonucleotide (aminoimidazole carboxamide ribonucleotide) is released to form an imidazole glycerol phosphate, which provides histidinol through transamination in the presence of glutamic acid. In the final phase, the reduction of histidinol leads to histidine. The overall course of histidine biosynthesis is described in members of the *Salmonella* and *Neurospora* genera.

5.6.6. Biosynthesis of nucleotides

Biosynthesis of pyrimidine nucleotides. This process is based on carbamoyl phosphate and aspartic acid. In the first step, aspartate transcarbamoylase mediates the condensation of aspartic acid with carbamoyl phosphate to form *N*-carbamoyl aspartic acid (**Fig. 5.81**).

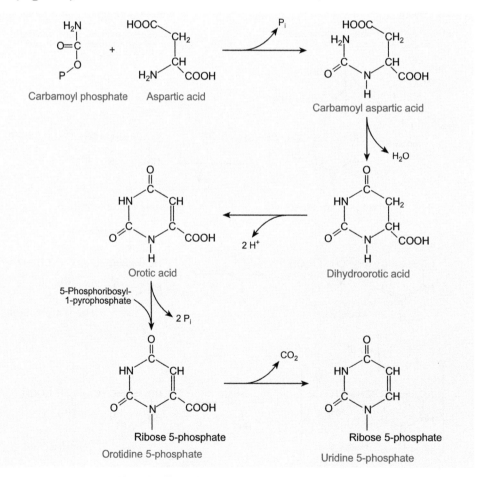

Fig. 5.81. Biosynthesis of uridine monophosphate.

In the next reaction, after dehydration, the ring closes to form dihydroorotic acid which is dehydrogenated to orotic acid. The enzymes catalyzing these reactions are dihydroorotase and dihydroorotate dehydrogenase, which is of flavin type. In the presence of phosphoribosyl pyrophosphate and pyrophosphorylase, orotic acid produces orotidine 5′-monophosphate, which is, after decarboxylation, converted by orotidyl 5′-monophosphate to uridine 5-phosphate (UMP), also known as 5′-uridylic acid.

Other pyrimidine nucleotides are formed by the phosphorylation of uridine monophosphate with the participation of ATP. The conversion to cytosine triphosphate is accompanied by transamination, which involves the attachment of the amide group of glutamine to the uridine triphosphate carbon.

Biosynthesis of purine nucleotides. Unlike pyrimidine bases, the purine nucleotide biosynthesis first produces a sugar and a phosphate component, which is linked to the formation of the purine nucleotide from small units of atomic groups as shown in **Fig. 5.82**.

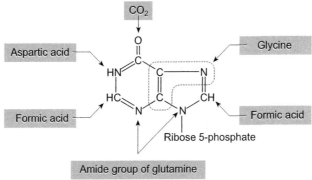

Fig. **5.82.** Origin of constituents in purine nucleotide biosynthesis.

The starting compound for the synthesis of purine bases, which is similar in animals and bacteria, is phosphoribosylamine formed during the transamination reaction between 5-phosphoribosyl-1-pyrophosphate and glutamine. In the presence of ATP, phosphoribosylamine condenses with glycine to form glycinamide ribonucleotide, which is the basis of the purine nucleotide. By attaching a single-carbon graft via formyltetrahydrofolic acid and subsequent amination mediated by glutamine, a formyl-glycinamidine ribonucleotide is formed, which is converted to 5-aminoimidazole ribonucleotide in the presence of ATP after the ring closure. The reactions are catalyzed by the respective ligases. These enzymes are also involved in further reactions, in which the carboxylation of 5-imidazole ribonucleotide is carried out on the corresponding acid, which, after transamination mediated by aspartic acid in the presence of ATP, provides 5-aminoimidazole-4-carboxamide ribonucleoside. With another formyl group, this compound provides 5-formamidoimidazole-4-carboxamide ribotide, which is converted to inosinic acid by closing a six-membered ring. The transfer of the formyl group occurs via formyl-THL; the ring closure is catalyzed by the enzyme. The overall course of inosinic acid biosynthesis is shown in **Fig. 5.83**.

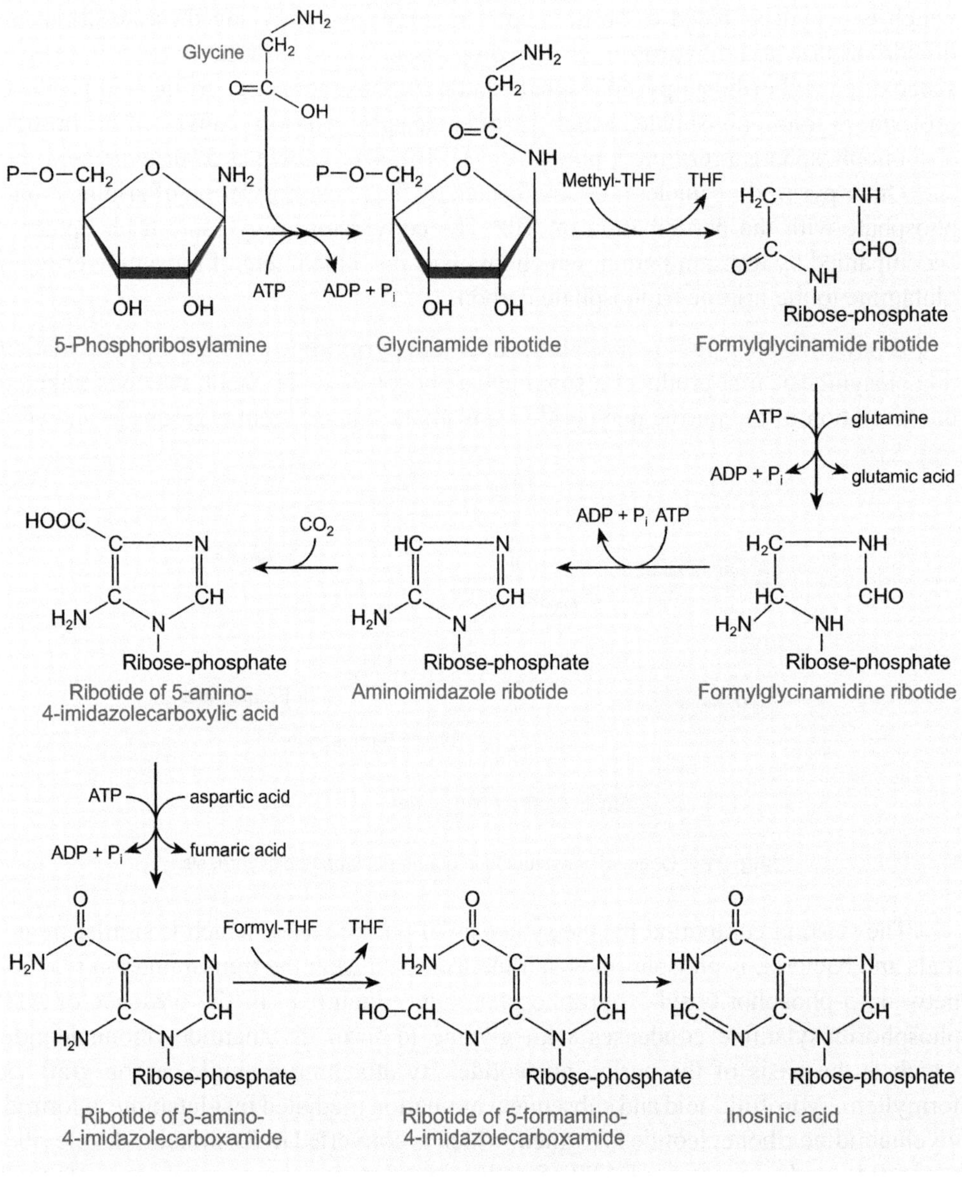

Fig. 5.83. Biosynthesis of inosinic acid.

Inosinic acid is a nucleoside monophosphate. It is the ribonucleotide of hypoxanthine and the first nucleotide formed during the synthesis of purine.

In the next phase, inosinic acid provides both adenylic and guanylic acids. The adenylic acid formation is accompanied by transamination with the participation of aspartic acid and GTP (**Fig. 5.84**). Guanylic acid is formed by the amination of an oxidized intermediate, such as xanthylic acid. The amino group is provided by glutamine.

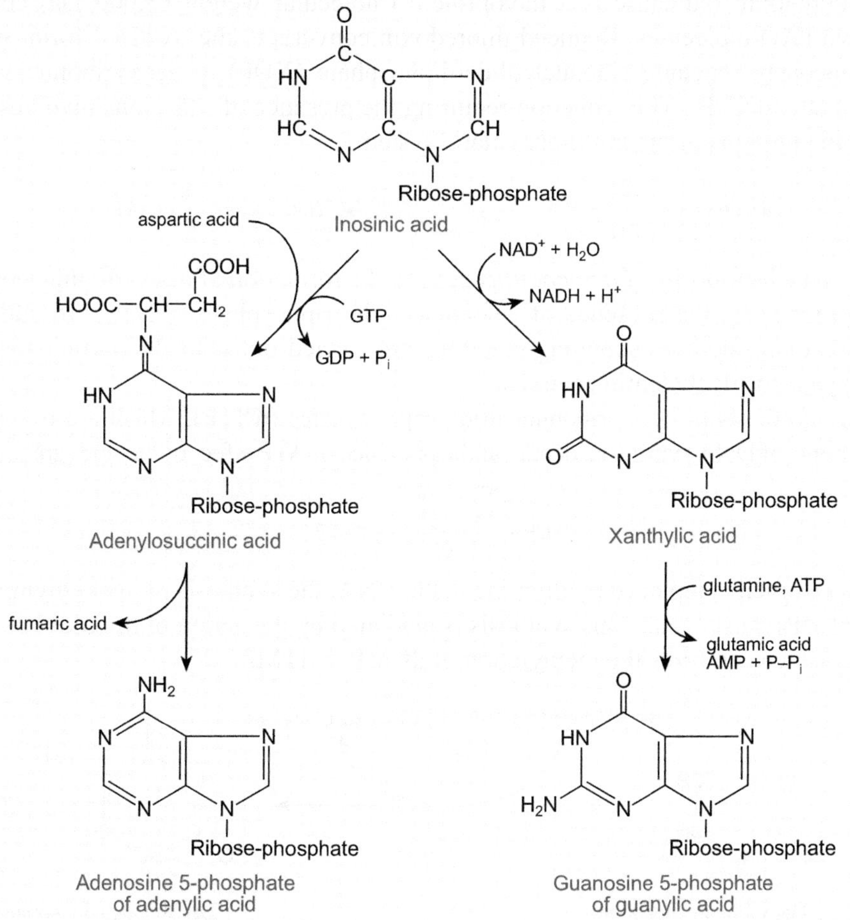

Fig. 5.84. Biosynthesis of adenylic and guanylic acids.

Reduction of ribonucleotides and deoxyribonucleotides. All four nucleoside diphosphates (ADP, GDP, UDP, and CTP) can be directly reduced to the corresponding deoxy analogs. This reduction is carried out with the participation of a multi-enzymatic system comprising four different enzymatic proteins. To reduce ribose to deoxyribose, a pair of hydrogen atoms is also required, the donor of which is NADPH + H$^+$ in the final stage. However, the electron donor is not this coenzyme, but a reduced form of thioredoxin. Its molecule (molecular weight 12,000 Da) is a thermostable protein containing 108 amino acid residues and two free HS groups that can be converted into the oxidized form and vice versa. The reduction of the disulfide form takes place in the presence of NADPH + H$^+$ in the reaction catalyzed by thioredoxin reductase, which is the second of four proteins of the previously mentioned multi-enzymatic system:

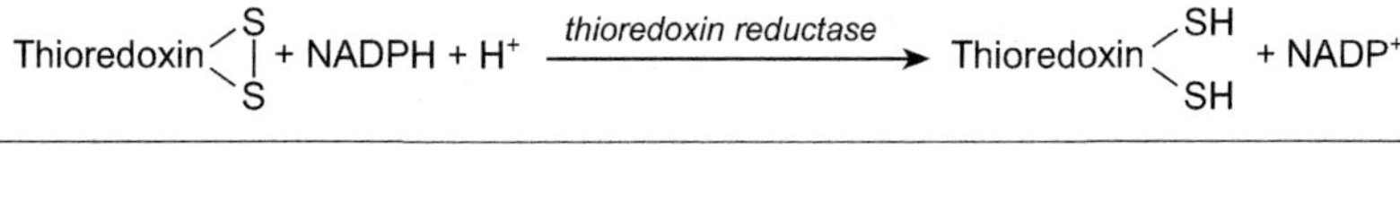

Thioredoxin reductase is a flavoprotein (molecular weight 68,000 Da) containing two FAD molecules. Reduced thioredoxin converts to the oxidized form, while simultaneously reducing ribonucleotide-diphosphate (XDP) to deoxyribonucleotide diphosphate (dXDP). The reduction requiring the presence of Mg^{2+} ions involves two enzymatic proteins of the multi-enzymatic system:

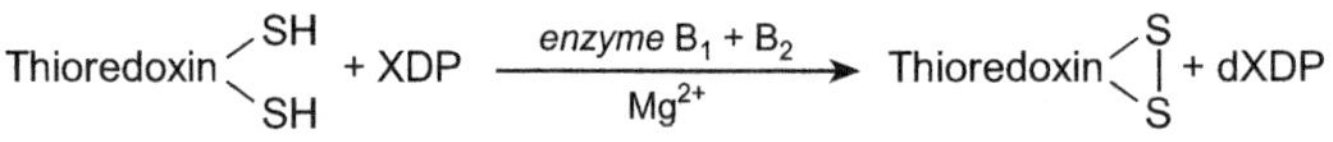

$$\text{Thioredoxin}\begin{array}{c} SH \\ SH \end{array} + XDP \xrightarrow[Mg^{2+}]{enzyme\ B_1 + B_2} \text{Thioredoxin}\begin{array}{c} S \\ S \end{array} + dXDP$$

In some lactobacilli (*Lactobacillus* genus), the formation of deoxyribonucleotides is conditioned by the presence of ribonucleoside triphosphates (XTPs). In addition to thioredoxin, the corresponding reactions are carried out with the participation of 5′-deoxyadenosylcobalamin coenzyme.

The synthesis of deoxyribonucleotide triphosphates (dXTP), which are the initial components of DNA synthesis, occurs at the presence of ATP effect of the relevant kinase:

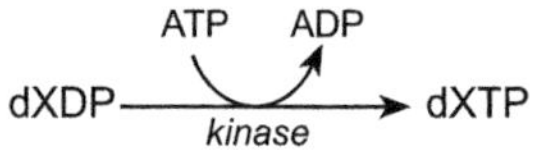

$$dXDP \xrightarrow[kinase]{ATP\ \ \ \ ADP} dXTP$$

Since uracil is replaced by thymine in the DNA, the synthesis of deoxythymidylic acid (dTMP) must occur. This synthesis is mediated by the action of thymidylate synthetase, which catalyzes the methylation of dUMP to dTMP:

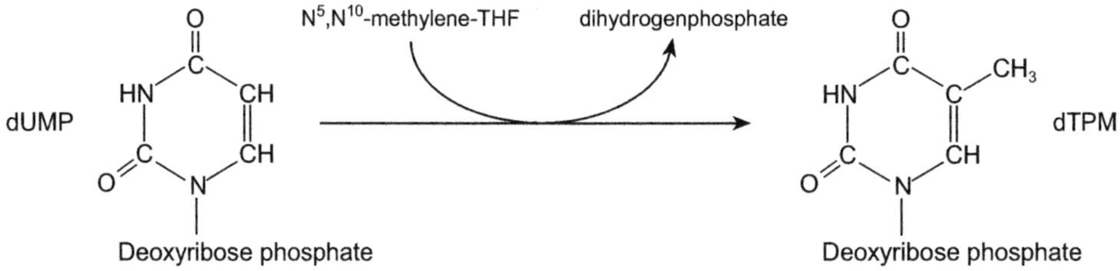

The coenzyme methylene-THF does not only serve as a donor of the methyl group but also as a donor of hydrogen in the transformation to dihydrofolic acid.

5.6.7. Biosynthesis of nucleic acids

Replication of DNA. The biosynthesis of DNA is happening in a semiconservative way, during which a new polynucleotide chain made by parental DNA replication in an exact copy. In the process of replication, the following enzymes take part:
- **nucleases** catalyzing the hydrolysis of the phosphodiester bond in positions 3′–5′ between neighboring nucleotides in the polynucleotide chain. Endonucleases are the most important for the replication of DNA; they split the bonds inside the chain and thereby determine a place for the initiation of replication;
- **DNA-polymerases** catalyze the formation of a phosphodiester bond between the α-phosphate group on 5′C of the nucleotide, entering into a polymer reaction, and the hydroxyl group on 3′C of the nucleotide in the polynucleotide chain, which serves as a primer (**Fig. 5.85**). Three polymerases were found in *E. coli*:

- **Pol I** catalyzing the addition of nucleotides to primers, which is a fragment of DNA;
- **Pol II**, which function is still unknown;
- **Pol III** catalyzing the addition of nucleotides to RNA-primers.

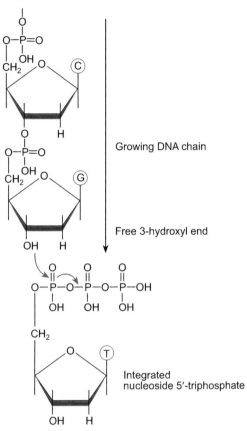

Fig. 5.85. Effect of DNA polymerase mechanism.

DNA ligase is connecting the ends of two chains in DNA by making phosphodiester bonds between the 3'-hydroxyl group at the end of the first chain and the 5'-phospho group at the end of the second chain. The reaction takes place with the occurrence of NAD^+, which appears in the reaction as a donor of the adenyl group, and by releasing the adenyl group, the needed energy is obtained.

$$NAD^+ + ligase \rightarrow ligase–AMP + NMN \text{ (nicotinamide mononucleotide)};$$

$$ligase–AMP + 5'P \text{ end of DNA} + 3'OH \text{ end of DNA} \rightarrow$$
$$phosphodiester \text{ bond} + ligase + AMP.$$

The typical replication of double chains of chromosomal DNA occur on individual chains after their separation. The process is initiated by the initial point of the replication, when the double helix opens by the effect of untangled proteins. The emerging gap is called ***replication eye***, which passes into a replication fork. Simultaneously with the separation of DNA chains in the place of their disconnection, the synthesis of a short chain of RNA takes place by the effect of RNA polymerase, which connects to the 5′ end of the polynucleotide chain of DNA. These RNA chains serve as primers, on which deoxyribonucleotides are connected step by step by the effect of DNA pol III and ATP with the creation of the phosphodiester bond 5′–3′. Deoxyribonucleotides are complementary to the nucleotides of template DNA. Because polymerization just happens in the direction 5′ → 3′ and chains of DNA are antiparallel, the replication of each chain runs in the opposite direction. The creation of short sequences of DNA chains, which contain around 1,000 nucleotides are called **Okazaki fragments (Fig. 5.86)**.

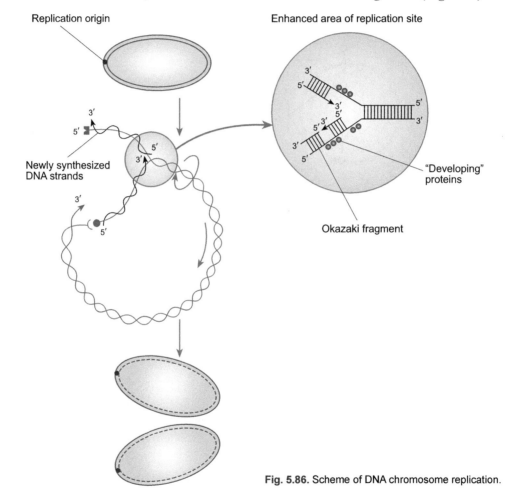

Fig. 5.86. Scheme of DNA chromosome replication.

RNA primers are hydrolase by the effect of pol I. The same enzyme catalyzes the connection of other nucleotides, which are complementary to template chains of DNA on the Okazaki fragments until the gaps are filled after the elimination of RNA primers. Simultaneously, the synthesis of new RNA primers begin, which determines the creation of new Okazaki fragments. This process repeats until the whole molecule of DNA is replicated. In the last phase, the neighboring ends 5′–3′ of the Okazaki fragments are connected by the effect of ligase and create a new polynucleotide chain. During the measurement of replication speed, it was found out that the polymerization of 9×10^4 nucleotides occurs in 1 minute.

Circular chromosomal double helix DNA of bacteria hold the form for the whole time of replication, which runs by the shift of the replication fork step by step along the whole chromosome. Both chromosomes contain one original (parental) fiber and one new fiber of DNA.

Biosynthesis of ribonucleotide acids occurs in the same way regardless of their difference in function. The process of transcription runs thanks to relevant ribonucleotides (ATP, GTP, CTP, UTP), complicated enzymatic system of RNA polymerase, and DNA which serves as a template. The polymerase that was isolated from cells of *E. coli* is formed from five subunits, which are polypeptide chains of different molecular weight: two α-chains (with molecular weight 41,000 Da), two β-chains (155,000 and 165,000 Da) and σ-chain (86,000 Da). It is presumed that the β′-chain contains a place, with which RNA polymerase bonds to DNA. The activity of RNA polymerase affects some other factors, which participated in ending the transcription and form proteins with molecular weight 200,000 Da. For example, the factor called Ψ factor regulates the synthesis of individual types of RNA depending on their amount.

The initiation of transcription happens on a specific place on DNA with an unusual sequence of bases. To that place, which is called ***promotor***, after branching, the double helix of DNA connects RNA polymerase, which catalyzes the polymerization of ribonucleotides and is complementary to the template DNA chain. The polymerization runs during the creation of phosphodiester bonds between the 3′-OH group of the growing RNA chain and the 5′α-phosphate group of free nucleoside triphosphate in the direction 5′ → 3′. Simultaneously, pyrophosphate is released:

$$n \, (ATP + GTP + CTP + UTP) \xrightarrow{\text{template DNA}} RNA + n \, P\text{–}P_i.$$

After the molecule of RNA is made, the synthesis is stopped, and RNA polymerase is released. Simultaneously, the newly synthesized RNA chain cuts off from the template chain. Thanks to that, there could be a restoration of hydrogen bonds between both DNA fibers, which return to the original structure of double helix. Different sizes of RNA molecules enable their synthesis on more places in the polynucleotide chains of DNA.

5.6.8. Biosynthesis of proteins

The process of biosynthesis of proteins is a complicated metabolic process, which runs in tight cooperation with nucleic acids. The most significant phase is the translation of genetic information, which was transferred during transcription from DNA to messenger RNA (mRNA). The result of this translation is the specification of the sequence of amino acids in a new polypeptide chain.

During translation, these processes are realized:

(1) amino acid activation;

(2) initiation of biosynthesis;

(3) elongation, when the lengthening of the peptide chain is happening;

(4) termination, which means the end of biosynthesis.

Amino acid activation runs with the participation of amino acid, ATP, and enzyme called aminoacyl synthetase. This enzyme has three bonding places, one of which is used for connecting with amino acids, the second for a bond with ATP, and the third for connecting with appropriate transmitted amino acid to tRNA. In the process of biosynthesis of proteins with 20 amino acids, the same number of specific enzymes — aminoacyl transferase and tRNA molecules — is used.

In the process of activation, a bond between amino acid and ATP is formed, firstly through aminoacyl adenylate (aminoacyl–AMP) by the effect of aminoacyl synthetase. Simultaneously, pyrophosphate is realized (**Fig. 5.87**).

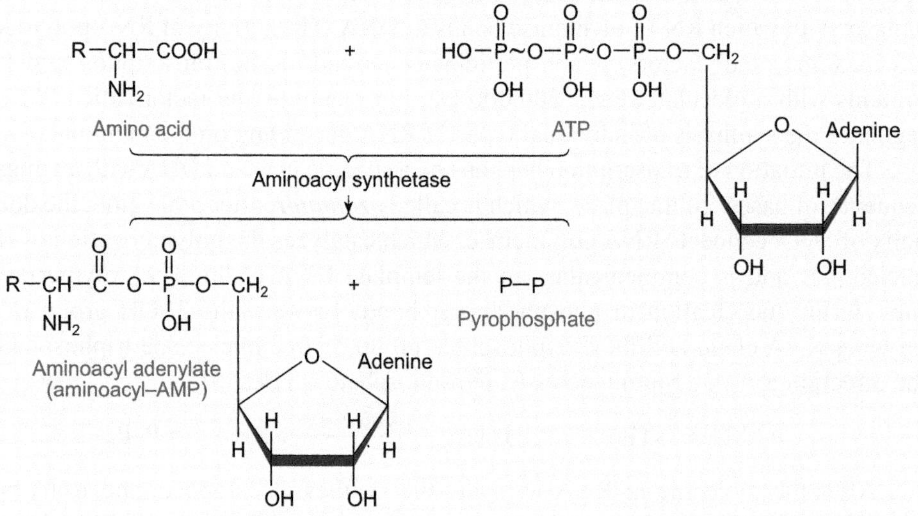

Fig. 5.87. Amino acid activation process.

The formed aminoacyl–AMP connects to the tRNA of the relevant amino acid with the participation of the same enzyme. The binding is between the carboxyl group of the amino acid and the 3′OH group of ribose. Simultaneously, the aminoacyl–tRNA complex is formed, and AMP is released according to this equation:

$$\text{Aminoacyl–AMP} + \text{tRNA} \xrightarrow{\text{aminoacyl synthetase}} \text{Aminoacyl–tRNA} + \text{AMP}.$$

The structure of the aminoacyl–tRNA is shown in the scheme in **Fig. 5.88**.

The biggest difference between individual tRNAs is observed in their inner structure, in the sequence of nucleotides, whose number is around 60. On the 3′ end of the nucleotide chain of each tRNA nucleotides are situated in the CCA sequence (cytidylic acid — cytidylic acid — adenylic acid). The shape of the tRNA molecule reminds a clover leaf, which is formed by four loops or branches signified by the character of the carried component. The components are an amino acid, pseudouridine, dihydrouridine, and anticodon. In some tRNAs, also a side branch occurs. Anticodon can be introduced as a triplet of nucleotides, which are complementary to a specific codon on messenger RNA. On the 3′ end, there is an anticodon closed by a purine nucleotide and on the 5′ end by a pyrimidine nucleotide.

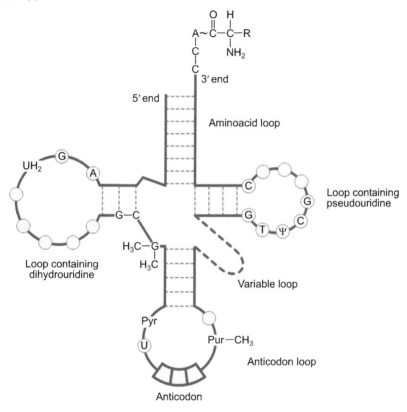

Fig. 5.88. General structure of aminoacyl–tRNA.

Initiation means the beginning of biosynthesis of proteins, which runs on ribozymes. The whole process starts with the dissociation of active ribozyme 70S into two subunits 30S and 50S. mRNA connects to the particle 30S via an initiation (start) codon. The initiation codon is sections on mRNA with an unusual sequence of nucleotides. Simultaneously, 30S also connects with the anticodon, tRNA, which transmits formylmethionine (tRNA$_{fMet}$). The connection of mRNA with subunit 30S and the attachment of tRNA$_{fMet}$ run with the participation of GTP and three unusual proteins called initiation factors IF 1–3. To the initiation complex formed, 50S ribosomal subunit is connected, which leads to the formation of active 70S ribosome. At the same time, GTP hydrolysis occurs, and the initiation factors, which can be shared at the next initiation, are released. The initiation process is shown in **Fig. 5.89**.

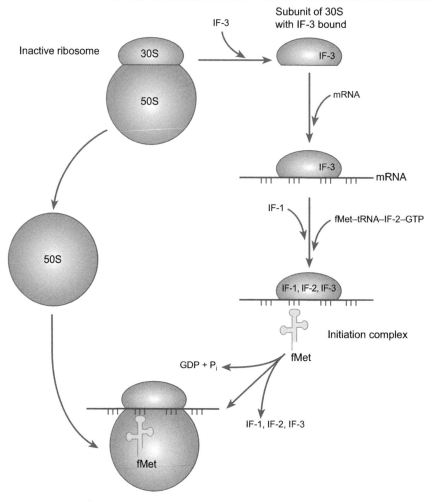

Fig. 5.89. Scheme of initiation of protein synthesis process.

Elongation represents an increase in the polypeptide chain on active 70S ribosome. The process is occurring on two binding sites: peptidyl (**P**) and aminoacyl (**A**). In the first stage after the transfer of tRNAfMet to the place, **P** connects with another molecule at site **A**. The codon of aminoacyl–tRNA is complementary to the codon of mRNA, which is located in the same place. Simultaneously, as a charge to GTP, a peptide bond is formed between the carboxyl group of the first amino acid and the amino group of the second amino acid. The connection of the two amino acids with the formation of a dipeptide is followed by the separation of the initial tRNA and with the release of tRNA from place P. By the effect of another elongation factor G and peptidyl transfe-rase, peptidyl–tRNA passes from place A to place P. Simultaneously with the movement of mRNA, her new codon joins place A. To the new codon with the complementary connection of an anticodon, another molecule of aminoacyl–tRNA is connected, and whole process repeats until the formation of a polypeptide chain. The scheme of the process of elongation is presented in **Fig. 5.90**.

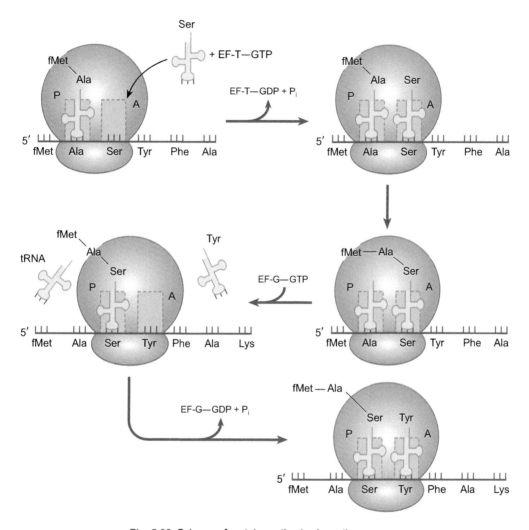

Fig. 5.90. Scheme of protein synthesis elongation process.

Termination the ending of proteosynthesis. The signal, which indicates the stopping of the process, is made by the termination codon with a nonsense sequence of nucleotides, for which no complementary anticodon aminoacyl–tRNA exists. The identification of codons with the nonsense sequence is possible because of relaxing factors R1–R3, which have proteinous character (molecular weight is around 44,000–47,000 Da). In the duration of the process of termination by the effect of these factors, peptidyl–tRNA is moved from place A to place P with the simultaneous hydrolysis of the bond between the end amino acid of the polypeptide chain and the relevant tRNA. Peptidyl transferase catalyzes the hydrolytic cleavage of that bond. At the same time, by relaxing polypeptide and tRNA, mRNA cuts off from ribosome. mRNA goes to an inactive form. Inactive ribosome 70S dissociates to subunits 30S and 50S with the participation of the initiation factor and enters the new process of proteosynthesis (**Fig. 5.91**).

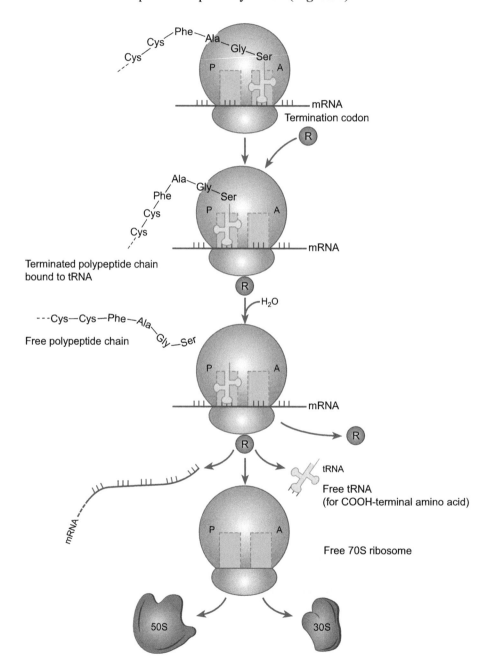

Fig. 5.91. Scheme of protein synthesis termination process.

Biosynthesis of proteins runs usually on several ribosomes simultaneously. Ribosomes are connected with the molecule of mRNA in the shape called polyribosome or polysome. The formation of polysome is good for using mRNA more effectively, because there is a possibility of the simultaneous synthesis of several polypeptide chains. Because 80 nucleotides are located in one ribosome. The size of the synthesized molecule of protein depends on the number of ribosomes, which are connected to mRNA. For example, a peptide chain contains around 150 amino acids. The peptide chain is encoded by mRNA, which is formed from 450 (3×150) nucleotides, which corresponds to 5–6 ribosomes.

The linear layout of amino acids in the polypeptide chain determines its linear structure. The grouping of ribosomes into a polysome enables the formation of a secondary or tertiary structure already during the process of synthesis. Polysomes contain six and more ribosomes, which form helical shapes, in which 30S subunits are oriented inside. Before the detachment from a ribosome, there is a possibility of the formation of spatial three-dimensional protein structures, which are characterized by covalent and non-covalent bonds. The further grouping of the polypeptide through noncovalent bonds allows the formation of the quaternary structure of the proteins.

5.7. Regulation of metabolism process

Allosteric proteins. The coordination of metabolic activity is ensured by two regulatory mechanisms. The result of their activity is:
- regulation of enzyme synthesis;
- regulation of enzymatic activity.

In the regulation, low-molecular compounds called **effectors** are involved. Cells can find them in the environment, or intermediates of their own metabolism can play this role. Components of both systems are specific **allosteric proteins**. Their molecule is made up of several polypeptide subunits called protomers and contains substrate binding sites (a catalytic center) and effector binding sites (an allosteric center).

The existence of multiple sites for the binding and activation of the substrate molecule is manifested in a change in the kinetics of the enzymatic reaction. The graphical expression of this change is a sigmoid (S) curve instead of hyperbole (H), which shows the course of the reaction of the monovalent enzyme with the substrate (**Fig. 5.92**).

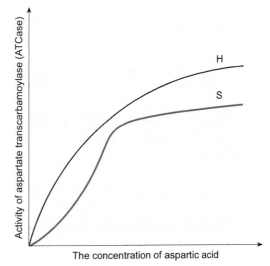

Fig. 5.92. Differences in kinetics of enzymatic reaction of native ATCase.

There are two types of allosteric proteins:

- *Allosteric enzymes*: the conformation and activity of enzymes vary upon association with an effector which can act as an activator or inhibitor. In the first case, the rate of the enzymatic reaction increases, in the other decreases;
- *Regulatory allosteric proteins*, which lack catalytic activity but may regulate the synthesis of enzymes in a particular metabolic pathway. Regulation is made possible by binding these proteins to the bacterial chromosome near structural genes, allowing control of their actions — DNA synthesis.

The schematic of allosteric enzyme binding with the substrate and the effector, including the change of conformation, is shown in **Fig. 5.93**.

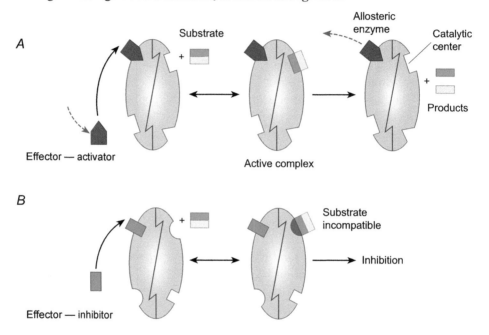

Fig. 5.93. Allosteric activation pattern of positive effector binding and allosteric inhibition by negative effector binding: effector — activator (*A*) and effector — inhibitor (*B*).

Thus, two types of allosteric proteins such as allosteric enzymes and regulatory allosteric proteins play an important role in the regulatory mechanisms of bacterial metabolism.

5.7.1. Regulation of enzyme synthesis

The synthesis of enzymes is regulated by *induction* or *repression*.

Enzyme induction. Bacteria carry genetic information for the synthesis of enzymes necessary for the use of various substrates as carbon and energy sources. The phenotypic expression of some genetic information is conditioned by the composition of the external environment, because an inducible enzyme can be synthesized only in the presence of a specific substrate called an inducer. If no substrate is present, the inducible enzyme concentration in the cell is very low.

The most common example of enzymatic induction is β-galactosidase formation. This enzyme hydrolyzes lactose, and its production is limited by the concentration of this saccharide. The resulting galactose induces the formation of other enzymes (galactokinase, transferase, epimerase) necessary for its further conversion to glucose-6-phosphate. This induction of enzymes necessary for the subsequent series of reactions is referred to as sequential.

The lactose fulfills two functions:
• It has a specific effect on the synthesis regulatory system.
• It provides energy and building material.

The inducible synthesis of enzymes must be distinguished from constitutive synthesis, in which the rate of production and the amount of enzyme are constant and independent of the presence of a specific substrate or inducer. Examples of constitutive enzymes are enzymes of the glycolytic pathway.

Enzymatic repression is the opposite of enzyme induction and means a relative reduction in the rate of formation of any enzyme or group of enzymes catalyzing the sequential biochemical reaction of a particular metabolic pathway. Repression is carried out by substances present in the cell — intruder repressors. Depending on the nature of the repressor, two types of repression are distinguished:
• repression of the end product;
• catabolic repression.

The repression of the end product is particularly characteristic of biosynthetic pathway enzymes. The end product D (corepressor) becomes part of the regulatory mechanism at a certain concentration and blocks further enzymatic synthesis (a, b, c) of the entire metabolic pathway as seen in the scheme:

$$A \xrightarrow{a} B \xrightarrow{b} C \xrightarrow{c} D$$

The result of this repression is stopping the synthesis of the individual intermediates of the path, including the final product. Once the intracellular level of the end product falls to a certain concentration limit, the production of biosynthetic enzymes and thus the biosynthesis of the end product are restored.

As an example of the regulation of synthesis of enzymes by repression of the final product, arginine biosynthesis in *E. coli* can be mentioned. When grown in a mineral environment, the cells produce all enzymes catalyzing the production of arginine from glutamic acid via ornithine, citrulline, and arginine succinic acid. However, after the addition of arginine to the cells, enzyme synthesis stops and resumes only after the cells have been transferred to a medium that does not contain the amino acid. Experimentally it has been demonstrated that the repression of the formation of enzymes involved in arginine synthesis occurs when the concentration of this amino acid exceeds the critical level of about 10 μ/mL.

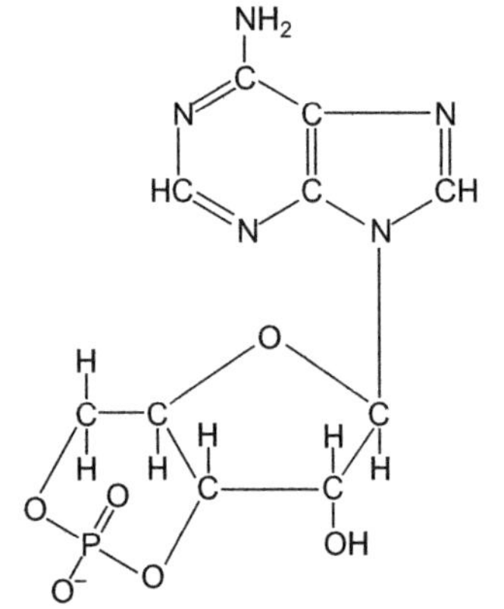

Fig. 5.94. Cyclic adenosine 3′,5′-monophosphate (cAMP).

Catabolic repression is used to regulate the synthesis of catabolic pathway enzymes. Catabolic repression occurs in *E. coli* cells cultured in a glucose- and lactose-containing environment. Glucose as a more usable carbon source is preferably converted and suppresses the synthesis of inducible enzymes that allow the use of lactose, that is, galactoside permease, β-galactosidase, and transacetylase. This phenomenon is referred to as the glucose effect. It means that glucose degradation products reduce the level of cyclic AMP (cAMP), which is the key metabolite for the transcription of the respective operon. This function is filled with cAMP in conjunction with the catabolic protein activator (CAP) (**Fig. 5.94**).

The induction of enzymes necessary for lactose hydrolysis is only initiated after complete glucose utilization. The time between glucose depletion and the onset of lactose utilization results in a marked slowdown in the growth of bacterial culture. The resulting characteristic double growth curve is known as diauxia.

Genetic model of enzymatic induction and repression. Inducible and repressible enzyme systems are genetically determined. Information about the primary structure of proteins, that is, enzymes, is fixed in DNA. Genes carrying this information are referred to as structural (**S**). The transcription of these genes produces mRNA. The initiation of mRNA synthesis is driven by the operator gene (**O**), which controls the transcription of adjacent structural genes. Next to the operator, there is the promoter (**P**), from which the transcription starts. Together, these genes form the so-called operon, which is a functional control unit. Beside the operon, the regulator (**R**) inducing allosteric protein formation, referred to as a repressor (**i**), is located. This substance contains binding sites that can bind to either the effector (**E**) or operator and induce or suppress the transcription initiation in the operon region.

The activity of an inducible system can be shown to synthesize enzymes that catalyze the conversion of lactose to *E. coli* cells. Structural genes that control the synthesis of the respective enzymes are referred to as **z** for galactosidase, **y** for galactoside permease, and **a** for transacetylase.

If there is an inductor in the environment that is identical to an effector, it will be linked to the repressor, which goes into an inactive form of incompetent connection with the operator. At the same time, cAMP with CAP is ligated to the promoter, which allows the RNA polymerase to be joined. This allows for the initiation of transcription of mRNA encoding the synthesis of the respective enzymes. In the absence of an inductor, the repressor binds to the operator, resulting in stopping the transcription process. This control method is referred to as the lac operon negative control in *E. coli* and is performed according to the scheme in **Fig. 5.95**.

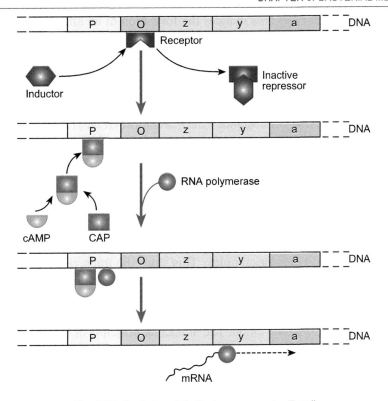

Fig. 5.95. Control model of lactose operon in *E. coli.*

An example of positive control is the regulation of the arabinose operon, which is also carried out in *E. coli* (**Fig. 5.96**). In this case, the operon forms genes encoding the synthesis of the enzymes epimerase (**D**), isomerase (**A**), and kinase (**B**).

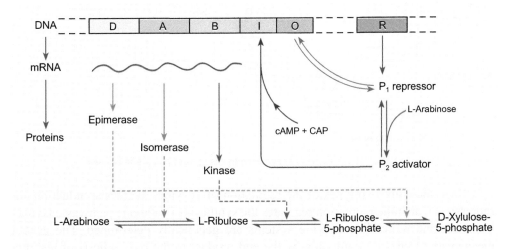

Fig. 5.96. Positive control of arabinose operon in *E. coli.*

The gene regulator (*R*) also induces the formation of an allosteric protein, but it can act as either a repressor (**P**$_1$) or an activator (**P**$_2$) (by modifying the conformation) of the operon transcription depending on the presence of arabinose.

If the arabinose acting as an inducer is present, the conformation of the protein from **P**$_1$ to **P**$_2$ changes, allowing its association with the initiator (**I**), thereby initiating the transcription of the structural genes and the synthesis of the respective enzymes. In the absence of arabinose, this substance exits in a conformation of **P**$_1$ to allow its binding to the operator, which results in stopping the transcription of the operon. The cAMP and catabolic protein activator CAP are also used to regulate the *E. coli* ara-operon.

The genetic mechanism of enzyme repression by the end product can be shown in the biosynthesis of histidine in *Salmonella enterica* serovar Typhimurium. This anabolic pathway is catalyzed by about ten enzymes. The synthesis of all these enzymes is suppressed by the end product of histidine. The apoptosis-producing regulator gene (*R*) is involved in the repression, which, in the presence of histidine, binds to the operator as a histidine–aporepressor complex. This stops the transcription of the genes of the entire operon. Thus, histidine acts as a compressor, which, in conjunction with an inactive aporepressor, causes the formation of an active repressor. In the absence of histidine, there is no correlation of the aporepressor with the operator, which can then control the initiation of transcription. The molecular principle of enzyme repression by the end product is illustrated in **Fig. 5.97**.

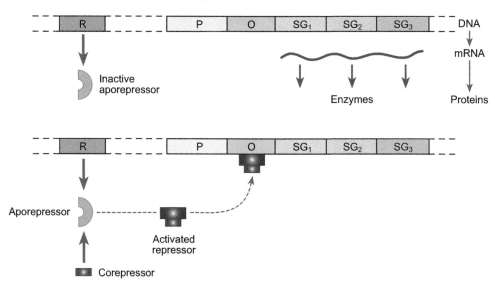

Fig. 5.97. Scheme of enzyme synthesis repression by end product.

Thus, there are two types of the regulation of enzymes synthesis, induction and repression. Catabolic repression regulates the synthesis of catabolic pathway enzymes. Inducible and repressible enzyme systems are genetically determined. The genetic mechanism of enzyme repression by the end product in the biosynthesis of histidine was studied in *Salmonella enterica* serovar Typhimurium.

5.7.2. Regulation of enzymatic activity

The regulation of metabolism by changing enzymatic activity occurs through either an end product or the modification of the enzyme structure.

When the enzymatic activity is controlled by the end product, this product only acts on the first enzyme of the metabolic pathway. The course of inhibition or also the negative feedback in this case can be expressed by the following scheme:

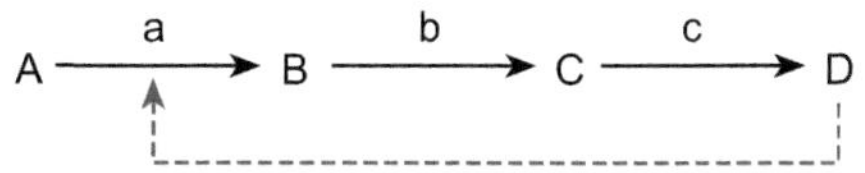

$$A \xrightarrow{\quad a \quad} B \xrightarrow{\quad b \quad} C \xrightarrow{\quad c \quad} D$$

The final product **D** inhibits enzyme activity according to this scheme. This results in the elimination of the formation of intermediate **B** and thus the disruption of the sequence of the entire metabolic pathway. This mechanism of regulation is particularly applicable to the biosynthesis of amino acids and nucleotides.

As a specific example, the inhibition of isoleucine synthesis from threonine in bacteria according to the scheme below can be used.

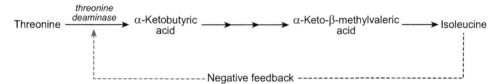

The formation of isoleucine has fully adapted to the need for the cell. If there is enough isoleucine in the nutrient medium, biosynthesis does not occur. Isoleucine can be dissimilated and used in energy metabolism. It is believed that the first enzyme of the isoleucine biosynthesis chain, threonine deaminase, is an allosteric protein, whose spatial arrangement allows the cell to bind either to the substrate, that is, threonine, or to the end product of biosynthesis, isoleucine. As a result of the reaction with the end product, the enzyme loses the ability to react with threonine so that the biosynthesis process stops.

In addition to inhibition or negative control, a positive control can also be carried out in the presence of an activator to inactivate the first enzyme of the metabolic pathway. The role of the activator is not played by a metabolite of the first pathway, but of the other pathway. The cell requires the final products of both pathways in a certain ratio.

Both methods of regulating enzymatic activity are employed in nucleotide biosynthesis. The first step of this synthesis is the formation of carbamoyl aspartic acid, which is catalyzed by transcarbamylase. This reaction suppresses the end product of CTP synthesis, which, upon reaching a certain concentration, acts as an inhibitor of that enzyme.

However, in the presence of ATP, this inhibition may be abrogated and the transcarbamylase reactivated (**Fig. 5.98**). This means that CTP formation is regulated by the relative ratio of CTP and ATP representing the final products of two distinct metabolic pathways.

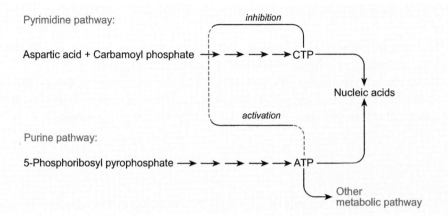

Fig. 5.98. Activation of nucleic acids synthesis via pyrimidine and purine pathways.

The regulation of activity by the chemical modification of the enzyme structure lies primarily in the change (binding or cleavage) of functional groups. This may be the process of methylation, acetylation, or phosphorylation.

For example, the catalytic properties of glutamine synthesis occur only when the enzyme is completely adenylated. The fully adenylated form, unlike the nonadenylated form, requires the presence of Mg^{2+} and a different pH optimum and has a different affinity for the end products of glutamine metabolism. Bacteria with the ability of this chemical modification synthesize at the same time a deadenylation enzyme, whose activity is dependent on various factors, such as α-ketoglutarate.

In addition to changes in the nature of functional groups, enzyme activity may be affected by the modification of its molecular structure in the presence of cofactors or low molecular substances and ions.

5.7.3. Specifics of regulation mechanisms

Regulation of branched biosynthetic pathways. Inhibition by the end product is usually regulated by the first specific enzyme in a certain sequence of biosynthetic reactions. Special problems arise when two or more end products are involved in the first stage of the metabolic pathway. If the substrate is converted to the final product E and also to the final product G and H in the path, the first step will be inhibited even if only one final product is accumulated, for example, H. As a result, the synthesis of the other two products E and G is also inhibited.

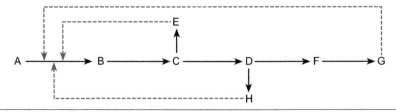

So far, several types of regulation have been observed for enzymes catalyzing branched or parallel metabolic pathways.

- The first stage is catalyzed by parallel (isofunctional) enzymes, each of which is regulated by a particular end product.
- The first step is catalyzed by only one enzyme, the inhibition of which produces the end product of each branch independently. This inhibition is called cumulative.

An example of regulating parallel acting enzymes is the inhibition of aspartokinase in *E. coli*. These enzymes catalyze the first stage of biosynthesis of threonine, isoleucine, methionine, and lysine from aspartic acid. Aspartokinase occurs in three isoforms (aspartokinase I, II, and III), which differ in sensitivity to threonine (aspartokinase I) and lysine (aspartokinase III) and other properties: the synthesis of these two amino acids as well as isoleucine is stopped by the action of lysine and threonine as end products. The biosynthesis of methionine catalyzed in the first stage by aspartokinase II is not affected by this inhibition (**Fig. 5.99**).

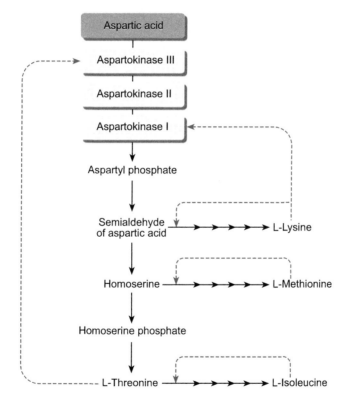

Fig. 5.99. Regulation of biosynthesis of amino acids derived from aspartic acid.

The biosynthesis of aromatic amino acids is regulated in a similar manner. It is also a branched run; its scheme is shown in **Fig. 5.99**. Inhibited regulatory enzymes, possibly activated by more than one regulator, are referred to as multivalent.

5.7.4. Regulation of energetic metabolism

In the regulation of energy metabolism in anaerobic glycolysis, about 20 times less energy is released from the same amount of substrate than in the oxidation process. In addition, the aerobic process is accompanied by increased biomass production, while reducing the consumption of the substrate. This phenomenon is referred to as the Pasteur effect. It can be explained by the action of regulatory mechanisms acting on key enzymes of glucose metabolism. Basically, there are three sections of the glucose cleavage pathway.

The first stretch involves enzymes of active glucose transport across the cytoplasmic membrane to glucose 6-phosphate. This intermediate acts as an inhibitor in allosteric inhibition of active glucose transport by feedback.

The second segment encompasses enzymes catalyzing the glucose 6-phosphate transfer to ATP and carbonaceous product in the fermentation or conversion of glucose 6-phosphate to ATP and citrate in anaerobic respiration. The inhibition by ATP-induced feedback is subject to phosphofructokinase formation. In contrast, AMP acts as a positive effector and inhibits ATP-induced inhibition. According to current knowledge, the allosteric sensitivity of phosphofructokinase to ATP and AMP represents the primary mechanism of the Pasteur effect.

The third stretch follows the conversion of citrate under anaerobic conditions, which increases the yield of ATP. At higher levels of ATP, isocitrate is hydrogenated by feedback. Its activity is dependent on NAD^+. Its allosteric activator is, for example, AMP. If the energy level is high and the AMP : ATP ratio is very small, then the enzyme activity decreases and the amount of citrate increases. The citrate can then serve to inhibit the preceding section by feedback. The affinity of citrate synthase for acetyl-CoA also decreases the higher concentrations of ATP, which prevents the accumulation of metabolites and reduction systems of this pathway.

The ATP : AMP ratio plays an important role in the regulation of some other metabolic processes. For example, in the autotrophic fixation of CO_2 in the Calvin cycle, AMP suppresses the reaction, during which ribuloso-1,6-diphosphate is formed due to phosphoribulokinase activity. This regulation allows to maintain a certain amount of ATP in the cell that provides cellular synthesis processes to the necessary extent.

These examples demonstrate that adenylates have important regulatory functions in the cell. The intracellular content of ATP, ADP, and AMP and their relative ratios determine the rates of a series of biochemical reactions and thus the course of catabolic and anabolic processes. The principle of this regulation is to activate or inhibit allosteric enzymes by the appropriate adenylate.

In studying the mechanisms of regulation of metabolic processes as well as in the practical application of acquired knowledge, strains, whose regulatory functions have been altered or eliminated by mutation, are used.

In terms of isolation and selection of mutants with impaired regulation, the mutant strains characterized by depression are at the forefront of interest. This is the elimination of repression of the synthesis of anabolic enzymes. Optionally, these are strains characterized by loss of ability to undergo allosteric inhibition. In both cases, the mutation results in an increased production of the end product of the biosynthetic pathway. For a mutant cell, the overproduction of the end product means that the metabolite displaces the relevant antimetabolite from the reaction and thus ensures cell growth. When comparing mutants with those properties, it was found that the removal of repression had less effect on the rate of synthesis of the final product than a change in the ability of allosteric inhibition.

This difference is due to the fact that in the case of strains that have lost the mechanism of inhibiting a certain pathway of biosynthesis, the end product of this pathway is accumulated irrespective of the presence of repression mechanisms. On the contrary, in mutants with apparent depression and disturbed allosteric inhibition by the end product, there is only limited formation of this metabolite.

These phenomena indicate that repression has a prime record for ensuring the economic mRNA and protein synthesis, while metabolite synthesis is regulated by allosteric inhibition.

In addition to a considerable theoretical benefit, allowing deeper knowledge and illumination of the mechanisms of regulation of metabolic processes, including their interconnection, the choice and use of bacterial mutants also have a great practical significance. This is mainly the possibility of obtaining not only active producers of active substances, such as antibiotics but also a number of other economically important products.

5.8. Metabolism of phototrophic bacteria

The main characteristic of phototrophic bacteria is the ability of conversion of light energy to chemical energy that is further used in the metabolism of the bacterial cell. Plants have the same ability. But there are some differences between bacterial and plant photosynthesis:
- the donor of hydrogen and electrons is not water, as it is in plant photosynthesis, but an inorganic or organic compound, for example, molecular hydrogen;
- the bacterial photosynthesis of bacteria is mostly under anaerobic conditions;
- bacterial photosynthetic pigments differ by structure and light spectrum;
- in addition, the localization of these pigments is different;
- the bacterial photosynthetic apparatus consists of only one center, the plant photosynthetic apparatus is made up of two centers: one secures photophosphorylation and the formation of reductive systems and the second one secures the photolysis of water.

5.8.1. Photolithotrophs

Photolithotrophs obtain energy from light and therefore use inorganic electron donors only to fuel biosynthetic reactions (e.g., carbon dioxide fixation). In photolithotrophic bacteria, NADH and ATP can be generated by cyclic and noncyclic photophosphorylation.

The physiological group of photosynthetic *photolithotrophic bacteria* is comprised three families:

- *Chlorobacteriaceae* (**green sulfur bacteria**) are represented by genera *Chlorobium, Pelodyction*, etc. They are strictly anaerobic and photolithotrophic. Reduced inorganic sulfur or hydrogen compounds of sulfur or hydrogen serve as donors of hydrogen and electrons. The source of carbon is CO_2, which is reduced through the *Calvin cycle*. As a precursor of cell material, acetate can be assimilated as well, but only through the reductive synthesis of pyruvate from acetyl-CoA and CO_2.

The **green nonsulfur bacteria** are often classified as a group within the green sulfur bacteria. It is made up of one genus, *Chloroflexus*, which consists of thermophilic microorganisms growing in neutral to alkali hot springs with the temperature around 45–70 °C. They have less chlorophyll that is covered by a number of carotenoids. They are photoorganotrophic, facultatively anaerobic. It is assumed that they gain organic matter from symbiosis with cyanobacteria.

- *Thiorhodaceae* (**purple sulfur bacteria**) consist of many genera, the most known are *Thiospirillum* and *Chromatium*. The dominant metabolic pathway is connected to H_2S utilization or different reduced sulfuric compounds as donors of hydrogen and electrons. They are strictly anaerobic. The reduction of CO_2, the source of carbon, occurs through the Calvin cycle.

Some species in this group are not *obligately photolithotrophic organisms*, because they can assimilate organic compounds, such as acetate. There are few species that require B_{12} as a growth factor.

- *Athiorhodaceae* (**purple nonsulfur bacteria**), including *Rhodospirillum, Rhodopseudomonas*, and *Rhodomicrobium*, are photoorganotrophic microorganisms, that use organic compounds (for example succinate, fumarate) as a source of hydrogen and electrons. Besides CO_2, they are capable of using organic compounds as carbon source with the simultaneous formation of ATP by cyclic phosphorylation.

Phosphorylation by this photolithotrophic bacterial group consists of two simultaneous reactions:

- photophosphorylation leading to the conversion of light energy to chemical energy;
- reduction of CO_2 as a carbon source for cell material.

Photophosphorylation is the process of light energy conversion into the energy of the macroergic bond in ATP with the simultaneous formation of the reductive system $NADH + H^+$. ATP can be created by *cyclic* or *noncyclic phosphorylation*.

The **reduction of CO_2** and its conversion to cell building material usually follow the metabolic pathway called *Calvin cycle* (**Fig. 5.100**). This pathway is characteristic for plants as well as for *photolithotrophic* and *chemophotolithotrophic* bacteria. Even photoorganotrophs can use this way when growing in the light.

The first step of CO_2 conversion is binding CO_2 to pentose ribulose diphosphate, catalyzed by the enzyme carboxydismutase. The unstable intermediate product splits through 3-phosphoglyceric acid to glutaraldehyde 3-phosphate, which serves as material for the biosynthesis of cell matter, as well as for the resynthesis of ribulose diphosphate.

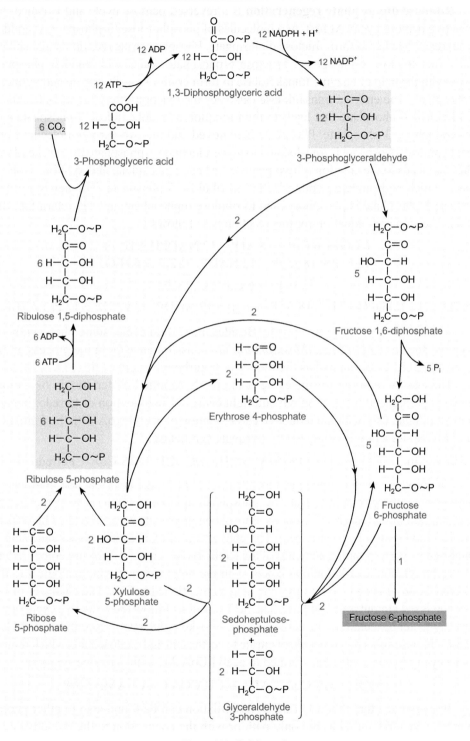

Fig. 5.100. Calvin cycle.

Ribulose diphosphate regeneration is a reversed pentose cycle and is called the reductive pentose cycle. At first, two molecules of phosphotriose (3-phosphoglyceraldehyde) react together to form fructose diphosphate. Phosphatase cleaves off the phosphate and forms fructose 6-phosphate. Fructose 6-phosphate reacts with another phosphotriose with the help of enzyme transketolase to form erythrose 4-phosphate and xylulose 4-phosphate. The enzyme transaldolase catalyzes the reaction of erythrose 4-phosphate and fructose 6-phosphate that leads to the formation of sedoheptulose 7-phosphate and glyceraldehyde 3-phosphate. Phosphate is removed and the two-carbon molecule from sedoheptulose phosphate is translocated to glyceraldehyde 3-phosphate by transketolase. This step leads to the creation of two pentoses, ribose 5-phosphate and xylulose 5-phosphate, which both undergo epimerization to ribulose 5-phosphate. The newly formed ribulose 5-phosphate is phosphorylated to ribulose diphosphate and can return into the cycle. The summary equation for this reaction is as follows:

$$6\ CO_2 + 6\ C_5P + 18\ ATP + 12\ NADPH + H^+ \rightarrow$$
$$12\ C_3P + 18\ ADP + 12\ NADP^+ + 12\ P_i + 6\ H_2O;$$

$$12\ C_3P \rightarrow 6\ C_5P + C_6P + P_i;$$

$$6\ CO_2 + 12\ NADPH + H^+ + 18\ ATP \rightarrow C_6P + 12\ NADP^+ + 18\ ADP + 17\ P_i + 6\ H_2O.$$

Reductive carboxylic acid cycle. Besides the Calvin cycle, some photosynthetic bacteria, such as *Chlorobium thiosulphatophilum*, assimilate CO_2 in the process of the reversed Krebs cycle, named as the reductive tricarboxylic acid cycle (**Fig. 5.10**1).

The energy required for the cycle is indirectly supplied by ferredoxin oxidation, which is prereduced by chlorophyll and light energy. The oxidation of ferredoxin happens simultaneously with the reductive carboxylation of acetyl-CoA with the formation of pyruvate. Reaction is catalyzed by pyruvate synthetase:

$$Ferredoxin_{red} + CO_2 + CH_3COS\text{-}CoA \rightarrow CH_3COCOOH + HS\text{-}CoA + Ferredoxin_{ox}.$$

The production of pyruvate is dependent on light. In the next reaction, pyruvate is converted to phosphoenolpyruvate by phosphoenolpyruvate synthase with the energy consumption of two macroergic bonds. Phosphoenolpyruvate is carbonylated by respective carboxylase to oxaloacetic acid. The conversion of oxaloacetate to succinic acid is catalyzed by corresponding enzymes of the Krebs cycle. Succinyl-CoA undergoes reductive carboxylation to α-ketoglutaric acid with the help of ferredoxin, analogical to the carboxylation of acetyl-CoA to pyruvic acid as mentioned before. The catalyzer of the reaction is α-ketoglutarate synthase. The next reactions follow the reductive cycle of tricarboxylic acids, including the carboxylation of α-ketoglutarate to isocitrate. In the last phase, citratase splits the molecule of citrate to acetic acid and oxaloacetate, which returns to the cycle. Acetic acid is transformed to acetyl-CoA due to the enzyme synthase.

$$CH_3COOH + HS\text{-}CoA + ATP \rightarrow CH_3COS\text{-}CoA + AMP + 2\ P_i.$$

It is assumed that this way of CO_2 assimilation can be found also in other photosynthetic bacteria, either as the only path or with the connection to the reductive pentose cycle.

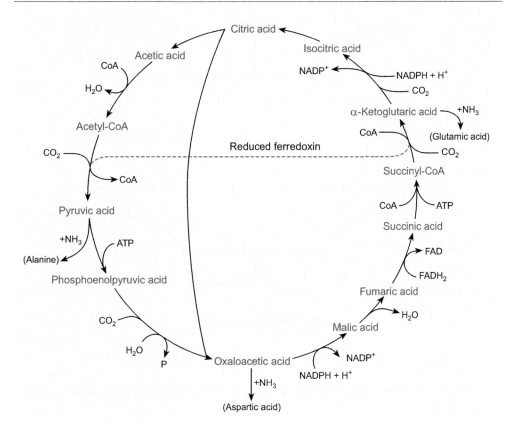

Fig. 5.101. Reductive cycle of carboxylic acids.

Thus, the reductive carboxylic acid cycle plays an important role for photolithotrophic bacteria where CO_2 is assimilated in the process of the reversed Krebs cycle. Ferredoxin oxidation, which happens simultaneously with the reductive carboxylation of acetyl-CoA with the formation of pyruvate, is also important in this process.

5.8.2. Photoorganotrophs

The dominance of the photoassimilation of organic compounds over the reduction of CO_2 in the group of photoorganotrophs is not clear and depends on the level of substrate oxygenation and growth conditions. If an organic substrate exists in the oxidized form, it can be partially dissimilated, and the gained energy can be used for the synthesis of reduction systems for the reduction of remaining substrate molecules, for instance, CO_2. If the substrate is highly reduced, it undergoes oxidation associated with CO_2 reduction and assimilation. An example is photometabolism of acetic and butyric acids, which are utilized by purple bacteria and mostly converted to poly-β-hydroxybutyric acid following the scheme in **Fig. 5.102**.

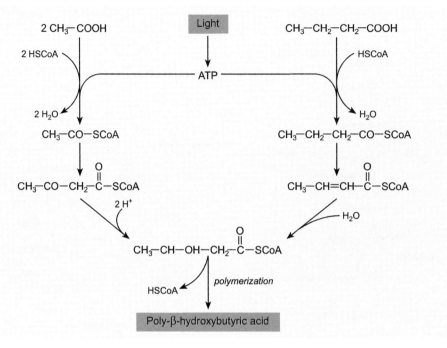

Fig. 5.102. Photometabolism of acetic acid and butyric acid.

The conversion of acetic acid to the polymer of β-hydroxybutyric acid is a reductive process:

$$2n\ CH_3COOH + 2n\ [H] \rightarrow (C_4H_6O_2)_n + 2n\ H_2O.$$

Most photoorganotrophs contain enzymes of the TCA cycle that enable the synthesis of reductive systems. The reductive conversion of acetic acid to poly-β-hydroxybutyric acid goes according to the summary equation:

$$9n\ CH_3COOH \rightarrow 4\ (C_4H_6O_2)_n + 2n\ CO_2 + 6n\ H_2O.$$

The process of acetic acid assimilation is highly effective; about 90% of the substrate is converted to cell material. This could be caused by the fact that enough ATP is synthetized in the process of cyclic phosphorylation to activate acetate to acetyl-CoA.

On the other hand, the conversion of butyric acid is an oxidative process and occurs in the presence of an appropriate acceptor of hydrogen.

$$n\ CH_3CH_2CH_2COOH \rightarrow (C_4H_6O_2)_n + 2n\ [H].$$

The role of acceptor is played by CO_2 that is assimilated in the Calvin cycle and converted into cell material. The process is summarized by the equation:

$$2n\ C_4H_8O_2 + n\ CO_2 \rightarrow 2\ (C_4H_6O_2)_n + (CH_2O)_n + n\ H_2O.$$

The anaerobic photoassimilation of butyric acid is necessarily connected to the reduction of CO_2, and the energy for both reactions is provided by ATP formed in the process of cyclic phosphorylation.

The conversion of acetic acid to itaconic acid and glutamic acid happens among the species of *Chromatium* genus:

Acetic acid ⟶ Acetyl-CoA ⟶ Citramalic acid

Glutamic acid ⟵ Itaconic acid

The scheme of the conversions, including the synthesis of cell material, is shown in **Fig. 5.103**.

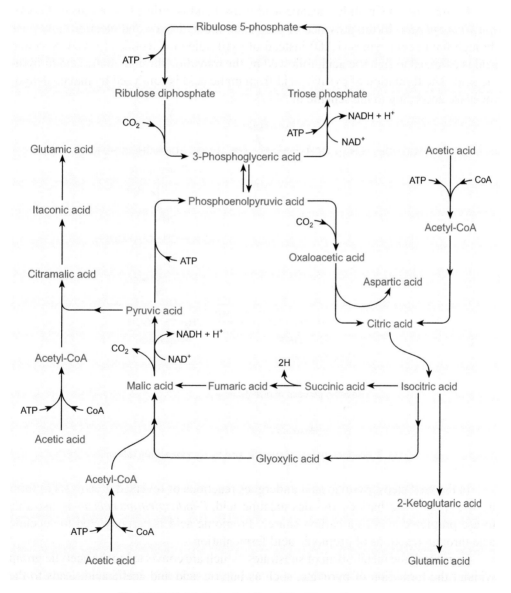

Fig. 5.103. Metabolism of acetic acid by *Chromatium* genus.

Purple nonsulfur bacteria *Rhodopseudomonas sphaeroides* convert acetic acid anaerobically in the presence of light to pyruvic acid instead of poly-β-hydroxybutyric acid. The conversion happens through acetyl-CoA followed by the formation of malic acid that is catalyzed by malate synthetase.

Acetic acid → Acetyl-CoA → Malic acid → Pyruvic acid.

Aside from acetic and butyric acids, photoorganotrophs are capable to utilize various organic compounds, for example, dicarboxylic acids, alcohols, and aromatic compounds.

As the substrate, mainly succinic acid is used and is utilized by strains of *Rhodospirillum rubrum*. In the presence of succinate, the hydrogen and electron flow goes through flavin coenzymes (FAD) instead of pyrimidine nucleotides (NAD). Succinic acid is oxidized to fumaric acid, followed by the transition through malic acid to pyruvic acid. The formation of pyruvic acid from malic acid is catalyzed by malate dehydrogenase according to the scheme in **Fig. 5.104**.

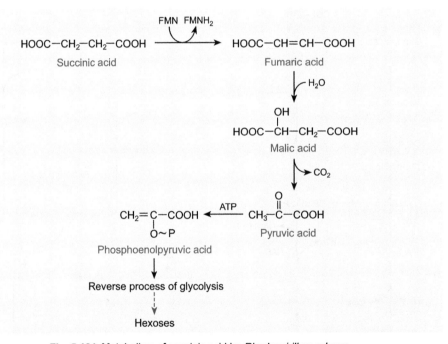

Fig. 5.104. Metabolism of succinic acid by *Rhodospirillum rubrum*.

In the next steps, pyruvic acid undergoes reactions of reversed glycolysis to form hexoses or polysaccharides. Besides succinic acid, *Rhodospirillum rubrum* is also able to use propionic acid as a carbon source. Propionic acid is carbonylated to succinic acid through reactions of propionic acid fermentation.

In general, the metabolism of substrates, which are converted to the acetylic group without the formation of pyruvate, such as butyric acid and acetic acid, leads to the formation of poly-β-hydroxybutyric acid. On the other hand, the substrates, whose conversion goes through pyruvic acid, such as succinate, malate, propionate, and in some bacterial groups acetate, provide saccharides and polysaccharides in the processes of biosynthesis.

Alcohols and acetone can also serve as a substrate for the photosynthetic reduction of CO_2, mostly for the bacteria from the family *Athiorhodaceae*. Some species use primary alcohols, other species can use secondary alcohols. The reduction of CO_2 requires the presence of the induced enzyme. The reaction process can be described by equations:

$$CH_3CH_2OH + 3\ H_2O \rightarrow 2\ CO_2 + 12\ H^+;$$

$$CH_3CH_2CH_2OH + 5\ H_2O \rightarrow 3\ CO_2 + 18\ H^+;$$

$$n\ CO_2 + 4n\ H^+ \rightarrow (CH_2O)_n + n\ H_2O.$$

The photometabolism of acetone, leading to the formation of cell material through various products, was discovered in *Rhodopseudomonas gelatinosa*. Acetone and carbon dioxide condensate with the formation of acetoacetic acid. If the intermediate products do not accumulate, the mechanism of cell material creation in *R. gelatinosa* is described by scheme in **Fig. 5.105**.

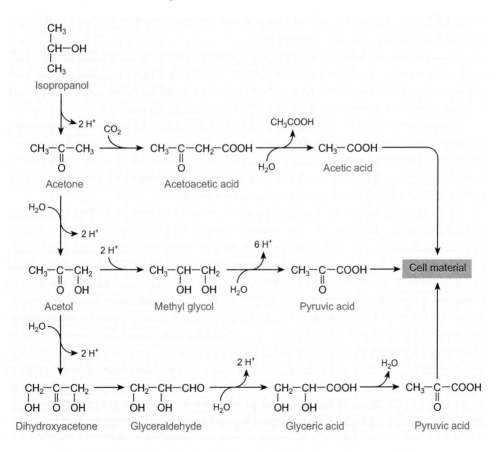

Fig. 5.105. Mechanism of cell material creation in *Rhodopseudomonas gelatinosa*.

From aromatic compounds, *Rhodospirillum palustris* can utilize benzoic acid as a substrate anaerobically in the light. The photometabolism of the acid occurs through a special reductive pathway, in which first cyclohexanecarboxylic acid is formed, then the acid is hydrated to 2-hydroxycyclohexanecarboxylic acid; after that, the newly formed acid is dehydrogenated, and the cycle is cleaved and opened to pimelic acid following the scheme:

The reduction of aromatic compounds is catalyzed by respective reductases with cooperation of ferredoxin.

***

To summarize, it should be noted that bacterial metabolism is a complex process in the cell. The bacterial cell is an open system characterized by exchange with the environment (matter, energy, information).

The metabolism is the flow of the mass, energy, and information through the cell. It includes processes of chemical transformation, which serve for gaining basic building material and energy for cell compound synthesis and for other vital processes of bacterial cells. Microbial metabolic activity is extremely diverse.

The metabolism of the cell (the flow of matter and energy through the cell) includes two opposing categories: catabolism (exergonic reactions) and anabolism (endergonic reactions). These two processes are coupled and interconnected. Some compounds can be reduced and others are oxidized. Catabolism and anabolism are two counter-processes that allow mutual existence in space and time.

The energetic metabolism takes place when the exergonic and endergonic reactions are coupled. To carry out the synthesis, the sum of the free energy changes of both coupled reactions must be negative or at least zero. Synthetic reactions usually take place with the participation of an energy-rich compound (ATP, a substance capable of forming a macroergic bond). ATP is a universal energy carrier and a universal phosphate donor in anabolic reactions. ATP accumulation can be archived at the substrate level or at the membrane level (in prokaryotes on the cytoplasmic membrane). Pyruvate is a central and important compound in metabolic processes, in particular, in fermentation as well as anaerobic and aerobic respiration.

Metabolic processes in the bacterial cells are strictly regulated, and the efficiency of cellular regulatory mechanisms is very high. The metabolic processes ensure the maximum use of environmental nutrients and prohibit the excessive synthesis of intermediates and terminal metabolites, responsible for rapid adaptation to environmental changes.

A photophosphorylation is the process of transforming the energy of a light quantum into the energy of a macroergic bond (ATP).

CHAPTER 6

GROWTH OF MICROORGANISMS IN NATURE

In this chapter, microorganisms as a part of the ecosystem are characterized. Symbiotic relationships between microbial organisms are presented. The physiological role of microorganisms in various ecosystems, including autochthonous and allochthonous, and oligotrophic microbial populations are described. The special attention is paid to the intracellular and internal population interactions and the quorum-sensing regulation of gene expression. Quorum sensing is a special type of the regulation of bacterial gene expression in conditions of high density of populations.

Luminescent bacteria occur in marine, freshwater, and terrestrial ecosystems. They can be free-living saprophytes or symbionts that live in the digestive tract or in the light organs of bony fish and squid. That is why the characterization of this physiological group of bacteria and the process of bioluminescence finalizes this chapter.

6.1. Microorganisms as part of the ecosystem

Under natural conditions, microorganisms never exist in the form of pure cultures. They do not only live in a mixed culture, but they are also part of the ecosystem, which includes organisms of other systematic groups. Any ecosystem contains microorganisms that perform two basic functions:

(1) synthesis of a new organic substance from CO_2 and other inorganic compounds, and

(2) destruction of accumulated organic material.

The ecological role of microorganisms lies in the fact that they can function at all levels of the ecosystem: fixing carbon as primary producers, using the energy of light and chemical bonds; being main reducers of organic substances; serving as a power source for many other organisms. The microbial cell contains an average of 50% carbon, 14% nitrogen, 4% phosphorus, and other elements.

The basic functions of microorganisms in natural places of distribution:
- mineralization of organic substrates to CO_2, NH_3, H_2, CH_4, and H_2O;
- provision of nutrients to chemoheterotrophic microorganisms;
- provision of nutrients to the simplest, nematodes, and soil insects;
- modification of complex compounds that become available to other organisms;
- conversion of compounds into soluble or gaseous form;
- production of compounds that suppress the activity of other organisms (antibiotics, toxins, bacteriocin).

Microorganisms coexist in nature in the form of populations of the same type or different types. The distribution of microorganisms depends on the availability of nutrients, toxic compounds, and limiting factors (temperature, water, pH, light, etc.). Usually, microorganisms in nature are not in optimal conditions. In some cases, a competitive relationship develops, and then microorganisms release substances (antibiotics, toxins, alcohols, acids) that suppress competitive microbiota. Rarely, there is a cooperative relationship. An example of close interactions is the anaerobic microbiota of the rumen. In the decomposition of polymer compounds, anaerobes capable of fermentation (e.g., clostridia), which split polymers, reduce hydrogen. They use fatty acids and alcohols to form acetate and hydrogen. Molecular hydrogen is used by methanogens, acetogens, and sulfate- and sulfur-reducing bacteria. Under these conditions, there is a *syntrophy*, which is a complete interdependence of microorganisms in terms of nutritional needs.

Symbiosis is a relationship between organisms not only due to the nutritional needs. Symbiosis differs in the components, the placement of organisms, and the nature of their relationship. Mutually beneficial relations are called *mutualism*, depressing ones are *parasitism*, and indifferent are *neutralism*. If the conditions change, the nature of the relationship may change. For example, a normal human microbiota is a mutualistic symbiosis, but if the immune status of a person decreases, some microorganisms can become pathogenic. Symbiosis is called *obligate*, when one organism cannot exist without another (e.g., rickettsia is a parasite of animal cells). An example of *facultative symbiosis* is bulbous bacteria of the *Rhizobium* genus that can exist without a plant. Another type of symbiosis is *commensalism*, when one partner of the association benefits, and for the other benefit is not important. In associations of aerobic and anaerobic bacteria, aerobics use oxygen and provide conditions for the development of anaerobes, but they do not benefit from it themselves. Exosymbiosis can be found in the rhizosphere, that is, the space around the roots, where the number of microorganisms is much larger than in the soil. Bacteria in the soil have weak metabolic activity due to insufficient supply of nutrients. With root growth, the nutrients necessary for microorganisms are released, and due to microbial growth, available forms of nitrogen compounds are made available to the plant. Microorganisms also release enzymes that contribute to the mineralization of complex organic compounds to simple, available to plants. The concept of the symbiotic relationship of microorganisms has undergone significant changes today, due to a change in the perception of the number of partners needed to create persistent symbiosis and to ascertain complex relationships between them.

An example of the interaction of microorganisms in nature may be an impassable lake. In the aerobic zone, where light is available, cyanobacteria and algae function, which can fix carbon dioxide during photosynthesis, form organic substances, and release oxygen. In this zone, chemoorganotrophs develop, which are capable of decomposing organic matter with the formation of final products, CO_2 and H_2O. In the anaerobic bottom zone, methanogens and homoacetogens use carbon dioxide to form methane or acetate. Sulphidogens reduce sulfate and elemental sulfur; denitrificators reduce nitrates to free nitrogen. Mixed bacterial populations can reduce Fe(III) to Fe(II) ions in anaerobic conditions. There is fermentation of alcohols, organic acids, and amino acids. In natural conditions, various groups of microorganisms constantly interact with each other. The predominant development of certain groups of microorganisms depends on changes in environmental factors, in particular, the presence of oxygen, light, and hydrogen sulfide, the corresponding temperature, and pressure.

6.2. Physiological role of microorganisms in ecosystems

Under conditions of high water content, sufficient nutrients, and a neutral pH value in soil or water, a large number of types of microorganisms develop. The more extreme physical and chemical characteristics of the environment are, the smaller the variety of species is, but the number of individuals belonging to the same species increases. Such interdependence between the number of species and number of individuals and the degree of extreme conditions is observed in many ecosystems: in hot springs, salt lakes, acid waters, intestines, and soils.

Sergei Winogradsky proposed to divide all microorganisms into two groups regarding their reaction to the effects of environmental factors. The first group includes *autochthonous microorganisms* that are constantly present, regardless of the amount of nutrients. Such microorganisms grow slowly (*low growth rate*). The second group is a group of *allochthonous microorganisms* that are present in wastewaters, where organic matter is sufficient. This group of microorganisms grows fast (*high growth rate*) and requires high concentrations of nutrients, while much energy dissipates. Later, another group of microorganisms that were resistant to environmental changes and able to survive adverse conditions was isolated.

In various ecosystems, there are also *oligotrophic microorganisms* that grow at low concentrations of nutrients. Cell survival strategies in the lack of energy substrate are:

(1) increased absorption of the substrate;

(2) reducing energy requirements of the cell.

The absorption efficiency of the substrate may increase due to the induction of increased transport systems or high-affinity transport systems. Another strategy for survival in the lack of substrate is an increase of the cell surface, for example, through the formation of prosthecae (stalk). Increasing the cell surface gives the advantage of starvation.

One of the ways to survive in a hunger strike is to reduce energy costs by creating spores or cysts. Some Gram-negative bacteria reduce the need for energy by reducing cell sizes. The formation of such dwarf forms is detailed in *Vibrio* sp. In this case, the degradation of proteins increases; the respiratory activity of the cells decreases; starvation is induced by new proteins (*Sti* proteins, from Starvation-induced); the composition of the fatty acids of the membranes changes; the system of high affinity transport of the substrate is induced; the metabolic activity is gradually reduced; and the cell becomes a resistant form, able to survive in unfavorable conditions.

Under conditions of starvation, a number of bacteria form extracellular mucus that supports the membrane potential and retains the proton gradient, protects the cells from drying out, and ensures the attachment of cells to various surfaces.

Sometimes microorganisms form ***biofilms***. In the simplest case, this is a monolayer of cells of the same type. The structure of the biofilm can be complicated depending on the presence of nutrients, light, and diffusion rate, while layers of different microorganisms can be formed, for example, the outer layer in photosynthetic bacteria, the middle layer in facultative chemoorganotrophs, and the inner layer in sulfate reducers. Investigations of such complex associations are carried out using a confocal scanning laser microscope, which allows receiving layer–layer images of the film at different depths of the biofilm. The general appearance of the film is reproduced by computer analysis. Biofilms can expand to large sizes, forming a microbial mat. Such groups are formed on stones, in freshwater lakes, and in marine coastal zones. Microbial moths contain cyanobacteria, which determine the structure of this cluster. In the lower part of the mat, sulfate reducers live. They form sulfide, which diffuses into anaerobic photosynthesis zone, where green and purple bacteria grow.

Some microorganisms can exist under different conditions either as independent single cells or as multicellular colonies of differentiated cells. One of the best characterized is *Dictyostelium mucus* mushroom. Under favorable conditions, this fungus lives in the form of single cells. Under conditions of fasting, cells form aggregates (**Fig. 6.1**). These aggregates (mucus) can move toward a more favorable environment, where they are transformed into a multicellular structure (the fetal body), on the surface of which the cells grow. The wind can move them to a place with more favorable conditions, where they grow, multiply, and live as individual cells until they have enough nutrition.

The photo from the scanning microscope (**Fig. 6.1**) shows how hundreds of thousands of single cells of the mucous membrane aggregate to form mucus (bottom left). In the appropriate place, they extend out to form a fruit body.

The transition from single-cell to multicellular growth is provided by intercellular interaction and indicates signaling processes between and within the cells. In *Dictyostelium*, when starving, the signal-molecule cAMP is released (**Fig. 6.2**).

Fig. 6.1. Transformation of *Dictyostelium* single cell into multicellular organism.

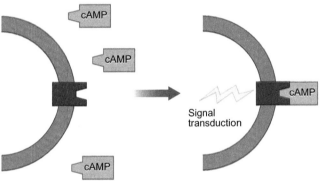

Fig. 6.2. Intracellular signaling.

This signal is transmitted to the cell by binding to the membrane-bound protein receptor on the cell surface. This binding causes several responses, which include movement toward higher cAMP concentrations as well as the formation and release of additional cAMP molecules. The cyclic AMF, which is recognized by receptors on the surface of the cell, initiates the formation of aggregates in *Dictyostelium*.

Cells aggregate with gradient cAMP. The cells then exchange additional signals and differentiate into a distinct cell type. Each of these cells expresses the corresponding genes. It is important to note that cAMP signals are characteristic of many organisms, including humans.

6.3. Intercellular and internal population interactions and quorum-sensing regulation of gene expression

Prokaryotic cells are capable of intercellular intrapopulation contacts and coopertive interaction. This behavior of bacteria is provided by the autoinduction sensory system, which is called quorum sensing (QS). *Quorum sensing* is a special type of regulation of bacterial gene expression, which functions in conditions of high density of populations. Compulsory components of the QS system are low-molecular-weight signaling molecules (autoinducers, AIs) that are readily diffused through the cell membrane and receptor proteins, to which the autoinducers bind. The concentration of an autoinducer in the medium is proportional to the number of bacteria present. At low densities of the population, bacteria produce the autoinducer at the basal level. As the population grows to a critical level, the number of AIs increases, and they accumulate in the medium. When reaching a certain threshold, the AI interacts with the corresponding receptor proteins; complexes of a **receptor protein** and the ***autoinducer*** bind to promoter regions of genes/operons, which cause the induction of expression of certain genes in bacteria. Bacterial communication with the AI is also carried out; it is the intercellular transmission of information in the population between bacteria belonging to different species, genera, and even families. Due to communication, bacteria can coordinate the control of gene expression, which promotes the survival of bacteria in ever-changing environments.

QS regulation was first described 30 years ago in bacteria *Vibrio fischeri*. Later, it turned out that this type of regulation is inherent in the bacteria of different taxonomic groups. QS regulatory systems play an important role in many bacterial cell processes, in particular, in the interaction of bacteria with higher organisms, animals, and plants, in the formation of biofilms, in the regulation of the synthesis of antibiotics, toxins, and enzymes. Autoinducers regulate the following processes:

1. ***Bioluminescence*** and "swarming" in bacteria of the genera *Vibrio*, *Proteus*, and *Serratia*. The bacterium *Vibrio fischeri*, which is a symbiont of some marine organisms, acquires its ability to glow only after reaching a high concentration of cells (10^{10}–10^{11} cells/mL), when an autoinjection of luminescence occurs. Separate cells of this bacterium in seawater do not exhibit shine. Similarly, the regulation of swarming on dense surfaces is typical of *Proteus mirabilis* and *Serratia marcescens*.

2. ***Virulence*** in *Erwinia carotovora* and *Pseudomonas aeruginosa*. For many microorganisms that are pathogenic to plants and animals, high population densities are necessary for the local accumulation of exoenzymes (cellulases, proteases, polygalacturonidases) that lyse the protective membranes of the cell, or for overcoming the host protective mechanisms. The activity of the genes is regulated by the AI, which is found in most Gram-negative bacteria.

3. ***Conjugation*** in *Agrobacterium tumefaciens*. Using the methods of transcriptional and proteomic analysis, it was established that QS systems function as global regulatory factors. As autoinverters, QS systems use a variety of signaling molecules. In Gram-negative bacteria, the most studied systems are those that function with autoinducers *N*-acyl homoserine lactones (**Fig. 6.3**).

There are more than 40 species of *N*-acyl homoserine lactones, which differ in length of the acyl chain (from C_4 to C_{18}) and the presence of different groups in the acyl chains. There are no strict species-specific AIs, the same AI can be formed and recognized by different species and bacterial genera.

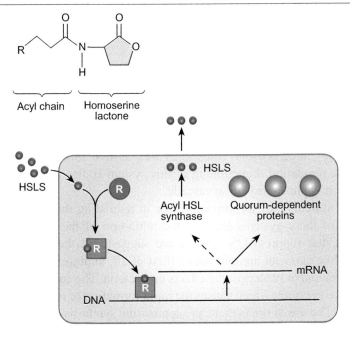

Fig. 6.3. Quorum sensing in Gram-negative bacteria: HSLS is homoserine lactone synthase.

The receptor proteins, which *N*-acyl homoserine lactones interact with, and their synthases, are homologous to the LuxR and LuxI proteins in *V. fischeri,* respectively. The most accurate QS regulation was studied in the *lux* operon of *V. fischeri.* The synthase of the autoinducer, *LuxI* protein, is responsible for the formation of *N*-hexanoyl-L-homoserine lactone. The *LuxR* receptor protein complex with the autoinducer binds to the lux operon promoter and activates its transcription. This leads to the synthesis of luciferase and the emission of light. With an increase in the population of cells, the concentration of the AI increases and becomes sufficient for the activation of the lux operon and the induction of this operon. It is revealed that the syntax gene of AI *LuxI* is included in the lux operon; therefore, the amount of the AI increases sharply with the induction of the operon. Many genes, including genes responsible for the synthesis of virulence factors, are activated by two QS systems in the pathogenic bacterium *Pseudomonas aeruginosa*, which causes severe respiratory infections. It has been found that more than 600 genes of *P. aeruginosa* are regulated by QS systems. It is very important to establish the fact that the AI itself, without bacteria *P. aeruginosa*, can affect the eukaryotic organism, interacting with components of the immune system (interleukins) and inhibiting the proliferation of lymphocytes and the formation of cytokines.

The QS regulation of virulence control in the Gram-positive bacterium *Staphylococcus aureus* and the control, competence, and sporulation in *Bacillus subtilis* were studied. In these systems, autoinducers of another type of peptides and peptides containing a thiolactone ring are used. The mechanism of functioning of QS systems in Gram-positive bacteria consists in the interaction of the AI with specific receptors —

two-component sensory histidine kinases. QS regulation involves a cascade of phosphorylation–dephosphorylation reactions. The phosphorylated regulator binds to DNA and activates the transcription of the target gene.

In other Gram-positive bacteria, streptomyces, QS systems are involved in morphological differentiation and the formation of secondary metabolites. Autoinducers in this case are γ-butyrolactones.

The QS system of *Vibrio harveyi* includes two types of autoinducers that interact with each other and participate in the regulation of the *lux* operon. Many other proteins, receptor proteins, three sensory kinases, as well as five small regulatory RNAs are involved in the regulation. Functioning of this system is also carried out through the phosphorylation/dephosphorylation cascade.

Since QS systems play an important role in regulating the virulence of bacteria, the inhibition of these systems leads to the inhibition of the synthesis of virulence factors. Drugs that suppress QS systems are suggested to be called antipathogenic inhibitors. Unlike classical antimicrobials (first of all, antibiotics), QS regulations do not have bactericidal or bacteriostatic effects on bacteria. The consequence of this is the lack of selective pressure, which causes the rapid formation of resistant forms of pathogenic bacteria. The use of drugs that can suppress the synthesis of virulence factors can help solve one of the most important problems of modern medicine. Another serious problem of antimicrobial therapy is the ability of pathogenic bacteria to form biofilms, leading to increased bacterial resistance to antibacterial drugs and disinfectants. Since it was shown that QS regulation plays an important role in the formation of biofilms, the use of drugs, which can suppress QS, is promising to inhibit such formation.

The inhibition of the functioning of QS systems can be achieved in various ways: inhibition of the synthesis of autoinducers, inhibition of binding of autoinducers to corresponding receptor proteins, and degradation of autoinducers by special enzymes. The strategy of antibiotic therapy, which is based on the inhibition of QS regulation, can also be used to combat phytopathogenic bacteria. The promising direction of the use of QS inhibitors is to prevent the formation of biofilms in the case of tube overgrowth.

The study of QS regulation is important for finding out the communication of bacteria. It is shown that autoinducers of one type of bacteria can significantly increase the pathogenicity of another species. An example of interspecific communication is the interaction between *P. aeruginosa* and *Burkholderia cepacia*. With a consistent infection, the pathogenicity of *B. cepacia*, which synthesizes the autoinducer at a very low level, is increased at the expense of the autoinducer of *P. aeruginosa*. This example indicates that the possible interaction of nonpathogenic bacteria (producers of autoinducers) and weakly pathogenic bacteria can lead to the development of infection.

QS systems carry out the global regulation of many cellular processes in bacteria. By this time, they have been studied in a few species, and the molecular mechanisms of different types of QS regulation are not sufficiently explored. The importance of studies of such a system of regulation of cellular processes in bacteria is shown by the proteomic analysis data on changes in the production of more than 150 plant proteins under the influence of bacterial autoinducers. Autoinducers of *P. aeruginosa* and *Sinorhizobium meliloti* can cause plants to form compounds that inhibit or stimulate *P. aeruginosa*. These data indicate a possible important role of QS regulation not only in bacteria but also in higher eukaryotes.

6.4. Luminescent bacteria and bioluminescence

Bioluminescence is a phenomenon that is quite widespread among living organisms. There are more than 800 species of organisms capable of bioluminescence, among which there are representatives of bacteria, fungi, algae, protozoa, crustaceans, mollusks, worms, and fish. The enzymes, substrates, and mechanisms of reactions that determine the luminescence of these organisms differ significantly, but in all cases, the reaction leading to the emission of light requires the participation of oxygen.

Luminescent bacteria occur in marine, freshwater, and terrestrial ecosystems (**Fig. 6.4**). They can be free-living saprophytes or symbionts that live in the digestive tract or in the light organs of bony fish and squid. Most luminous bacteria belong to the genera of *Photobacterium*, *Alteromonas*, and *Beneckea* (*Vibrio*). Representatives of *Photorhabdus* genus (formerly identified as *Xenorhabdus*) were also from the soil. The best-studied group of luminous organisms are bacteria. A catalog of luminous bacterial cultures provides a description of strains, substrates for their selection, storage conditions, and important biotechnological properties. The appendix contains a description of the preparations for bioluminescent analysis.

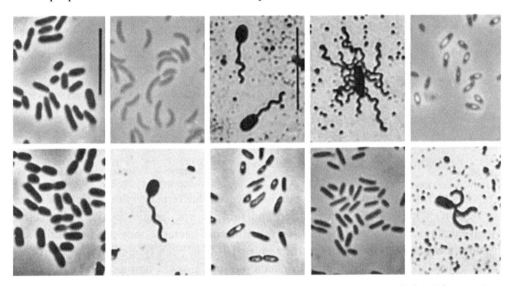

Fig. 6.4. Shapes of luminescent bacteria of *Vibrio* and *Photobacterium* genera isolated from various ecosystems.

Most prokaryotes capable of luminescence are Gram-negative facultative anaerobes, which move with the help of one to eight flagella (**Fig. 6.5**).

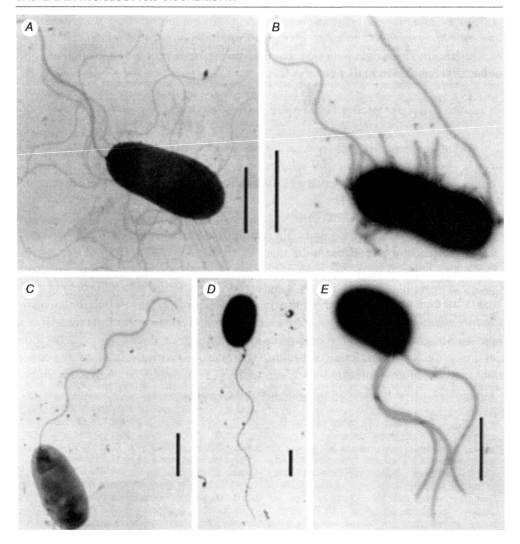

Fig. 6.5. Luminescent bacteria: *Photobacterium leiognathi* (A), *Vibrio fischeri* (B), *Photobacterium phosphoreum* (C), *Vibrio harveyi* (D), *Photobacterium fischeri* (E).

In anaerobic conditions, almost all luminescent bacteria carry out formic acid fermentation or mixed-type fermentation and form formic, acetic, lactic, and succinic acids, alcohol, CO_2, and acetoin. Like most marine bacteria, they are halophilic and require a concentration of salts in the range of 1–2%. Each type of luminescent bacteria differs according to its growth needs, optimal growth temperature, and kinetic properties of luciferase, involved in the generation of light. However, all luminous bacteria have similar mechanisms of light formation.

The *luminescence* is the process of aerobic oxidation, which leads not to the formation of ATP, but to the excitation of an intermediate product, which then produces light.

The *luciferin* and the enzyme *luciferase* are involved in the luminosity reaction in the mollusc *Pholas dactylus*. Bioluminescence was studied best in the American firefly *Photinus pyralis*. Luciferase (L) catalyzes the reaction of reduced luciferin (L-H_2) with ATP; the product of this reaction is adenylate which produces light when oxidized.

$$L\text{-}H_2 + ATP + L \rightarrow L \times L\text{-}H_2 - AMP + PP_i,$$

$$L \times L\text{-}H_2 - AMP + O_2 \rightarrow \text{light} + \text{products}.$$

In bacterial luminescence, several components are also involved: the reconstituted flavin mononucleotide (FMN), O_2, and long-chain aldehyde *tetradecanal*. Luciferase belongs to monooxygenases and carries out the conjugate oxidation of NADPH and the long-chain aldehyde. The reaction is accompanied by the emission of blue-green light (wavelength 490 nm).

$$NADP + H^+ + R\text{--}CHO + O_2 \rightarrow NADP^+ + H_2O + R\text{--}COOH + h\nu$$

Luciferase is localized in the cytoplasm; it contains a bound FMN, whose reduction is provided by NADFN. In the reduced form, $FMNH_2$ is oxidized with oxygen to form peroxiflavine and a stable intermediate product (*aldehyde complex*). After the oxidation of the substrates, the complex slowly decomposes with releasing the products. At this stage, there is an emission of light.

$$L\text{-}FMNH_2 + O_2 \rightarrow L\text{-}FMNH\text{--}OOH + R\text{--}CHO \rightarrow L + FMN + H_2O + R\text{--}COOH + h\nu$$

Aldehyde, which is involved in the luciferase reaction, is formed from the corresponding fatty acid under the action of a reductase complex consisting of synthetase, transferase, and reductase.

$$R\text{--}COOH + ATP + NADPH \rightarrow NADP^+ + AMP + PP_i + R\text{--}CHO$$

The intensity of the luminosity depends on the concentration of ATP and NADPH; so, this reaction is used to determine the low concentrations of NADPH or ATP using a luciferase complex of bacteria or the firefly *Photinus pyralis*, respectively.

The genes encoding the bacterial luciferase subunits (*luxAB*) and the polypeptides of the complex enzymes of fatty acid reduction (*luxCDE*) are cloned and sequenced from the *lux* operon of luminescent bacteria of three genera *Photobacterium*, *Vibrio*, and *Xenorhabdus*. LuxCDE genes flank *luxAB* genes in different types of luminescent bacteria; genes are read in the order of *luxCDABE*. In bacteria of the *Vibrio* genus, enzymes of the luciferase complex are encoded by the *luxCDABEGH* locus. The *luxAB* genes of this complex are structural luciferase genes; *luxCDE* genes encode a complex of enzymes for the reduction of fatty acids; *LuxGH* gene products are needed for the reduction of flavin.

Bacterial luciferase is a heterodimer consisting of two different polypeptides α and β (with molecular weights 40 kDa and 37 kDa, respectively). The active center is located in the α-subunit. These two polypeptides have 30% identity in the amino acid sequence. It is believed that the α-subunit function is to maintain the conformational changes of the α-subunit in the process of catalysis.

The *luxCDE* genes encode polypeptides required for the conversion of fatty acids into a long-chain aldehyde. Proteins (transferase, synthetase, and reductase) have been purified from *P. phosphoreum* as part of the multifencing reductase complex with a molecular weight of 500 kDa. The gene *luxG* encodes the protein associated with flavin reductase and other electron transport enzymes. Perhaps, the product of this gene is related to the formation of FMNH2 for the luminescence reaction. The gene *luxH* encodes the protein that has 60% identity with the protein that catalyzes the synthesis of the predecessor riboflavin, namely, 3,4-dihydroxy-2-butanone-4-phosphatase, in *Escherichia coli*. Along with this genome, there are three more *rib* genes that are involved in the conversion of riboflavin precursors.

All these genes have a common transcriptional terminator, indicating a possible relationship between lux and rib genes in luminescent bacteria. The *luxL* and *luxY* genes encode proteins that have 30% identity with enzymes that catalyze the conversion of **lumazin** to riboflavin. The blue fluorescent protein (**lumazin protein**) in *Photobacterium phosphoreum* and *Photobacterium leiognathi* changes the wavelength of light emission ($\lambda_{max} \sim 478$ nm), causing a blue light. The *LuxY* gene encodes a yellow fluorescent protein. The presence of it in *V. fischeri* causes a shift in the wavelength of luminescence ($\lambda_{max} \sim 545$ nm) and the emission of yellow rather than blue-green light. The molecular mechanisms of such changes have not been clarified.

Regulatory genes, *luxI* and *luxR*, that control the expression of genes involved in the luminescence process are cloned and sequenced in *V. fischeri* and *V. harveyi*. The *luxI* gene is required for the synthesis of an autoinducer that controls the expression of luminescence in *V. fischeri*. The *luxR* gene product acts as an autoinducer receptor and thus can activate the expression of the lux operon. In the lux systems of other types of luminescent bacteria, regulatory genes are not identified.

The ability to grow and bioluminescence depend on the composition of the medium and the density of cells in the culture. The study of the effect of culture density in a symbiotic bacterium *Vibrio fischeri* has shown that the regulation of luminescence, which depends on the density of the culture (effect "quorum sensing"), is carried out at the transcription level.

Cells of *V. fischeri* constitutively synthesize and secrete a small amount of an autoinducer (N-(β-ketocaproyl)-L-homoserine lactone or n-(3-oxohexanoyl)-l-homoserine lactone) in the medium. At high cell density, the concentration of the autoinducer reaches a threshold value, at which it is able to induce luminescence. The autoinducer easily penetrates through the cell membrane and binds to its receptor, the transcription regulator *LuxR*, which activates the expression of the *luxGDABEGH* genes in these conditions. A similar type of autoregulation has been found in free-living luminescent bacteria, whose population density in natural conditions does not reach the level required for the induction of a luminescent complex.

It is believed that **lux autoinducers** can be a part of a common family of signal transducers. The bacterial lux system can serve as a good model for investigating common signaling systems involved in developmental processes, intracellular communication, and even symbiosis.

The identification of lux autoinducers and regulatory proteins in *V. fischeri* and *V. harveyi* has become a biochemical and genetic basis not only for the elucidation of bioluminescence process mechanisms but also for the development of rapid and sensitive methods for the determination of various compounds. The basis of these methods is the change in the intensity of the luminescence of biologics after the action of the substance, the analysis of which is carried out. The concentration of the substance is determined by changing radiation parameters. Bacterial luminescence has a high sensitivity to various inhibitors of biological activity: anesthetics, drugs, industrial poisons, insecticides, pesticides, and medicinal preparations. Lack of specificity makes the application of luminous bacteria for ecological monitoring prospective. In this case, the total effect on luminous bacteria determines the toxicity of the sum of harmful substances. The ease of luminescence measurement, the speed of the method (analysis duration is 1–3 min), high sensitivity (from 10^{-3} to 10^{-12} M), the ability to automate the measurement, and statistical processing of results provide luminescent bacteria with significant advantages over other biological tests.

Recombinant bioluminescent strains containing *lux* genes isolated from natural luminescent bacteria are produced and successfully used. The *lux* genes that are placed under the appropriate promoter and transferred to a nonluminescent organism can be used as sensors for a number of processes and substances, whose capabilities are determined by the properties of the expression system. The use of recombinant *E. coli* luminescent strains containing a cloned luciferase gene is well established for the analysis of fresh water. Reporter strains that have high specificity to a specific toxic agent are obtained.

Another direction of the practical application of the luminescence phenomenon is the use of cloned *lux* operons to predict the stability of the expression of foreign genes in different conditions of the existence of the host organism. Such marker systems are designed for many microorganisms as well as for higher organisms and plants. As test objects, instead of luminescent bacteria, the reactions catalyzed by luciferase, the coupled enzyme system NADH:FMN-oxidoreductase and luciferase, can be used. Tests on the basis of coupling chains of enzymes with luciferases allowed to expand the amount of substances for bioluminescent analysis and to use those compounds that do not directly participate in bioluminescence. The use of enzymatic reactions as indicative antigen–antibody reaction systems resulted in the creation of various variants of homogeneous and heterogeneous immuno-enzymatic methods, in which both enzymes, along with horseradish peroxidase and alkaline phosphatase, use luciferases. Immunobioluminescence methods allow to selectively analyze substances that do not participate directly or indirectly in the process of bioluminescence, including xenobiotics, inhibitors of biological activity, and toxicants.

Highly purified bacterial luciferases are isolated from four types of luminescent bacteria: *P. leiognathi*, *P. phosphoreum*, *V. fischeri*, and *V. harveyi*. Among the luminescent bacteria, there are strains promising for many other enzymes: oxidoreductase, amino acids decarboxylase, restriction endonuclease, cellulase, and chitinase. Chitinase producers can be used to produce chitin derivatives that are used in the food and medical industry. A number of strains of luminescent bacteria are producers of neuraminidase and restriction enzymes, which are used for fundamental research as well as for the production of medical products.

***

Thus, bacteria are an integral part of the ecosystem and never exist in the form of pure cultures in the natural conditions. Not only do they live in a mixed culture, but they are also part of the ecosystem, which includes organisms of other systematic groups.

The extreme physical and chemical factors affect the environment. The interdependence between the number of species, the number of individuals, and the degree of extreme conditions is observed in many ecosystems.

Bacterial populations are capable of intercellular intrapopulation contacts and co-operative interaction (quorum sensing). The quorum sensing is a special type of regulation of bacterial gene expression, which functions in conditions of high density of populations.

There are more than 800 species of organisms capable of bioluminescence. Bioluminescence is a phenomenon that is quite widespread among living organisms. Luminescent bacteria occur in marine, freshwater, and terrestrial ecosystems. The luminescence is the process of aerobic oxidation, which leads not to the formation of ATP, but to the excitation of an intermediate product, which then produces light.

CHAPTER 7

EFFECT OF ENVIRONMENTAL FACTORS ON BACTERIAL CELLS

In this chapter, the influence of environmental factors on bacterial populations and their mechanisms are described. External environmental factors with antibacterial effect may be physical or chemical. Bacteriostatic and bactericidal effects are characterized. The effect of the physical factors, such as drought, temperature, pressure, ultrasound, radiation, sunlight, ultraviolet rays, X-rays, and γ-rays, on bacterial viability is presented. The chemical factors, including the effect of pH, redox potential, disinfectants, phenol and phenolic compounds, alcohols, halogens, heavy metals and their compounds, oxidants, dyes, soaps, and synthetic detergents, are also described.

A unique group of chemotherapeutics is antibiotics. Effects of chemotherapeutics and antibiotics, their mechanisms of action, and the toxic influence on bacterial cell populations are represented. A special attention is paid to antibiotics inhibiting cell wall synthesis, disturbing the function of the cytoplasmic membrane, the metabolism of nucleic acids, and protein synthesis.

7.1. Effect of external factors on bacteria

The life processes of bacteria strongly depend on the environment. These conditions are a reflection of the joint action of various factors. The general feature of most of them is the lawful character of their effect, which is manifested by the characteristic points of minimal, optimal, and maximal effects (**Fig. 7.1**). With regard to bacterial life, the minimal effect of the agent can be related to the onset of growth and the initiation of metabolic processes. The optimal effect corresponds to the maximum growth rate, the largest activity of substance, and energy exchange. At the maximum intensity of the effect of the relevant factor, the cells die.

Bacterial Physiology and Biochemistry
http://doi.org/10.1016/B978-0-443-18738-4.50007-3,

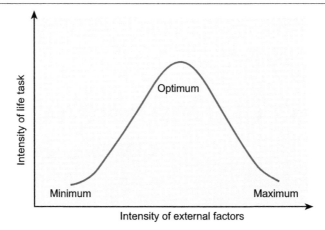

Fig. 7.1. Effect of external factors on bacteria.

The effect of environmental factors on bacteria can have positive and negative expressions. Positively acting factors are mainly used where they help accelerate growth or increase metabolic activity. On the other hand, external factors with adverse effects are found to be particularly useful in conditions where the presence of bacteria is undesirable or even harmful.

Bacteriostatic and bactericidal effects. When observing the adverse effect of factors on bacteria, we can see that some of them act bacteriostatically, others bactericidally. The bacteriostatic effect lies in the fact that the bacterial cells stop dividing so that their number does not increase while the factor is present (**Fig. 7.2**).

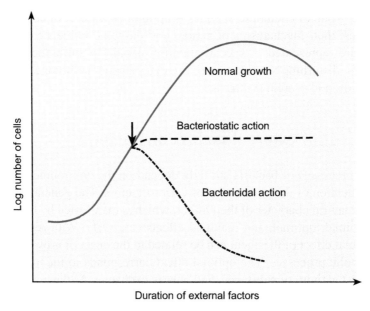

Fig. 7.2. Effect of external factors on bacteria.

If the factor is removed from the environment, the cells can continue living. In the presence of a bactericidal factor, the ability to reproduce and cell viability are lost. In some cases, lysis of the cell may occur. Death does not usually occur at the same time in the whole population but is progressive and can be expressed by the so-called logarithmic curve of death (**Fig. 7.3**). Certain chemicals may exhibit a concentration-dependent bacteriostatic effect. At extremely low concentrations, some toxic substances can even act as stimulants.

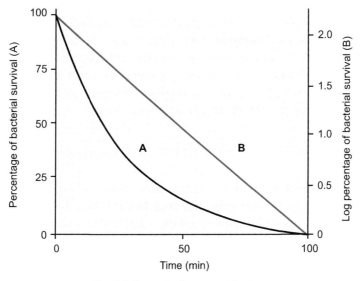

Fig. 7.3. Logarithmic curve of death.

Conditions affecting the effect of external factors. External environmental factors with antibacterial effect may be physical or chemical. In practice, they are mostly applied in bacterial disinfection by either sterilization or disinfection. When sterilizing, all bacteria, including other forms of living matter, are killed. The aim of the disinfection is to kill the pathogenic organisms of infectious diseases that occur mainly around the infection. In particular, chemotherapeutics and especially antibiotics are used to combat the bacterial infection of a macroorganism. The antibacterial effect of external factors can be affected by different conditions. These are in particular:

1. **Nature and intensity of factor action.** Most physical and chemical factors affect virtually all bacterial cells. In contrast, chemotherapeutics and antibiotics are characterized by a selective, that is, species or group-specific, effect on bacteria. In addition, the effect of physical factors is influenced by, for example, temperature and radiation dose. For chemical substances, the intensity of the action is dependent on the concentration; in addition, it often determines whether the effect of the substance is bacteriostatic or bactericidal.

2. **Duration of exposure.** The limiting factor is also the duration of action, the length of which increases the harmfulness of the effect of the respective factor on the cells. The duration of exposure is closely related to the temperature at which the microorganism

is exposed to the factor. For example, *E. coli* cells subjected to phenol die much more rapidly at 42 °C than at 30 °C when the phenol acts bacteriostatically for almost 10 hours.

3. **Nature of the organism.** The sensitivity of bacteria to the effect of external factors is conditioned by both the species and the physiological state of the cell. The rate of activity is also determined by the density of bacterial cell suspension and age, that is, the growth phase. In general, vegetative forms are relatively more sensitive than bacterial spores that are often highly resistant to adverse effects. Similarly, the cells in the early growth phases, in particular the lag phase and the accelerated growth phase, are more sensitive to the effects of external environment factors than the older cells.

4. **Nature of the environment.** The effect of a given factor may be amplified or weakened by the chemical composition and physical properties of the environment. For example, a highly viscous environment or a medium containing organic substances contributes to weakening the effect. The amplification of the effect can be achieved, for example, by changing the pH, raising the temperature, etc.

7.2. Mechanisms of the effect of environmental factors

Although the mechanism of destructive effect of external factors on bacteria is not yet fully known, it can be reasonably assumed that it is manifested in damage to vital systems, whose activity stops and the cell dies. The cell can survive if it is adjusted to the effect of the agent by means of physiological or genetic changes.

The disruption of vital systems and functions by external factors in bacteria can be caused by damage to cellular structures, cell walls, and cytoplasmic membranes and by the inhibition of enzymatic activity by stopping metabolic processes.

7.2.1. Physical factors

Physical factors that can cause bacterial cell damage or potentially death under certain conditions include drought, temperature, pressure, ultrasound, surface tension, and radiation.

Drought. Most bacteria belong to so-called hydrophilic organisms, that is, those that require freely accessible water in their environment. Only some actinomycetes can be included among *xenophilic* organisms that can also use hygroscopic water, which is bound in the form of a molecular layer on the surface of soil particles. In the absence of water, bacterial cells are dehydrated; therefore, they rapidly reduce their metabolic activity and die despite living in normal air and temperature conditions. Some Gram-negative cocci are particularly sensitive; for example, gonococci and meningococci die in a water-free environment within a few hours. Streptococci are more adaptable and can withstand these conditions for weeks. The viability of tuberculosis-inducing *Mycobacterium tuberculosis* (cells from dried sputum) is maintained longer. Other forms of bacteria, such as spores, cysts, and capsuled cells, are characterized by high resilience to drought.

Temperature. It can fulfill functions within a certain range of temperature, characterized by minimum, optimum, and maximum cardinal points. At the minimum temperature, all bacterial life stops. Although some cells die in these conditions, most of them retain viability. The optimum temperature is associated with the highest rate of reproduction and metabolic activity, although some metabolic processes may differ in optimum temperature requirements. Above the maximum temperature bar, life symptoms cease, and further increases in temperature affect the cells lethally.

The range between the minimum and maximum temperatures represents the temperature range of bacterial growth. Its range can vary considerably for different bacteria. While it is relatively wide in saprophytic bacteria, it is very narrow in pathogenic bacteria. Depending on the temperature range, bacteria can be divided into several groups. These are:

1. *Psychrophilic bacteria.* Psychrophilic bacteria grow best at temperatures below 20°C. They occur in oceans, deep lakes, and cold springs. They are characterized by slow reproduction and the reduced activity of metabolic processes. In laboratory conditions, many psychrophilic bacteria easily adapt to higher temperatures.

2. *Mesophilic bacteria.* Mesophilic bacteria require a temperature in the range of +20–+40 °C. These bacteria are the majority. In addition to saprophytic forms, they include parasitic, zoopathogenic, and human pathogenic bacteria with an optimum growth rate at about 37 °C, which corresponds to the human and animal body temperature.

3. *Thermophilic bacteria.* Thermophilic bacteria have optimal growth temperatures around +55 °C. Strictly thermophilic (stenothermophilic) bacteria do not grow at temperatures below +30 °C; optionally thermophilic (eurythermophilic) may grow below this threshold. The essence of thermophilia is probably the higher thermostability of enzymes conditioned by their structure. In addition, the ribosomes of thermophilic bacteria are more stable at higher temperatures than mesophilic ribosomes. The maximum thermophilic temperature is between +75 and +90 °C. Thermophilic organisms include actinomycetes and sporulating forms of bacteria. They are found in hot springs, manure, compost, etc.

The lethal effect of temperature lies primarily in protein denaturation and the inactivation of one or more enzymes associated therewith. Respiratory enzymes are particularly sensitive to higher temperatures. The thermal inactivation kinetics shows that specific proteins, including those of thermophilic bacteria, are considerably more stable than mesophilic forms (**Fig. 7.4**).

It is clear that resistance is an expression of mutational changes affecting the primary structure of most cellular proteins. The lethal effect, however, depends not only on the temperature and duration of action but also on the conditions, in which the bacteria are exposed to the effect of temperature, for example, wet heat is much more efficient than dry heat. The effect of temperature also increases the acidic and alkaline environment. Conversely, colloidal solutions containing proteins and polysaccharides reduce the effect of temperature.

The effect of elevated temperature is used in the sterilization of culture media, various devices, and so on.

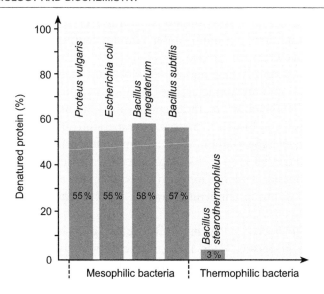

Fig. 7.4. Stability of cytoplasmic proteins in mesophilic and thermophilic bacteria at 60 °C.

In order to express the degree of resistance, the term "thermal death time" is sometimes used. This term expresses the shortest time required to kill the bacterial suspension at a given temperature while respecting other conditions. The experimental determination of thermal death time requires careful control of these conditions, which include, in particular, the environmental properties (composition, pH, etc.) and the number of cells, which has a significant role in the temperature effect. The thermal inactivation of the cells is most frequent according to the curve shown in **Fig. 7.5.**

Bacteria stored at or below the freezing point have no apparent metabolic activity. It turns out that at low temperatures, proteins are subject to gradual changes, which are conditioned by the weakening of their hydrophobic bonds, important especially for the tertiary structure. As to all other types of bonds, their strength increases when the temperature is lowered. The attenuation of hydrophobic bonds leads to the inactivation of mainly allosteric and ribosomal proteins, which are

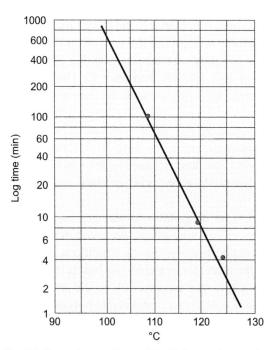

Fig. 7.5. Dependence of thermal death time on temperature.

especially sensitive to temperature reduction. The effect of low temperatures can also be seen in disrupting the transport of substances through the cytoplasmic membrane, leading to growth inhibition.

The effect of the low temperature may be bacteriostatic, if the cooling is done quickly enough. Otherwise, the cell structure will be broken down by ice crystals in the cell. If freezing is performed in a vacuum, the ice crystals sublimate, so dehydrated cells can be retained for a prolonged period without loss of viability. This method of drying bacteria cells is used in the so-called lyophilization. Only some bacteria, for example, meningococci and gonococci, are exceptionally sensitive to low temperatures.

Pressure. Changes in the structure of the bacterial cell can also be induced by pressure, both osmotic and hydrostatic.

The *osmotic pressure* in bacterial cells is greater because of their higher concentration of soluble substances than in the surrounding environment. Its value for Gram-positive bacteria is about 0.2 MPa and for Gram-positive, 0.05–0.1 MPa. However, due to the considerable rigid cell wall and osmoregulation, bacteria do not exhibit the plasmopathy phenomenon that is common in plant cells. The increasing osmotic pressure of the environment is manifested in hypertonic solutions and prevents the cells from receiving water with its dissolved nutrients from the environment because they create so-called physiological drought around them. Cells are subject to plasmolysis, which can be lethal if it lasts too long. The plasmolytic effect of high salt concentrations (10–15%) or sugar concentrations (5–70%) is used in practice in food preservation since most bacteria are highly susceptible to higher osmotic pressure.

The most bacteria grow well in environments containing less than 2% of salts. Higher concentrations are harmful to bacteria. However, there are bacteria that grow best only in higher salt environments (up to 30%). They are the so-called halophilic forms, among which there are mainly marine or salt lake bacteria. Previous studies of halophiles indicate that the mechanism balancing osmotic pressure lies mainly in the high concentration of salts in the intracellular solution.

Most bacteria are influenced by changes in *hydrostatic pressure*. As a result of higher pressure at normal temperatures, slowing or loss of movement, stopping growth, weakening of virulence, metabolic changes, or death occur in many cases. The effect of increased pressure also depends on the heat and the duration of the action. Hydrostatic pressure greater than 100 MPa inactivates the vegetative cells of most bacteria. Sporulating bacteria tolerate the pressure increase much more than nonsporulating bacteria. They can survive under pressure conditions up to 12, 100 MPa. Sea-bed or oil-bearing bacteria also tolerate hydrostatic pressure very well. These bacteria are referred to as barophilic. The mechanism of action of hydrostatic pressure consists primarily in decreasing cell volume and increasing the viscosity of the cell content, resulting in the inactivation of enzymes and hence in the reduction of the rate or stopping of biochemical reactions, especially of protein synthesis.

Ultrasound. Sound waves, especially at frequencies above 20 kHz, have a harmful effect on bacteria. Vegetative cells are quickly killed, with younger cells less resistant than older cells. Fibrous bacteria are the most sensitive to ultrasound, bacilli are less sensitive, and cocci are the most resistant. Particularly high sensitivity was found in bacterial spores and acid-resistant bacteria. The lethal effect of ultrasound is attenuated by the increased media viscosity, possibly by the presence of proteins and surface tensioning agents.

The harmful effect of ultrasound is manifested mainly by gas cavitation, consisting in the formation of small gas bubbles in the protoplasm. This phenomenon causes mechanical damage to the cell and its disintegration.

Ultrasound is now used to make so-called cell-free preparations or to release intracellular enzymes for their isolation and biochemical studies. This method can isolate cellular components, for example, ribosomes, cell wall, and cytoplasmic membrane fractions.

Radiation. The bacteria are vulnerable to any radiation that can be absorbed by the cells and cause chemical changes. The harmfulness of the effect depends on the amount of energy contained in the absorbed light quantum, whereas the quantum energy content depends indirectly on the wavelength of the radiation (**Fig. 7.6**).

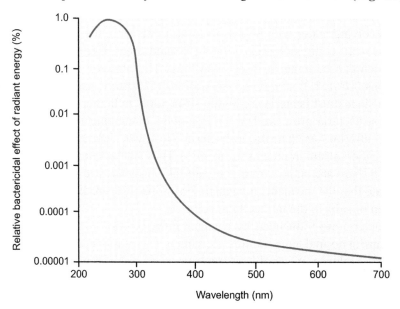

Fig. 7.6. Dependence of radiation effect on light wavelength.

Chemical changes of molecules or atoms cause the bacterial cell to absorb the amount of radiation with a wavelength of about 1000 nm. This range includes sunlight, ultraviolet rays, X-rays, gamma rays, and cosmic rays (**Fig. 7.6**). The energy of light quanta of higher wavelength is too small to be enough to cause chemical changes. The effect of this radiation is mainly manifested by heat (infrared radiation).

Sunlight. Sunlight is the most natural source of radiation that acts destructively on bacteria. Its effect is weakened by the presence of pigments or slime layers. An exception is photosynthetic bacteria for which sunlight is a source of energy.

Sunlight can also act indirectly by changing the environment. For example, staphylococci do not grow on agar plates exposed to sunlight for several hours. It is believed that irradiated soil produces toxic waste products such as peroxides, which act bactericidally. The effect of sunlight on bacteria is substantially increased by some dyes, such as methylene or toluidine blue, that bind to the cells. In the presence of these dyes, the cells exhibit increased sensitivity to that portion of the visible light that is absorbed by the dye. It is induced by the dye transitioning into a highly active, the so-called triplet state.

A subsequent reaction with diatomic oxygen produces O_2^- — active oxygen, which is a strong oxidizer and has a rapid lethal effect on bacteria. This phenomenon is referred to as photodynamic effect of light or also photosensitization (**Fig. 7.7**).

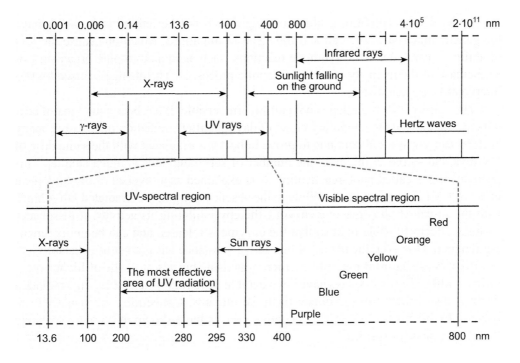

Fig. 7.7. Biologically significant types of radiation by wavelength.

Ultraviolet rays. Ultraviolet radiation has a considerable bactericidal effect. It acts most strongly in a 265-nm wavelength region, where it is maximally absorbed by nucleic acids and nucleoproteins. Although sunlight is partly composed of UV rays, most of the radiation with shorter wavelengths is retained by the atmosphere (ozone, clouds, and smoke), so the impact of UV rays on Earth is very limited. For this reason, sunlight has a less bactericidal effect than the UV rays used in the laboratory.

The lethal effect of UV rays, which are mainly absorbed by the pyrimidine bases (thymine and cytosine), lies primarily in the ionization and excitation of atoms, resulting in structure damage or breakdown of the molecules. Thus, DNA-inhibiting dimers of thymine can form. Thymine dimer is formed as a covalently bonded complex of two adjacent thymines on a single strand of DNA. This damage is very frequent but almost 90% of thymine dimers are repaired within a short time of the order of minutes, and only a few are experimentally observable and originate future changes at the cell level.

Thymine   Thymine dimer

Under certain conditions, ultraviolet light may not be lethal but it can damage the genetic material of the cell and thus lead to mutations. Mutations cause the loss of ability to carry out some synthetic reactions, such as in auxotrophic organisms, or an increase of the intensity of some metabolic processes. The changes caused by this intervention are hereditary.

The effect of UV radiation is not entirely irreversible. It has been found that if bacteria are exposed to the temporary effect of daylight after irradiation of UV rays, many bacteria can survive and continue to grow. In bacteria irradiated with the same dose of UV rays, but stored in the dark, the number of surviving cells is much smaller. This phenomenon is called photoreactivation. It is explained as a reversal of the biological effect of UV radiation on DNA. While in the absence of light, the irradiated DNA binds with the so-called photoreactive enzyme, thereby inhibiting its activity, it dissociates with the effect of visible light so that the enzyme is released and can begin its repairing function. However, the nature of this active substance has not yet been established.

X-rays are harmful not only to microorganisms but also to cells of higher organisms. Unlike UV rays, X-rays have considerable penetration capability. This radiation effect is also widely used to obtain bacterial mutants. Concerning the sensitivity of bacteria to the X-ray effect, Gram-negative species have shown to be more sensitive than Gram-positive species.

γ-**Rays** have a considerable ability to penetrate matter and are generally lethal for microorganisms. Bacteria are sensitive to this radiation, similarly to UV radiation and X-rays. Particularly high resistance is seen in sporulating and pigmented bacteria, particularly *Micrococcus radiodurans* strains. Due to its considerable penetration capacity and strong bactericidal effect, this radiation is mainly used to sterilize packaged food. According to previous results, it can be concluded that γ radiation induces the ionization of intracellular water molecules under the formation of hydroxyl radicals, which then react with other components, causing damage to the cell (**Fig. 7.8**).

In addition to the above-mentioned forms of electromagnetic radiation, corpuscular radiation has a wide harmful effect on the bacteria. Strong bactericidal action is known especially for cathode ray, which is used in medical practice to sterilize surgical instruments, drugs, and other materials.

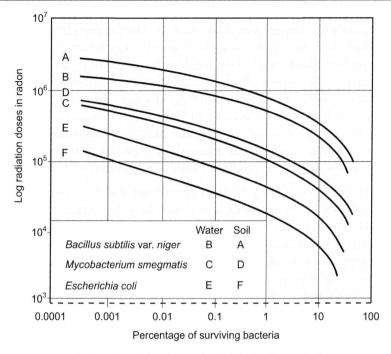

Fig. 7.8. Sensitivity of some bacteria to the X-ray effect.

Nowadays, many microorganisms are used as suitable models to study the mechanism of radiation effect on living cells of higher organisms. In addition to this fundamental research in radiation biology, the possibilities of using radiation for the sterilization of biological material are also monitored. This so-called cold sterilization can be applied even when disinfecting substances are sensitive to temperature, especially some foods and pharmaceuticals.

7.2.2. Chemical factors

Chemical factors, which influence bacterial life, are mainly pH and E_h of the environment and disinfectants. In the therapy of disease caused by bacteria, chemotherapeutics, including antibiotics, are used.

Effect of pH on the environment. Successful bacterial growth is only possible with a suitable reaction environment, that is, a suitable concentration of hydrogen ions. Like the temperature, the pH of the environment for each bacterial strain has its minimum, optimum, and maximum values, allowing cell growth and multiplication as well as other life processes. These values correspond approximately to the pH values required for the activity of vital enzymes. However, the temperature at which they were determined must be mentioned at the same time. pH limits for bacteria range from 4 to 10. Most bacteria grow best in a neutral environment, that is, around pH 7.0. Neutral culture mediums are required primarily by pathogenic bacteria, occupying the blood or lymph of the animal with pH around 7.4. On the other hand, acidophilic bacteria grow very well even in acidic environments, for example, *Thiobacillus thiooxidans*, oxidizing sulfur to sulfuric acid, also grows at a pH of about 1. On the other hand, alkaliphilic bacteria (urobacteria, denitrifying and proteolytic bacteria) grow best in alkaline media. Buffers play an important role in pH regulation. They help in keeping the environment at the required levels and preventing abnormal pH changes during the growth of bacteria that may affect their biochemical activity. In biological systems, amino acids, polypeptides, and proteins are used as buffers. A regular component of conventional cultivation environments are salts of weak acids, especially phosphates, acetates, and carbonates. To neutralize media, which are rapidly acidified, $CaCO_3$ is added.

Redox potential (E_h). The growth of aerobic and anaerobic bacteria depends not only on the presence or absence of O_2, but also on the redox conditions of the environment. The indicator of these conditions is redox potential, expressed by the value of E_h. Redox potential is a measure of the tendency of a chemical species to acquire electrons from or lose electrons to an electrode and thereby be reduced or oxidized, respectively. For most aerobic bacteria, an environment with an E_h value ranging from +0.2 to +0.4 V (at pH 7.0) appears to be suitable. In contrast, the obligatory anaerobic forms usually do not grow in a medium with E_h above -0.2 V. Decreasing of the E_h value can be achieved by the addition of thioglycolic acid, cysteine, ascorbic acid, etc. To reduce the redox potential, Na_2S may also be used but due to its toxicity in a limited amount only (below 0.1 g/L). These substances contain reactive groups that can be converted as needed to a reduced or oxidized form and thus change the redox potential to the desired value. In this way, strictly anaerobic organisms, such as *Clostridium tetani* and *Clostridium perfringens*, also grow in the presence of oxygen. Anaerobe enzymes are believed to contain groups that can only be active in a reduced environment.

The redox potential of the medium is significantly influenced by changes in pH and metabolism products that have the character of redox systems such as H_2O_2.

Disinfectants. Certain chemical compounds act harmfully not only on bacteria but also on most animals or plants. They are therefore primarily used to remove germs of infection from the outside environment, that is, to disinfect.

Factors influencing the selection and use of disinfectants include:

1. **The nature of the material to which the substance is to be applied**. For example, a chemical commonly used to disinfect contaminated equipment is not suitable for skin applications where it can cause serious damage to cellular tissue along with the removal of microbes.

2. **The properties of the organism exposed to the action of the disinfectant**. Different organisms are unequally sensitive to the effect of disinfectants. For bacteria, this different sensitivity may be conditioned by the age of the cell or by its composition (**Fig. 7.9**). Vegetative and capsule-free cells are therefore more sensitive in this respect than spores and encapsulated cells.

3. **The conditions under which the active ingredient is applied**. These include in particular the temperature, concentration, and duration of application of the disinfectant concerned. An important role is also played by the nature of the environment,

in which the bacteria are exposed to the effect of the substance. Experimentally, it was found that the relationship between the concentration of the disinfectant and the rate of bacterial death is not linear but exponential.

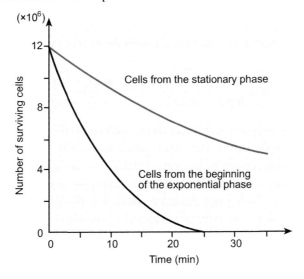

Fig. 7.9. Comparison of the sensitivity of bacterial cells from different growth phases and the effect of the disinfectant on time of action.

Dilution coefficient. Mathematically, the relationship between the concentration of the disinfectant and the time required to kill the cells can be expressed by the following expression:

$$C^n \cdot t = A \quad \text{or} \quad n\log C + \log t = \log A.$$

C is the concentration of disinfectant, n is the dilution coefficient, concentration exponent depending on the nature of the substance, and t is the time required for the lethal effect of the disinfectant (**Fig. 7.10**).

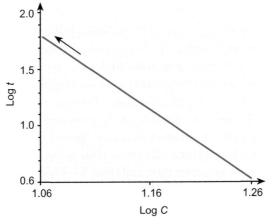

Fig. 7.10. Relationship between the concentration and the time required for lethal effect.

If the time required for the lethal effect of a disinfectant for a given bacterial population is determined for a series of concentrations and the results are expressed in the appropriate equations

$$n\log C_1 = \log A - \log t_1; \quad n\log C_2 = \log A - \log t_2, \text{ etc.,}$$

we can deduce the values n_1, n_2, and others from the pairs of equations according to the relationship below:

$$n_1 = \frac{\log t_2 - \log t_1}{\log C_1 - \log C_2}; \quad n_1 = \frac{\log t_3 - \log t_2}{\log C_2 - \log C_3}.$$

The value of the dilution factor (n) varies with each disinfectant and also with the temperature, at which the disinfection takes place. Generally, a substance having a high n value acts at a given concentration very lethally but, in dilution, it quickly loses its efficacy. As a measure of lethal effect, the concentration of a given substance used to kill 50% of individuals in the population is used. It is designated by the LD_{50} symbol and is used to compare the activity of various disinfectants.

Conditions of disinfectant applicability. Active substances must meet certain essential requirements to be used as disinfection. These requirements include:
- The substance must be toxic to microorganisms at low concentrations and at room temperature and must not be toxic to humans and animals.
- The substance must be well soluble in water or other commonly available liquids.
- It must be stable, that is, its antibacterial properties cannot be changed during treatment.
- The substance should not have any adverse effects on surrounding objects exposed to its effect (e.g., it does not cause corrosion of metals).
- The substance should be free of unpleasant odor.

Among the most important and most commonly used chemical compounds with disinfecting effect, the following substances should be mentioned: phenol and phenolic compounds, alcohols, halogens, heavy metals and their compounds, oxidants, dyes, soaps, synthetic quaternary ammonium salt detergents, and some gases.

Phenol and phenolic compounds. Even though there are many more efficient chemicals, which are also active at lower concentrations, phenol and phenolic compounds belong to the most important and most commonly used disinfectants (**Fig. 7.11**). In microbiology, phenol is often used to compare the effectiveness of other disinfectants. This is done using the so-called phenolic coefficient, which expresses the ratio between the lethal concentration of the compared substance and the phenol at the highest dilution after 10 minutes of action against the same strain of bacteria. The effectiveness of the investigated substance is the greater, the higher the coefficient value is. Depending on the concentration and time, for which the cells are exposed, the phenol acts bactericidally or bacteriostatically.

The phenol activity decreases at low temperatures, in an alkaline environment, and in the presence of an organic material. On the contrary, the presence of salts increases its effectiveness. Phenol is less efficient against spores. In the form of an aqueous solution, phenol is also used to disinfect contaminated instruments, sputum, urine, etc. The effect of phenol lies primarily in a reduction of the surface tension of the environment. This manifests itself in cell wall damage, which results in a gradual destruction of the cell.

Some phenol derivatives, such as cresol, orthonitrophenol, hexylresorcinol, and hexachlorophene, are much more effective than phenol alone. Especially effective is hexachlorophene, which acts bacteriostatically at high dilution, especially against Gram-positive bacteria. In combination with soap, it can be used to disinfect the skin.

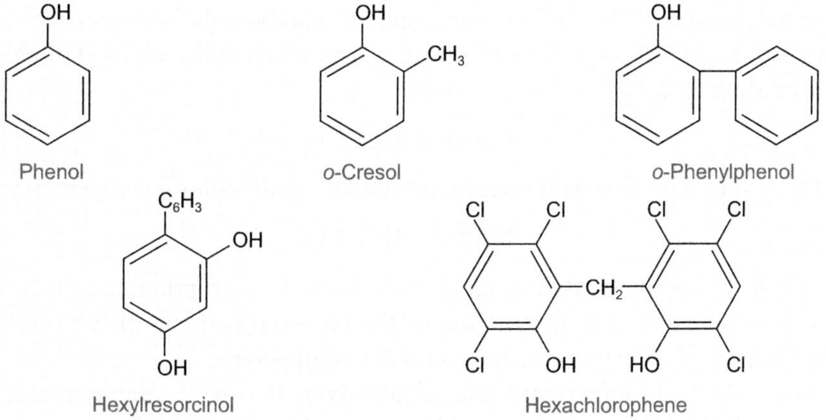

Fig. 7.11. Structures of phenol and some phenolic compounds.

Alcohols. The most commonly used disinfectant is 50–70% ethanol. However, it cannot be used for sterilization because it does not practically work on spores in the concentration effective on vegetative cells. Methanol is less bactericidal than ethanol. On the other hand, higher alcohols (propyl alcohol, butyl alcohol, and amyl alcohol) are more efficient. The bactericidal effect of alcohols increases with increasing molecular weight. Propyl alcohol and isopropanol at a concentration of 40–80% are used to disinfect the skin. High-molecular-weight alcohols mix very poorly with water and are therefore usually used for local disinfection.

Alcohol causes coagulation of proteins. However, high concentrations of alcohol have a strong dehydration effect, as it removes the water from the cell and cannot enter the cell. In this case, it acts rather bacteriostatically. This explains the relatively low efficiency of absolute ethanol on bacteria. In addition to gross disinfection of the skin, alcohol is also used to disinfect clinical oral thermometers.

Halogens are toxic to most bacteria. Their effect is inversely proportional to their atomic mass. One of the most widespread disinfectants in this group is chlorine. Compressed chlorine gas is widely used to purify water supplied to water mains. Like chlorine, also other compounds have a disinfecting effect. The most commonly used are chlorates (chlorine lime) and chloramines. The advantage of chloramines (chloramine-T and azochloramide) is that they are more stable than chlorates (**Fig. 7.12**).

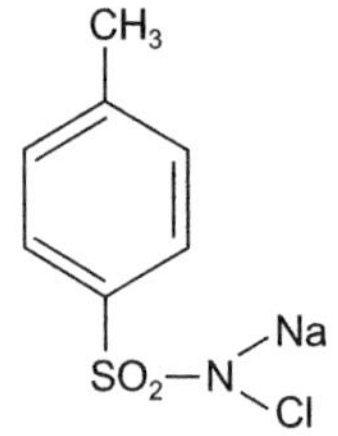

Fig. 7.12. Structures of chloramine-T and azochloramide.

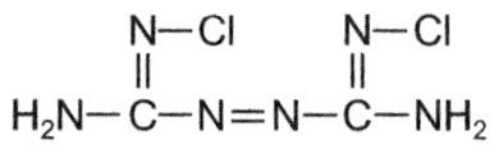

The bactericidal effect of chlorine and its compounds is associated with the decomposition of hypochlorous acid, which is formed upon the addition of chlorine to water according to the following equation:

$$Cl_2 + H_2O \rightarrow HCl + HClO.$$

At the same time, this acid is decomposed into hydrochloric acid and oxygen:

$$HClO \rightarrow HCl + O.$$

Oxygen is a strong oxidizing agent and causes the disruption of cellular components and whole cells. The destruction of the bacterial cell can also be induced by direct clustering of chlorine with some cellular components.

Iodine belongs to important bactericidal agents. It is well soluble in alcohol and in aqueous solutions of potassium iodide (tincture of iodine). Iodine is also bound in organic compounds with a surfactant effect. The advantage of these substances is that they do not stain the relevant object, do not irritate the skin, and are practically odorless.

Iodine acts bactericidally on almost all kinds of bacteria and on spores. Its effect is reduced by a larger amount of organic material in the environment. The mechanism of iodine action has not yet been fully elucidated. It is believed to bind to specific proteins, probably enzymes, and, by inhibiting their activity, induce destruction of cellular components. It is used mostly to disinfect the skin but is also suitable for disinfecting water and air.

Heavy metals and their compounds. Most heavy metals, whether alone or in the form of compounds, have harmful effects on bacteria. The most effective are silver, mercury, copper, and arsenic.

Silver has a strong bactericidal effect due to the release of metal ions into the environment. This so-called oligodynamic effect of silver lies in the high affinity of some cellular compounds to these ions. It is used for cleaning drinking water and for the production of antiseptic preparations (dressings, ointments, etc.). As a disinfectant, $AgNO_3$ and some organic silver salts (e.g., lactate, citrate) are used. The effect of silver is explained by the inactivation of the sulfhydryl groups in enzyme systems and the formation of insoluble salts.

Mercury is the most effective of all heavy metals. For disinfection, it is most often used in the form of a sublimate ($HgCl_2$), which kills vegetative cells at a dilution of

1:1000. Spores are affected after a longer period. Organic mercury compounds also have a strong bactericidal effect. Their advantage is relatively low toxicity for higher organisms. For this reason, they are used as antiseptics. Important mercury preparations include merthiolate, merbromin, and di(phenylmercuric) monohydrogen borate.

Mercury effect is based on the inhibition of enzymes that contain sulfhydryl groups. At the same time, a cellular protein precipitates. The binding process of the sulfhydryl group occurs according to the following equation:

$$\text{Enzyme} \begin{matrix} \diagup \text{SH} \\ \diagdown \text{SH} \end{matrix} + HgCl_2 \longrightarrow \text{Enzyme} \begin{matrix} \diagup S \\ \diagdown Hg \\ \diagdown S \end{matrix} + 2\,HCl$$

Active enzyme   Inactive enzyme

Copper and its salts also have a significant bactericidal effect. The presence of copper salts ($CuSO_4$, $CuCl_2$) induces the coagulation of proteins, which can be reversible under certain conditions. Just like silver, metallic copper acts oligodynamically.

For its bactericidal effect, *arsenic* is used in organic compounds, which are less toxic to higher organisms. Of more important trivalent arsenic preparations that are significantly more effective than pentavalent arsenic compounds, arsphenamine (salvarsan) and neosalvarsan are effective in syphilis treatment, caused by *Treponema pallidum*. Their mechanism of action is the inactivation of sulfhydryl groups by the formation of mercaptides.

Oxidants. Of the oxidizing agents, hydrogen peroxide and potassium permanganate are mainly used in disinfection. Hydrogen peroxide is usually used at a 3% concentration in which it is a mild antiseptic. Its effect lies in strong oxidation properties. Similarly, potassium permanganate ($KMnO_4$), another strong oxidizer, is also used as an antiseptic. Both substances oxidize sulfhydryl and other functional groups of enzymes, causing the degradation of vital cell systems.

Dyes. Most of the substances used to color bacteria have an inhibitory effect. Acidic and basic dyes work similarly to anionic and cationic surfactants. Basic dyes are usually more effective under normal pH conditions. They have a greater affinity for phosphate groups of nucleoproteins than for carboxyl groups of proteins. Some dyes, especially acridine dyes, are bactericidal, which is attributed to their strong ionization or to complexes with vital metals (chelation). The range of dyes has a selective effect on bacteria. For example, crystal violet inhibits most Gram-positive bacteria and some fungal species (fungistatic effect) but does not act on Gram-negative bacteria. Brilliant and malachite greens, like acriflavine, are selectively effective on Gram-positive bacteria.

The way the dyes act on bacteria is still unclear. It is likely that, in addition to chelation, their effect is associated with binding to cellular proteins, causing changes that disrupt the function of cellular enzymatic systems. This assumption is confirmed by the inhibitory effect of gentian violet on cell-wall peptidoglycan biosynthesis. The presence of the dye blocks the coupling of *N*-acetylglucosamine disaccharide and muramic acid with the peptide chain and interferes with the synthesis of murein (**Fig. 7.13**).

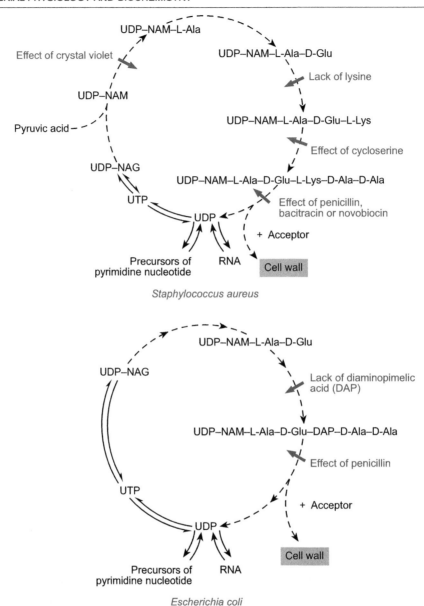

Fig. 7.13. Inhibitory effect of dyes and antibiotics on cell wall biosynthesis.

Soaps and synthetic detergents. Potassium or sodium salts of higher fatty acids, soaps, are slightly bactericidal. Pneumococci and some streptococci are relatively sensitive to soaps, whereas staphylococci and acid-resistant bacteria are highly resistant. The main importance of soaps lies primarily in the mechanical removal of microbes from the surface of the skin. In addition, soap reduces the surface tension of the environment and emulsifies and disperses oils with impurities, which also removes microorganisms.

Nowadays, synthetic surfactants — detergents, which are often more beneficial than soap, are used abundantly, as they do not coagulate and do not form sediments. Some are highly bactericidal. Their molecules typically contain both hydrophilic and hydrophobic groups. According to the charge, they are divided into anionic, cationic, and nonionic detergents.

1. **Anionic detergents.** Anionic detergents contain fatty acids with anionic groups formed by sulfates, phosphates, carboxyls, and sulfonamides. Unsaturated higher fatty acids and their salts, such as oleic acid, linoleic acid, and linolenic acid, lyse some cocci, mycobacteria, etc. They are less effective against Gram-negative enterobacteria, which are affected by salts of saturated fatty acids (sodium laureth sulfate).

2. **Cationic detergents.** Cationic detergents are quaternary ammonium salts with hydrophilic groups (cetrimonium bromide, cetylpyridinium chloride, etc.). Like soaps, they reduce surface tension and increase bacterial cell wettability. They are not only especially effective against Gram-positive bacteria but also act on Gram-negative bacteria. At lower concentrations, they act bacteriostatically. Because of their antibacterial effect, relatively low toxicity, solubility, and stability, quaternary ammonium salts are particularly useful in sanitation.

3. **Nonionic detergents.** Nonionic detergents contain, in addition to the hydrophobic aliphatic chain, a nonionized hydrophilic group formed by polymerized ethylene oxide or a polyhydroxy alcohol. They have only a slight bactericidal effect, but they reduce the surface tension of the environment. For these properties, Tween 80 (derived from polyethoxylated sorbitan and oleic acid) is often used. In its presence, the character of the growth of some bacteria changes, or bacterial growth is accelerated (*Mycobacterium tuberculosis*).

The mechanism of activity of detergents lies in the change of surface charge and surface structure of cells, leading to the disruption of the osmotic barrier and the function of the cytoplasmic membrane. At the same time, specific enzymes are inactivated, and cell proteins coagulate. Gram-positive bacteria are particularly sensitive to detergents. On the other hand, spores and acid-resistant bacteria appear to be resistant to the activity of these substances.

$$(C_{12}H_{25}OSO_3)^- Na^+ \qquad \left[\begin{array}{c} R_1 \\ | \\ R_4 - N - R_2 \\ | \\ R_3 \end{array} \right]^+ Cl^-$$

Detergents

Anionic	Cationic
(Quaternary ammonium salt)	(Sodium lauryl sulfate)

Disinfection with gas. When sterilizing sensitive biological materials, plastic containers, enclosed rooms, etc., which cannot be treated with conventional disinfectants, gaseous substances, such as formaldehyde, ethylene oxide, and β-propiolactone, may be used.

Formaldehyde (CH_2O). Formaldehyde starts evaporating only at higher concentrations and at elevated temperatures. At elevated temperature, it polymerizes and forms a rudimentary paraformaldehyde. In practice, formaldehyde is best known in the form of formalin, which is a 37–40% formaldehyde solution. Formalin and para-formaldehyde release gaseous formaldehyde (a proper disinfectant) by heating. It is also used to sterilize and inactivate toxins and viruses that retain their antigenic properties. The disadvantage of formaldehyde is the long time needed to kill vegetative forms of bacteria and low efficiency against spores. The best results are achieved by the application of formaldehyde at a temperature of about 22 °C and a relative humidity of 60–80%. The effect consists in the aldehyde group reacting with the anionic and hydroxyl groups of proteins and nucleic acids according to the following equation:

$$H-C{\overset{H}{\underset{O}{}}} + H_2N-CH_2-R \longrightarrow H-C{\overset{H}{\underset{N-CH_2-R}{}}} + H_2O$$

Ethylene oxide (CH_2-O-CH_2). Ethylene oxide is a disinfectant with a strong bactericidal effect not only on vegetative forms of cells but also on spores. At a temperature of less than 10.8 °C, it changes into a liquid. Increasing the temperature above this value quickly causes evaporation, and the substance becomes a flammable gas. For this reason, working with ethylene oxide itself is very challenging and requires the careful observance of safety measures, such as when working with other flammable substances. Ethylene oxide is also used for disinfection in a mixture with CO_2 or halogen kerosene. These mixtures are nonflammable but retain the disinfecting effect of free ethylene oxide. The strong bactericidal effect of ethylene oxide is due to its high penetration ability. For this property, it is used to disinfect packaged materials, bundles of substances, plastic containers, pharmaceutical preparations, and contaminated rooms.

The bactericidal effect of ethylene oxide lies in alkylation reactions with free carboxyl, amine, or sulfhydryl groups of organic compounds, including enzymes and other substances of a proteinaceous nature.

$$H_2C-CH_2{\underset{O}{\diagdown \diagup}} + Enzyme-SH \longrightarrow Enzyme-S-CH_2-CH_2OH$$

β-Propiolactone $\left(\overset{O}{\overset{\|}{\underset{H_2C-CH_2}{C}}}\right)$. β-Propiolactone is a liquid substance used at 20 °C to sterilize vaccines, tissue transplants, and other unstable biological materials. At higher temperatures, it changes into a nonflammable gas with a strong bactericidal effect on vegetative cells and spores. The greatest efficiency is achieved at 25 °C and relative humidity around 70–80%. It is more active than the previous gases. However, unlike ethylene oxide, it has lower penetration ability. It is mostly recommended to disinfect rooms instead of using formaldehyde. Its disadvantage is the irritating effect on the skin. Repeated usage and higher doses show carcinogenic properties. The mechanism of action of β-propiolactone is similar to that of ethylene oxide.

7.2.3. Chemotherapeutics

Chemotherapeutics are substances that have a toxic effect on bacteria, but unlike disinfectants they do not act harmfully on higher organisms. If their chemical composition is known, they can also be prepared artificially. A special group of chemotherapeutics are antibiotics.

The mechanism of action of many chemotherapeutic agents appears to be closely related to their structural relationship with compounds that are required by bacteria to produce coenzymes, proteins, and nucleic acids. As structural analogs of these compounds, chemotherapeutics take their place in biochemical reactions and inhibit vital processes. In general, chemotherapeutics are used in the treatment of various infectious diseases. Among the most important chemically known and chemically formulated chemotherapeutics, there are sulfonamides, which are 4-aminobenzoic acid (PABA) antagonists (**Fig. 7.14**).

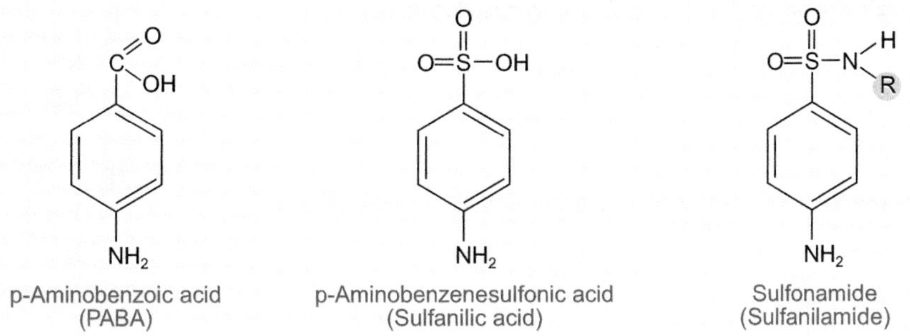

p-Aminobenzoic acid (PABA)   p-Aminobenzenesulfonic acid (Sulfanilic acid)   Sulfonamide (Sulfanilamide)

Fig. 7.14. Structure of some antagonists of 4-aminobenzoic acid.

The displacement of 4-aminobenzoic acid causes bacterial sulfonamides to interfere with the formation of folic acid, which is a precursor of coenzyme F (N^{10}-tetrahydrofolic acid), which is involved in the synthesis of certain amino acids and purines. The therapeutic properties of individual preparations differ according to the nature of the sulfonamide nitrogen-bonded radical (R) (**Table 7.1**). Sulfadiazine and sulfamerazine are the most widely used in medicine and have a broad spectrum of antibacterial activity without causing any toxicity to the patient. Sulfonamides are particularly preferred in the treatment of diseases caused by meningococci and *Shigella* species, staphylococcal and streptococcal diseases, and urinary tract infections caused by Gram-negative bacteria. They are also used preventively against postsurgery infections.

The effect of sulfonamides is bacteriostatic and can be abolished by excess 4-aminobenzoic acid. The application of sulfonamides in low doses leads to the rise of resistant strains of bacteria.

A similar competitive inhibitory effect is also seen in isonicotinic acid hydrazide (INH). It appears to be a structural analog of vitamin B_6, that is, pyridoxine, which is a transaminase component. In therapy, it is used in combination with antibiotics, predominantly against the causative agent of tuberculosis, *Mycobacterium tuberculosis*.

Table 7.1. Individual preparations with different sulfonamide nitrogen-bonded radicals (R)

Sulfonamide	Radical – R
Sulfapyridine (4-amino-*N*-pyridin-2-ylbenzenesulfonamide)	Pyridine
Sulfathiazole (4-amino-*N*-(1,3-thiazol-2-yl)benzenesulfonamide)	Thiazole
Sulfadiazine (4-amino-*N*-pyrimidin-2-ylbenzenesulfonamide)	Pyrimidine
Sulfamerazine (4-amino-*N*-(4-methylpyrimidin-2-yl)benzenesulfonamide)	4-Methyl-pyrimidine
Sulfamethazine (4-amino-*N*-(4,6-dimethylpyrimidin-2-yl)benzenesulfonamide)	4,6-Dimethylpyrimidine
Essential metabolite	Pyridoxine Salicylic acid
Structural analog	Isonicotinic acid hydrazide (INH) 4-Aminosalicylic acid (PAS)

para-Aminosalicylic acid (PAS) is used in the treatment of tuberculosis. It is a growth factor (salicylic acid) antagonist of *Mycobacterium tuberculosis*.

Nitrofurans are chemotherapeutics derived from furfural (furan-2-carbaldehyde) by the addition of nitro groups (**Fig. 7.15**). Individual substances differ in the nature of the side groups attached to the carbon atom of the aldehyde group. Nitrofurans act on Gram-positive and Gram-negative bacteria, some protozoa, and fungi. Their significance is limited in therapy due to mutagenic and carcinogenic effects.

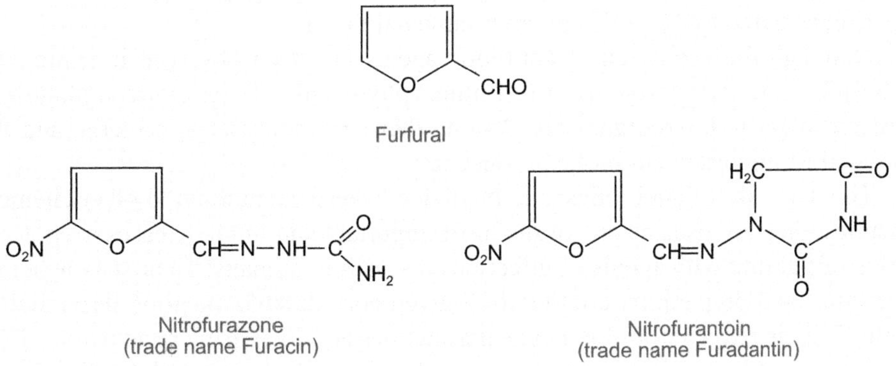

Furfural

Nitrofurazone
(trade name Furacin)

Nitrofurantoin
(trade name Furadantin)

Fig. 7.15. Structures of furfural and its derivatives.

Among newer antibacterial chemotherapeutics, nalidixic acid is known to be effective against some Gram-negative bacteria (*E. coli*). Trimethoprim combined with sulfonamides exhibits a synergistic antibacterial effect (**Fig. 7.16**).

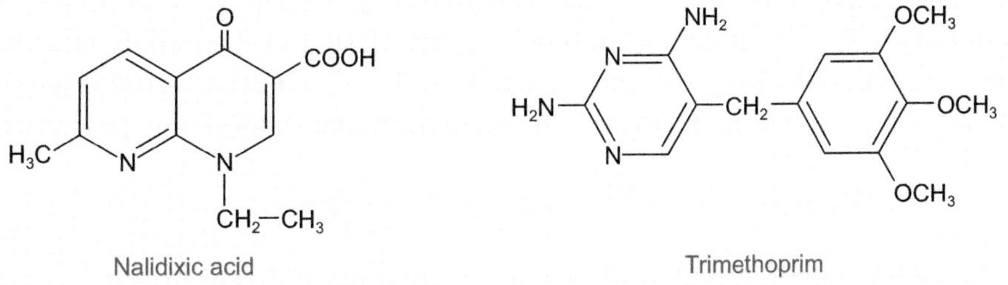

Nalidixic acid

Trimethoprim

Fig. 7.16. Structures of nalidixic acid and trimethoprim.

In addition to inhibiting the synthesis of coenzymes, the synthesis of proteins and nucleic acids may be impaired by the action of some structural analogs. In addition, the effect of these substances affects the replication of phage particles. Therefore, they are important in the chemotherapy of viral infections, that is, in the inhibition of virus synthesis in host cells. As chemotherapeutics, structural analogs of amino acids and pyrimidine and purine bases may be used in these cases.

7.3. Antibiotics and their mechanisms of action

Antibiotics are substances produced by living organisms. Some may also be acquired artificially or semisynthetically. Their action is mainly directed against vital functions of microorganisms, but they can also be toxic to plants and animal cells. The effect of antibiotics is selective and generally occurs in negligible concentrations. Their biological function is one of the forms of antagonistic relations between organisms that are gradually formed during phylogenetic development. In this respect, it can be considered as a significant environmental factor.

Although the formation of antibiotic agents has been observed in some higher organisms (e.g., basic proteins in fish mites, phytoneids of higher green plants), it is most prevalent in microorganisms. Among them, actinomycetes, bacteria, and fungi are the most important antibiotic producers.

The main use of antibiotics is in medicine, where, in addition to other chemotherapeutics, they are used as one of the most important and in many cases irreplaceable means of fighting with agents of infectious diseases in humans. Their search, study of properties, and preparation are therefore given considerable attention throughout the world. This is reflected in the increasing number and increasing production of these substances and in the expansion of their application to livestock and crop production.

The nature of the effect of most antibiotics has not yet been fully elucidated. In general, it can be assumed that, like other chemotherapeutics, antibiotic stimulation affects sensitive and vital systems in the cell, resulting in disruption of subsequent metabolic reactions. The induced changes may cause growth to stop or even death of the bacterial cell. Resistant strains of bacteria lack this sensitive system or can avoid it in the presence of an antibiotic using an alternative metabolic pathway. Some bacteria often cause an antibiotic to be directly destroyed by an enzyme, or prevent its penetration into the cell. The effect of antibiotics is manifested in the inhibition of various metabolic processes. They can interfere with cell wall synthesis, cytoplasmic membrane function, nucleic acid and protein metabolism, phosphorylation processes, etc.

7.3.1. Antibiotics inhibiting cell wall synthesis

Antibiotics inhibiting cell wall synthesis include penicillin, bacitracin, cycloserine, and vancomycin. **Penicillin** is a heterocyclic compound derived from 6-aminopenicillanic acid. Its molecule forms a thiazole core to which a β-lactam ring with a side aliphatic or aromatic group is attached (**Fig. 7.17**).

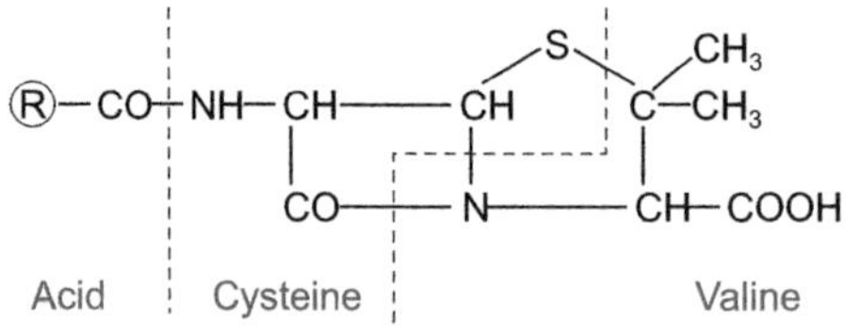

Fig. 7.17. Penicillin structure: acid, cysteine, valine.

The most important strains are *Penicillium notatum* and *P. chrysogenum*. In particular, penicillin G (benzylpenicillin), effective mainly against Gram-positive bacteria, spirochetes, and some actinomycetes, is widely used. Several other penicillins, which are synthesized using suitable precursors added to the culture medium during fermentation, are also important. It has been shown that this interaction can affect the appearance of the side chain, and thus the properties of antibiotics. Penicillin O (allylmercaptomethylpenicillin), propicillin ((1-phenoxypropyl)penicillin), etc. (**Table 7.2**) were obtained in this way.

Table 7.2. Semisynthetic derivatives of 6-aminopenicillanic acid

Penicillin antibiotics	Side chain character
Penicillin G (benzylpenicillin)	$-CH_2-\langle\rangle$
Penicillin V (phenoxymethylpenicillin)	$-CH_2-O-\langle\rangle$
Propicillin ((1-phenoxypropyl)penicillin)	$-CH-O-\langle\rangle$, CH_2-CH_3
Ampicillin (α-aminobenzylpenicillin)	$-CH-\langle\rangle$, NH_2
Methicillin (dimethopenicillin)	H_3C-O, H_3C-O
Penicillin O (allylmercaptomethylpenicillin)	$-CH_2-S-CH_2-CH=CH_2$
Pheneticillin (phenoxymethylpenicillin)	$-CH-O-\langle\rangle$, CH_3
Clometocillin (3,4-dichloro-α-methoxybenzylpenicillin)	$-CH-\langle\rangle-Cl$, $O-CH_3$, Cl
Oxacillin (5-methyl-3-phenyl-4-isoxazolylpenicillin), etc.	H_3C ... O ... N

Most of these substances are acid-resistant and in this respect more advantageous than penicillin G. Some of the prepared penicillins show even a wider spectrum of action. An example is ampicillin (α-aminobenzylpenicillin), against which both Gram-negative and Gram-positive bacteria are sensitive. An important feature of other penicillins is that they are not subject to penicillinase — β-lactam ring cleaving enzyme. Therefore, they can be used against bacteria, especially staphylococci, which have become resistant to other penicillin antibiotics for example, clometocillin (3,4-dichloro-α-methoxybenzylpenicillin), oxacillin (5-methyl-3-phenyl-4-isoxazolylpenicillin), etc.

The effect of penicillin is also inhibited by penicillin amidase cleaving the side chain from the penicillin molecule (**Fig. 7.18**). Unlike penicillinase, this enzyme does not participate in the development of resistance, but is of considerable importance in obtaining 6-aminopenicillanic acid prepared from natural penicillins and in the production of semisynthetic penicillins.

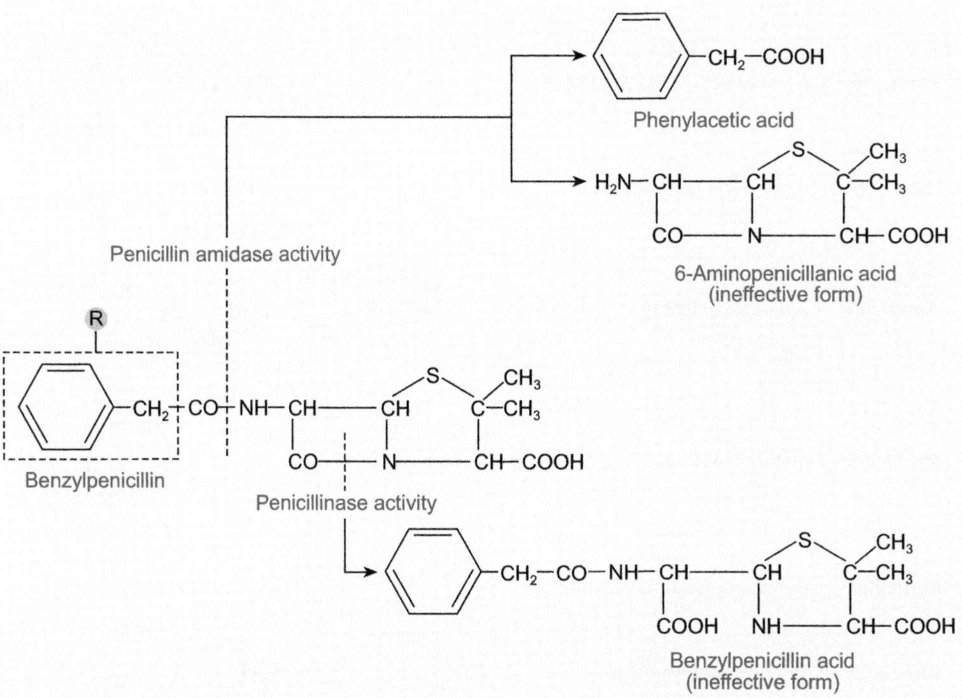

Fig. 7.18. Penicillinase and penicillin amidase effects on penicillin.

The effect of penicillin lies in the disruption of cell wall peptidoglycan synthesis. The connection of muropeptide fragments to peptidoglycan murein, that occurs in *Staphylococcus aureus* via the pentaglycine bridge, is inhibited by penicillin bound via the β-lactam ring to a specific transpeptidase, which leads to blocking of its activity. The realization of this bond is made possible by the structural relationship of penicillin to the terminal group of muramyl pentapeptide — D-alanyl dipeptide (**Fig. 7.19**).

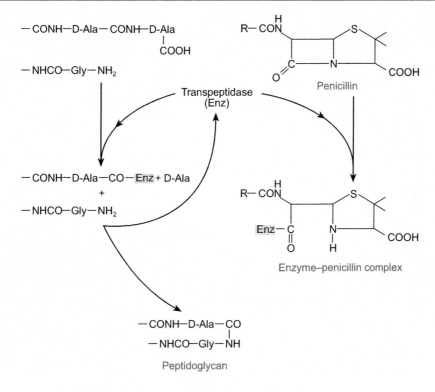

Fig. 7.19. Mechanism of penicillin's effect on *Staphylococcus aureus.*

The inhibitory effect of penicillin on cell wall synthesis is confirmed by the fact that it only acts on growing cells that produce spheroplasts, capable of growth and reproduction, in the hypertonic environment. However, in conventional culture media, penicillin acts bactericidally on bacterial cells; the cells crack and die.

Penicillin is less effective on Gram-negative bacteria, probably due to the more complex chemical structure of their cell wall, especially a low content (1–10%) of murein. Nevertheless, these bacteria can also produce spheroplasts in the presence of higher concentrations of penicillin and osmotic stabilizer or after the destruction of the outer lipopolysaccharide layer of the cell wall by EDTA.

Bacitracin is a neutral polypeptide antibiotic produced by *Bacillus licheniformis* and *B. subtilis* strains. The most effective one is bacitracin A, the molecule of which consists of amino acids, including D-isomers, pyrrolidine, and sulfathiazole (**Fig. 7.20**). The action of bacitracins is directed against Gram-negative bacteria and spirochetes.

The inhibitory effect of bacitracin antibiotics on cell wall synthesis is probably associated with the content of some D-amino acids in its molecule and with the presence of an amide bond between the ε-amino group of lysine and the carboxyl group of aspartic acid. The mechanism of action consists in disrupting the bonds of the peptidoglycan, at the position of lysine or diaminopimelic acid. Thus, the synthesis of peptidoglycan, whose fragments accumulate in the environment, is disrupted.

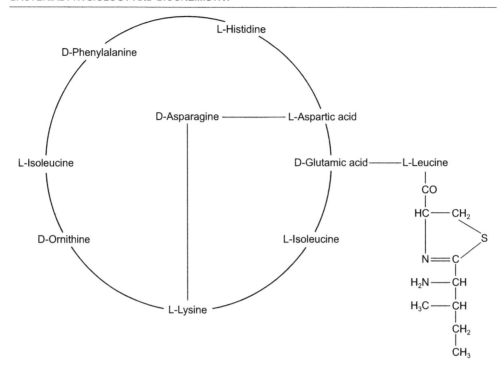

Fig. 7.20. Structure of bacitracin A

D-Cycloserine (oxamycin) is involved in the inhibition of D-alanyl-D-alanine and racemic alanine synthesis. Cycloserine, sold under the brand name Seromycin, is an antibiotic used to treat tuberculosis. Cycloserine was discovered in 1954 from a species of *Streptomyces* genus. For the treatment of tuberculosis, cycloserine is classified as a second-line drug, that is, its use is only considered if one or more first-line drugs cannot be used. In *E. coli* cells, D-cycloserine causes spheroplast formation. The antibacterial effect of cycloserine lies in its structural similarity to D-alanine.

Cycloserine   D-Alanine

Vancomycin is produced by the strains of *Streptomyces orientalis* and contains sugars and amino acids in its molecule. It acts on Gram-positive bacteria and spiro-chetes, in which it prevents the incorporation of amino acids into peptidoglycan and induces rapid accumulation of muramyl peptides in the environment. Vancomycin is an antibiotic used to treat a number of bacterial infections and recommended for the treatment of complicated infections and severe *Clostridium difficile* colitis.

7.3.2. Antibiotics disrupting the plasma membrane

Antibiotics disrupting the plasma membrane have inhibitory action by binding to proteins or lipids of the membrane, thereby interfering with its integrity and physiological function. As a result, growth and cell proliferation are stopped. Substances, whose effect is directed against the synthesis and function of the cytoplasmic membrane, include polymyxins, nystatin (trade name Fungicidin), and amphotericin.

Polymyxin is an antibiotic produced by *Bacillus polymyxa* strains. A strong bactericidal effect is seen in polymyxin B (**Fig. 7.21**). It is a basic polypeptide, whose constituent is 5-methylheptanecarbanic acid in addition to amino acids. Polymyxin acts against Gram-negative bacteria. In particular, it plays a role in the treatment of infections caused by *Pseudomonas pyocyanea* strains, against which other antibiotics are not effective. The effect of polymyxin lies in its specific binding to the membrane and disruption of its permeability, which leads to cell lysis.

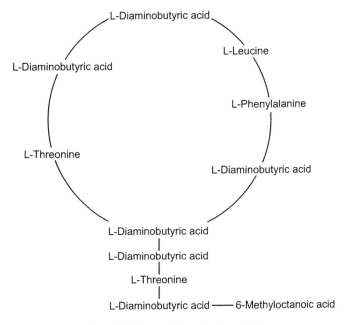

Fig. 7.21. Composition of polymyxin B.

Nystatin (Fungicidin) produced by *Streptomyces noursei* strains and amphotericin isolated from *Streptomyces nodosus* are polyene antibiotics effective against fungi. The site of action is the cell membrane, where these antibiotics cause binding to sterols to change physical properties. The bacteria are not affected due to the absence of sterols in their cytoplasmic membrane. This antibiotic increases the permeability of plasma membranes to small monovalent ions. It increases the basal level of transepithelial 36Cl flux approximately 1.5-fold and eliminates UTP stimulation of this flux.

7.3.3. Antibiotics disturbing the metabolism of nucleic acids

These antibiotics include primarily mitomycin, novobiocin, and actinomycin.

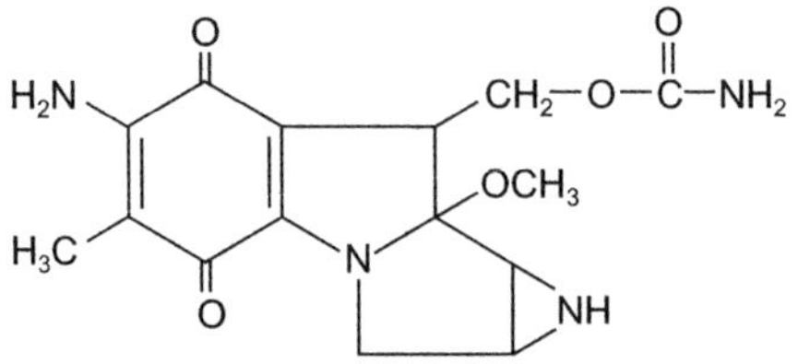

Mitomycin C

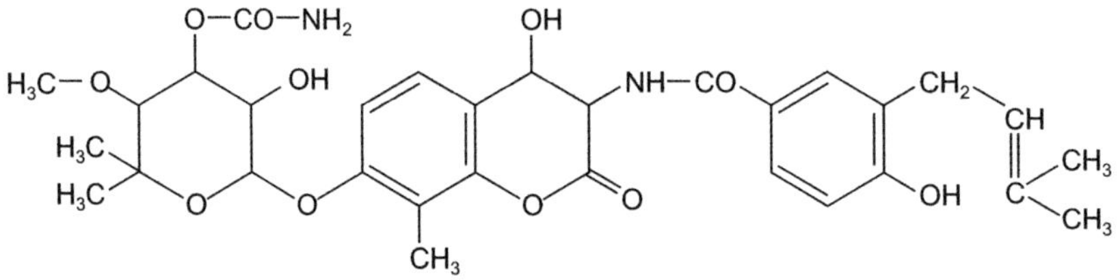

Novobiocin

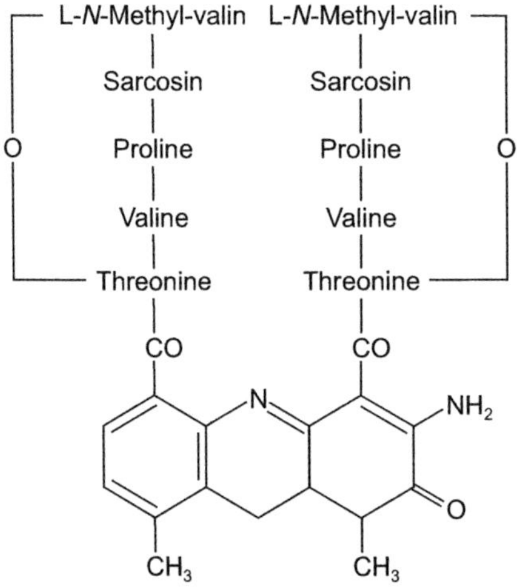

Actinomycin D

Mitomycin is formed by actinomycetes *Streptomyces caespitosus*. Mitomycin C has a broad spectrum of activity, including that against bacteria, protozoa, and other eukaryotic systems along with cancer cells. Therefore, it is widely used in the treatment of some forms of malignant growth. In the presence of mitomycin, the synthesis of cellular RNA is inhibited, and it is hydrolytically cleaved into acid-soluble fragments. However, it does not affect the synthesis of phage DNA; on the contrary, in lysogenic cells, it induces the development of the active phage from the prophage. This phenomenon shows a different pathway of phage DNA biosynthesis from cell DNA biosynthesis.

Novobiocin is a derivative of coumarin produced by *Streptomyces niveus* and *S. sphaerotilus* strains. It acts primarily on Gram-positive bacteria, in which it inhibits the cell wall synthesis and at the same time disturbs the integrity of the cytoplasmic membrane. However, the primary mechanism of action of novobiocin is the suppression of DNA polymerase activity, which results in the inhibition of DNA synthesis.

Actinomycin is a mixture of substances produced in cultures of *Streptomyces antibioticus*, *S. flavus*, *S. chrysomallus*, etc. It is particularly effective against Gram-positive bacteria. Actinomycin D, the molecule of which is formed by two identical amino acid chains linked by peptide bonding to substituted phenazinedicarboxylic acid, is best known. The effect of actinomycin lies in its binding to DNA, which is then deactivated. By complexation with DNA, actinomycin blocks the synthesis of DNA-dependent RNA polymerase and thus ribonucleic acids.

7.3.4. Antibiotics inhibiting protein synthesis

In the biosynthesis of proteins, antibiotics interfere most often with the translation process, namely, the elongation phase. The sites of action are usually ribosomes that bind antibiotics to either 30S or 50S subunits. The most effective antibiotics influenced on 30S ribosomal subunits are tetracyclines, streptomycin, and other aminoglycans.

Tetracycline antibiotics. Tetracyclines contain a naphthacene nucleus in their molecule. The production strains are *Streptomyces aureofaciens* (tetracycline and chlortetracycline) and *Streptomyces rimosus* (oxytetracycline). Tetracycline antibiotics act against Gram-positive and Gram-negative bacteria, some actinomycetes, spirochetes, protozoa, rickettsia, and large viruses. They do not affect *Proteus* species and *Pseudomonas pyocyanins*.

The mechanism of action of tetracyclines is the suppression of protein synthesis by inhibiting aminoacyl-tRNA binding to iRNA in 30S ribosomal subunit. In addition, they disrupt the formation of polyribosomes.

Tetracycline

Oxytetracycline

Chlortetracycline

Streptomycin is produced by strains of *Streptomyces griseus* and *S. bikinensis.* Its molecule consists of streptidine (guanidinoinositol), streptose, and *N*-methylglucosamine. (**Fig. 7.22**). In addition to streptomycin, streptonicozide produced by the reaction of the aldol group of streptose with the hydrazine group of isoniazid and dihydrostreptomycin resulting from the hydrogenation of this group are known. Streptomycin, in combination with penicillin, is used in a standard antibiotic cocktail to prevent bacterial infection in cell culture.

The effect of streptomycin, which acts on Gram-positive and Gram-negative bacteria, especially on *Mycobacterium tuberculosis*, is varied and is manifested by the disruption of the cytoplasmic membrane, inhibition of the synthesis of nucleic acids, proteins, etc. However, the primary mechanisms of action are irreversible binding of streptomycin to the ribosomal subunit 30S and proteosynthesis suppression. Other aminoglycoside antibiotics, such as neomycin produced by *Streptomyces fradise* and kanamycin, which is produced by *Streptomyces kanamyceticus*, have a similar effect.

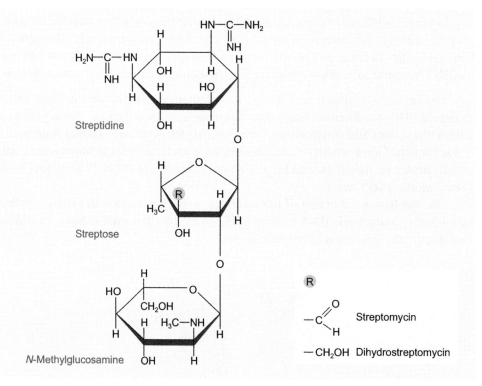

Fig. 7.22. Structure of streptomycin.

Thus, streptomycin is a protein synthesis inhibitor. It binds to the small 16S rRNA of the 30S subunit of the bacterial ribosome, interfering with the binding of formyl-methionyl-tRNA to this subunit.

Chloramphenicol, lincomycin, and a group of macrolide antibiotics affected proteosynthesis are primarily focused on 50S ribosomal subunits.

Chloramphenicol (Chloromycetin). Chloramphenicol is *p*-nitrophenyl-2-dichloroacetoamide-1,3-propanediol. Its producer is *Streptomyces venezuelae*. It acts on Gram-positive and Gram-negative bacteria, actinomycetes, spirochetes, rickettsiae, and some large viruses. Strains of *Proteus vulgaris*, *Pseudomonas pyocyanes*, and *E. coli* are insensitive to this antibiotic.

Cloramphenicol inhibits protein synthesis by binding to 50S ribosomal subunits. The bond is stereospecific. Since the formation of polyribosomes continues even after stopping the proteosynthesis, it appears that the antibiotic does not affect the initiation of this synthesis, but the formation of peptide bonds.

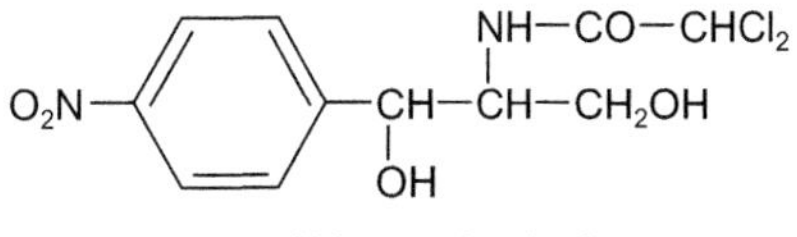

Chloramphenicol

Lincomycin has a relatively simple molecule containing octyl amidically bound to propylhygrinic acid. It is produced by *Streptomyces lincolnensis*. It acts only on Gram-positive cocci and bacteria.

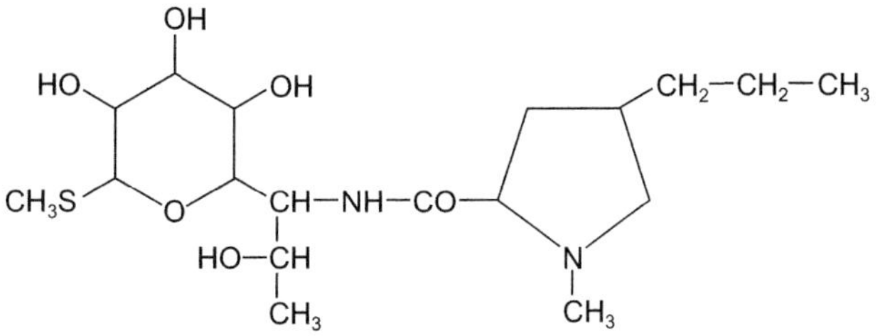

Lincomycin

Similarly to chloramphenicol, lincomycin binds to the 50S ribosomal subunit and inhibits proteosynthesis by disrupting aminoacyl-tRNA binding to iRNA. Unlike chloramphenicol, it induces the dissociation of the polyribosomes to 30S and 50S subunits.

Macrolide antibiotics contain a macrocyclic lactam ring (macrolide) to which a sugar component is attached (**Fig. 7.23**). As an example, *erythromycin* produced by the strains *Streptomyces erythraeus* can be used. It is effective against Gram-positive bacteria, rickettsiae, and large viruses. The mechanism of action is similar to chloramphenicol and consists in inhibiting the formation of peptide bonds. **Oleandomycin**, produced by *Streptomyces antibioticus*, and **spiramycin**, produced by *Streptomyces ambofaciens* strains, work alike. **Puromycin** is produced by *Streptomyces alboniger* strain. It is structurally similar to the terminal aminoacyladenine of tRNA. Its ability to interfere with proteosynthesis in prokaryotes and eukaryotes is the reason why it is not used clinically.

Inhibition is mediated by the binding of puromycin to the section of ribosomal 50S reserved for aminoacyl-tRNA. The following reaction with peptidyl-tRNA results in pepridylpromycin formation, which causes the disruption of the peptide chain elongation and its release from the ribosome. After a certain interval, depending on how the iRNA moves through the ribosome, the process may be restored. The effect of puromycin is the abortive synthesis of several oligopeptides with terminal amino acids.

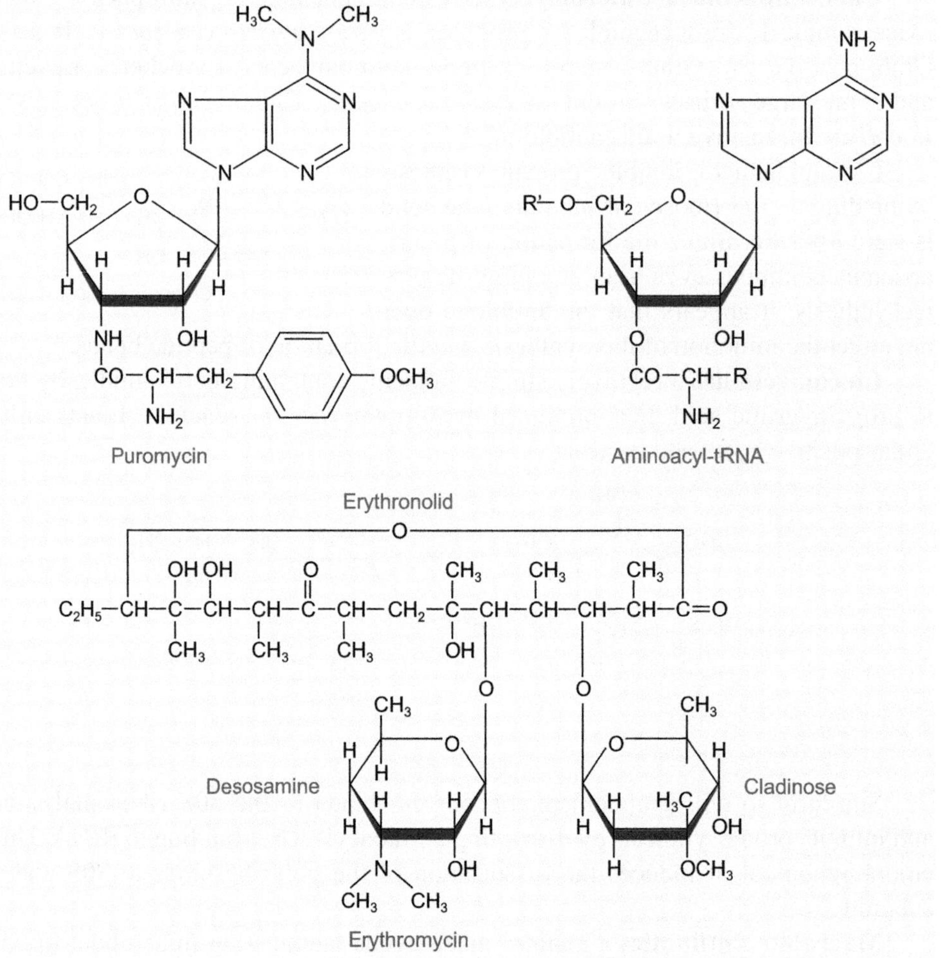

Fig. 7.23. Structures of some macrolide antibiotics.

Thus, the main types of macrolide antibiotics are erythromycin, oleandomycin, and puromycin. They have different structures and are produced by various species of the *Streptomyces* genus.

7.3.5. Antibiotics affecting phosphorylation process

Phosphorylation processes are affected mainly by **gramicidins (Fig. 7.24)**. Gramicidins are ionophores; their dimers form ion channel-like pores in cell membranes and cellular organelles. Inorganic monovalent ions, such as potassium (K^+) and sodium (Na^+), can travel through these pores freely via diffusion. Gramicidin, also called gramicidin D, is a mix of ionophore antibiotics, gramicidin A, B, and C, which make up about 80%, 5%, and 15% of the mix, respectively. Each has two isoforms, so the mix has six different types of gramicidin molecules. They are produced by *Bacillus brevis* strains. Gramicidins are cyclic peptides, in which D-isomers are present in addition to natural amino acids. They occur in mixtures, where individual substances differ in their amino acid composition and antibacterial activity. They act on Gram-negative cocci, corynebacteria, clostridia, and neisseria. Gramicidin S, a cyclic dipeptide, can be mentioned as an example.

Gramicidins act as antibiotics against Gram-positive bacteria like *Bacillus subtilis* and *Staphylococcus aureus*, but not well against Gram-negative ones like *E. coli*. This group of antibiotics is often mixed with other antibiotics like tyrocidine and antiseptics. It acts mainly by blocking phosphorylation reactions and electron transfer.

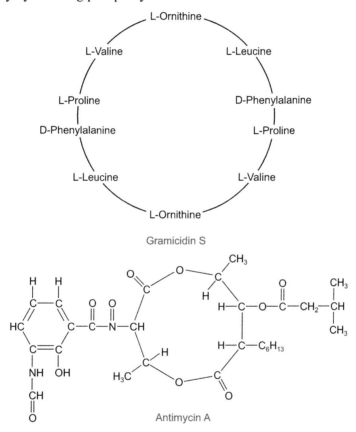

Fig. 7.24. Composition of gramicidin S and antimycin A structure.

A similar effect is shown by the antibiotic **antimycin**, which interrupts electron transport between cytochrome *b* and cytochrome *c*. Antimycin A was first discovered in 1945 and registered for use as a fish toxicant in 1960.

Antimycin A is a secondary metabolite produced by *Streptomyces* bacteria and a member of a group of related compounds called antimycins. Antimycin A is classified as an extremely hazardous substance in the United States.

Antimycin A is an inhibitor of cellular respiration, specifically oxidative phosphorylation. This antibiotic binds to the Qi site of cytochrome c reductase, inhibiting the oxidation of ubiquinone in the Qi site of ubiquinol, thereby disrupting the Q cycle of enzyme turn over. It also causes the disruption of the entire electron transport chain. Due to this, there can be no production of ATP. Cytochrome c reductase is a central enzyme in the electron transport chain of oxidative phosphorylation. The inhibition of this reaction disrupts the formation of the proton gradient across the inner membrane of the mitochondria. The production of ATP is subsequently inhibited, as protons are unable to flow through the ATP synthase complex in the absence of a proton gradient. This inhibition also results in the formation of the toxic free radical superoxide. The rate of superoxide production exceeds the ability of the cellular mechanisms to scavenge it, overwhelming the cell and leading to cell death. It has also been found to inhibit the cyclic electron flow within photosynthetic systems along the proposed ferredoxin quinone reductase pathway.

***

Thus, the bacterial life processes strongly depend on the environment. These conditions are a reflection of the joint action of various factors. External environmental factors with antibacterial effect may be physical or chemical. The physical factors affecting bacterial viability are drought, temperature, pressure, ultrasound, radiation, sunlight, ultraviolet rays, X-rays, and γ-rays. The chemical factors include the effect of pH, redox potential, disinfectants, phenol and phenolic compounds, alcohols, halogens, heavy metals and their compounds, oxidants, dyes, soaps, and synthetic detergents.

A unique group of chemotherapeutics is antibiotics. Antibiotics inhibit cell wall synthesis, disturbing the function of the cytoplasmic membrane, the metabolism of nucleic acids, and protein synthesis.

RECOMMENDED REFERENCES

Anderson D., Salm S., Allen D. Nester's Microbiology: A Human Perspective. *McGraw Hill*, 8th edition; 2015; 896 p.

Bartlett J.M.S., Stirling D. A short history of the polymerase chain reaction. *Methods in Molecular Biology*, 2003; 226: 3–6.

Barton L.L. Structural and Functional Relationships in Prokaryotes. *Springer-Verlag, New York*, 2005; 818 p.

Barton L.L., Fauque G.D. Biochemistry, physiology and biotechnology of sulfate-reducing bacteria. *Advances in Applied Microbiology*, 2009; 68: 41–98.

Bensimon A., Heck, A.J.R., Aebersold R. Mass spectrometry–based proteomics and network biology. *Annual Review of Biochemistry*, 2012; 81(1): 379–405.

Bertone P., Snyder M. Advances in functional protein microarray technology. *FEBS Journal*, 2005; 272(21): 5400–5410.

Black J.G. Microbiology: Principles and Explorations. *Wiley*, 8th edition, 2012; 968 p.

Blackstock W.P, Weir M.P. Proteomics: quantitative and physical mapping of cellular proteins. *Trends in Biotechnology.*, 1999; 17(3): 121–127.

Brown T.A. Gene Cloning and DNA Analysis: An Introduction, 7th edition. *Wiley-Blackwell Publisher*, 2016; 376 p.

Daffé M., Etienne G. The capsule of *Mycobacterium tuberculosis* and its implications for pathogenicity. *Tubercle and Lung Disease*, 1999; 79(3): 153–169.

Dijkhuizen L., Levering P.R., De Vries G.E. The physiology and biochemistry of aerobic methanol-utilizing Gram-negative and Gram-positive bacteria. In *Methane and Methanol Utilizers, Springer, Boston, MA*, 1992; pp. 149–181.

Friedrich C.G. Physiology and genetics of sulfur-oxidizing bacteria. *Advances in Microbial Physiology*, 1997; 39: 235–289.

Gobbetti M., De Angelis M., Corsetti A., Di Cagno R. Biochemistry and physiology of sourdough lactic acid bacteria. *Trends in Food Science & Technology*, 2005; 16: 57–69.

Hallenbeck P.C. Recent Advances in Phototrophic Prokaryotes. *Springer*, 2010; 376 p.

Hastings J.W., Potrikusv C.J., Gupta S.C., Kurfürst M., Makemson J.C. Biochemistry and physiology of bioluminescent bacteria. *Advances in Microbial Physiology*, 1985; 26: 235–291.

Chawl S. Carbon and nitrogen metabolism. 2008; 54 p.
http://nsdl.niscair.res.in/jspui/bitstream/123456789/803/1/CarbonMetabolism.pdf

Chen X., Schauder S., Potier N., Van Dorsselaer A., Pelczer I., Bassler B.L., Hughson F.M. Structural identification of a bacterial quorum-sensing signal containing boron. *Nature*, 2002; 415: 545–549.

Jacob F., Brenner S., Cuzin F. On the regulation of DNA replication in bacteria. *Cold Spring Harbor Symposia on Quantitative Biology.*, 1963; 28: 329–348.

Kim B.H., Gadd G.M. Prokaryotic Metabolism and Physiology. *Cambridge University Press*, 2nd edition, 2019; 504 p.

Kushkevych I. Intestinal Sulfate-Reducing Bacteria, *MUNIPRESS*, 2017; 320 p.

Leaver M., Domínguez-Cuevas P., Coxhead J.M, Daniel R.A., Errington J. Life without a wall or division machine in *Bacillus subtilis*. *Nature*, 2009; 457: 849–853.

Levin P.A., Grossman A.D. Cell cycle: the bacterial approach to coordination. *Current Biology*, 1998; 8(1): 28–31.

Ma Z., Jacobsen F.E., Giedroc D.P. Coordination chemistry of bacterial metal transport and sensing. *Chemical Reviews*, 2009; 109(10): 4644–4681.

Madigan M., Bender K., Buckley D., Sattley W., Stahl D. Brock Biology of Microorganisms. *Pearson*, 15th edition; 2017; 1056 p.

Mager H.I., Tu, S.C. Chemical aspects of bioluminescence. *Photochemistry and Photobiology*, 1995; 62: 607–614.

Meighen E.A. Bacterial bioluminescence: organization, regulation, and application of the lux genes. *FASEB Journal*, 1993; 7: 1016–1022.

Moat A.G., Foster J.W., Spector M.P. Microbial Physiology. *Wiley-Liss. Ink., New York*, 2002; 733 p.

Nelson D.C., Hagen K.D. Physiology and biochemistry of symbiotic and free-living chemoautotrophic sulfur bacteria. *American Zoologist*, 1995; 35(2): 91–101.

Newton A., Ohta N. Regulation of the cell division cycle and differentiation in bacteria. *Annual Review of Microbiology*, 1990; 44(1): 689–719.

Nicholas M.W., Nelson K. North, south, or east? Blotting techniques. *Journal of Investigative Dermatology*. 2013; 133: e10.

Nisnevitch M. Prokaryotes: Physiology, Biochemistry and Cell Behavior (Biochemistry Research Trends). *Nova Science Pub Inc; UK ed. edition*, 2014; 273 p.

Patterson G.W., Nes W.D. Physiology and Biochemistry of Sterols. *AOCS Publishing*, 1992; 395 p.

Preiss J., Romeo T. Physiology, biochemistry and genetics of bacterial glycogen synthesis. *Advances in Microbial Physiology*, 1990; 30: 183–238.

Prescott Lansing M. Microbiology. 5th edition, 2002; 1147 p. (www.mhhe.com/prescott5).

Rosenberg E., DeLong E.F., Lory S., Stackebrandt E., Thompson F. eds., The Prokaryotes: Prokaryotic Physiology and Biochemistry. *Springer, Berlin, Heidelberg*, 2013; 662 p.

Sabidó E., Selevsek N., Aebersold R. Mass spectrometry-based proteomics for systems biology. *Current Opinion in Biotechnology*, 2012; 23(4): 591–597.

Southern E.M. Detection of specific sequences among DNA fragments separated by gel electrophoresis. *Journal of Molecular Biology*, 1975; 98(3): 503–517.

Tortora G., Funke B., Case C. Microbiology: An Introduction. *Pearson*, 12th edition, 2016; 960 p.

Van Houdt R., Michiels C.W. Role of bacterial cell surface structures in Escherichia coli biofilm formation. *Research in Microbiology*, 2005; 156: 626–633.

Venil C.K., Zakaria, Z.A., Ahmad W.A. Bacterial pigments and their applications. *Process Biochemistry*, 2013; 48(7): 1065–1079.

Wells T.J., Tree J.J., Ulett G.C., Schembri, M.A. Autotransporter proteins: novel targets at the bacterial cell surface. *FEMS Microbiology Letters*, 2007; 274(2): 163–172.

White D., Drummond J., Fuqua C. The Physiology and Biochemistry of Prokaryotes. *4th edition. Oxford University Press, New York*, 2012; 696 p.

Woese C.R., Kandler O., Wheeliss M.L. Towards a natural system of organisms: proposal for the domains Archaea, Bacteria and Eucarya. *Proceedings of the National Academy of Sciences of the United States of America*, 1990; 87: 4576–4579.

Wolgemuth Ch., Hoiczyk E., Kaiser D., Oster G. How myxobacteria glide. *Current Biology*, 2002; 12: 369–377.

INDEX

363